Engineering Materials and Processes

Series editor

Brian Derby, Manchester, UK

For further volumes:
http://www.springer.com/series/4604

Yury Mironovich Volfkovich
Anatoly Nikolaevich Filippov
Vladimir Sergeevich Bagotsky

Structural Properties
of Porous Materials
and Powders Used
in Different Fields of
Science and Technology

 Springer

Yury Mironovich Volfkovich
A.N. Frumkin Institute of Physical
  Chemistry and Electrochemistry
Moscow
Russia

Vladimir Sergeevich Bagotsky
Mountain View, CA
USA

Anatoly Nikolaevich Filippov
Department of Higher Mathematics (535)
Gubkin Russian State University
  of Oil and Gas
Moscow
Russia

Vladimir Sergeevich Bagotsky is Deceased.

ISSN 1619-0181
ISBN 978-1-4471-6912-3   ISBN 978-1-4471-6377-0   (eBook)
DOI 10.1007/978-1-4471-6377-0
Springer London Heidelberg New York Dordrecht

Printed on acid-free paper

Springer is part of Springer Science+Business Media (www.springer.com)

Foreword

Porous materials and powders with different pore sizes are used in many areas of industry, geology, agriculture (crops and livestock), pharmacy, and science. These areas include a variety of devices and supplies—filters and membranes, chemical power sources (batteries and supercapacitors), sintering, paper, pharmaceutical powders; thermal insulation and building materials; oil-bearing geological, gas-bearing and water-bearing rocks; biological objects—food materials, soils and roots of plants, mammalian skin, and others.

Until now, books and reviews that described the properties of the porous structure of various materials and products were focused on the materials used in certain relatively narrow areas such as powder metallurgy, water treatment, electrochemical energetics, oil and gas industry, leather industry, soil science, etc. However, various porous and powder materials and products have many properties in common. These are characteristics of the porous structure and transport processes in these objects. Therefore, the aim of the authors was to write a book that would amalgamate information about the porous structure of various materials and products used in various industries, geology, agriculture, pharmacy, and science. Each section of the book was written by well-known experts in their respective fields of expertise.

The book is divided into the following summarizing chapters on individual types of materials: technical materials (paper, components of chemical power sources (electrochemical energetics), powder metallurgy, thermal insulating materials, powdered medical substances used in the pharmaceutical industry, membrane materials for gas separation and sorbents), natural materials (oil and gas bearing rocks, groundwater rocks), and biological materials (food materials, soils and plant roots, hide and skin of mammals). The chapter devoted to modern methods of investigation of the porous structure and wetting (hydrophilic–hydrophobic) properties of any porous and powder materials as well as the chapter on mathematical modeling of filtration processes in porous media, demonstrating common approaches for different materials, combine all sections mentioned above. Mathematical models validated experimentally allow optimizing these processes, as well as the porous structure of different products. The principles and approaches of mathematical modeling are united and thus applicable to all porous objects such as artificial and created by nature during the process of evolution, i.e., inorganic and biological.

This book allows the reader (a) to understand the basic regularities of mass transfer occurring in a qualitatively different porous materials and products and (b) to optimize the functional properties of porous and powdered products and materials.

The book is written in the popular language and is intended for a wide audience, specializing in different fields of science and engineering: engineers, geologists, geophysicists, oil and gas producers, agronomists and other agricultural workers, physiologists, pharmacists, researchers, teachers, and students.

The authors express deep gratitude to the scientific employees of A.N. Frumkin Institute of Physical Chemistry and Electrochemistry of the Russian Academy of Sciences (Moscow, Russia), namely Valentin E. Sosenkin, Nadezhda F. Nikolskaya, Daniil A. Bograchev and also to Daria Yu. Khanukaeva and Sergey I. Vasin from Gubkin Russian State University of Oil and Gas (Moscow, Russia) for the performance of experimental and computational work. The authors also express their gratitude to Yulia S. Dzyazko (V. I. Vernadskii Institute of General and Inorganic Chemistry of the National Academy of Science, Kiev, Ukraine), who is the author of two chapters—for the great help in the scientific and organizational work related to writing the present book.

We announce with deep regret that our co-author, Vladimir Sergeevich Bagotsky, died in the process of writing this book in November 2012. He was an outstanding scientist in the field of electrochemistry, chemical power sources, and macrokinetics processes in porous electrodes. V. S. Bagotsky had initiated the publication of the book and greatly contributed to its content. We express our deepest condolences to his family and friends and all those who knew him.

Moscow, December 2013   Yury Mironovich Volfkovich
                                               Anatoly Nikolaevich Filippov

Contents

Contributors

Vladimir Sergeevich Bagotsky A.N. Frumkin Institute of Physical Chemistry and Electrochemistry of Russian Academy of Sciences, RAS, Moscow, Russia

Yuliya Sergeevna Dzyazko V. I. Vernadskii Institute of General and Inorganic Chemistry, National Academy of Sciences of Ukraine, Kiev, Ukraine

Mikhail L'vovich Ezerskiy Moscow, Russia

Anatoly Nikolaevich Filippov Department of Higher Mathematics, Gubkin Russian State University of Oil and Gas, Moscow, Russia

Boris Yakovlevich Konstantinovsky Kiev National University of Construction and Architecture, Kiev, Ukraine

Olena Romanovna Mokrousova Department of Leather and Fur Technology, Kiev National University of Technologies & Design, Kiev, Ukraine

Alexander Vladilinovich Rastorguev Department of Hydrogeology, Faculty of Geology, Lomonosov Moscow State University, Moscow, Russia

Mikhail Mikhailovich Serov «MATI»—Russian State University of Aviation Technology, Moscow, Russia

Mikhail Yurievich Sidorenko Moscow State University of Food Production, Moscow, Russia

Oksana Leonidovna Tonkha National University of Life and Environmental Science of Ukraine, Kiev, Ukraine

Yury Mironovich Volfkovich A.N. Frumkin Institute of Physical Chemistry and Electrochemistry, RAS, Moscow, Russia

Vladimir Vasilievich Volkov A.V. Topchiev Institute of Petrochemical Synthesis, RAS, Moscow, Russia

Yury Pavlovich Yampolskii A.V. Topchiev Institute of Petrochemical Synthesis, RAS, Moscow, Russia

Introduction

Porous materials and porous media with pores of different form and size, as well as powders with different particle size are used in different fields of science, technology, and everyday life per se or for manufacturing of industrial products. Structural and wetting properties of porous materials and powders have a significant influence on different processes and on the functional properties of these materials. The number of known methods for investing these properties is limited and most of them have specific limitations. During the 1980s, a new method was developed called Method of Standard Contact Porosimetry (MSCP). This method is devoid of the limitations of other methods and allows the quantitative investigation of a broad range of properties of porous materials.

Porous materials and porous media with pores of different form and size, as well as powders with different particle size, are used in different fields of science, technology, and everyday life per se or for manufacturing of industrial products. These fields include: (a) different devices and equipments, such as filters and membranes, chemical and electrochemical reactors, etc.; (b) building materials, paint-and-varnish items, cermet articles, fabrics, etc.; (c) geological structures—oil- and gas-bearing strata, agricultural soil; (d) biological items—human and animal skin, timber, and many others.

Structural and wetting properties of porous materials and powders have a significant influence on different processes and on the functional properties of these materials (rate of gas and/or liquid flows, electrical and thermal conductivity, sound permeability a.o.). The stability and aging rate of articles manufactured from these materials is often determined by their structural properties.

For optimization of devices and other items made from porous materials and powders and for modeling and optimization of their functional parameters, a detailed knowledge of all their structural and wetting properties is necessary. Such knowledge is also necessary for a correct interpretation of different natural and biological phenomena.

The number of methods for investigating structural properties of porous materials is very limited. The widely used Method of Mercury Porosimetry (MMP) has many drawbacks. It cannot be used for soft and frail materials since under the influence of high mercury pressures such samples are subjected to deformation or even demolition during measurement. Practically no existing methods for the determination of wetting properties of porous materials are known.

During the 1980s, in the A.N. Frumkin Institute of Physical Chemistry and Electrochemistry of the Russian Academy of Sciences in Moscow, a new method was developed called MSCP. This method is devoid of the limitations of MP and allows the quantitative investigation of a broad range of properties of porous materials: pore size distribution dependences, specific pore surface area, pore corrugation, wetting (liophilic/liophobic) properties, isotherms of capillary pressure and of binding energy, and much other information. At the beginning this method was used for investigating fuel cell components. In the following the application fields of this method were considerably extended.

Besides problems of experimental investigations in the field of experimental investigations there exist some theoretical problems connected with the interpretation and mathematical modeling of flow processes in porous media. The d'Arcy model which is often used for this purpose often leads to results contradicting experimental data. With the recent development of micro- and nanofluidicity concepts new approaches are needed. The Brinkman model for porous media gives the possibility to take into account detailed structural and surficial properties, particularly those that can be measured by the new highly informative MSCP.

Chapter 1
Experimental Methods for Investigation of Porous Materials and Powders

Yury Mironovich Volfkovich and Vladimir Sergeevich Bagotsky

Abstract The best known method for investigating the porous structure of different materials is the method of mercury porosimetry MMP, which has some serious drawbacks. The method of standard contact porosimetry is based on the laws of capillary equilibrium. If two (or more) porous bodies partially filled with a wetting liquid are in capillary equilibrium, the values of the liquid's capillary pressure p^c in these bodies are equal. In this method the amount of a wetting liquid in the test sample is measured and compared with the amount of the same liquid in a standard sample with a known pore structure. Using different working liquids the wetting properties of the test sample can be determined.

1.1 Classical Methods

1.1.1 Methods for Investigation of Structural Properties

The geometric structure of porous materials can be characterized by porograms, viz., the integral pore size distribution function (PSDF) describing the distribution of pore volume versus pore radii V, r, or the differential PSDF (dV/dr), r.

Using the equation,

$$S_{\text{pore}} = 2 \int_0^r (1/r)(dV/dr)dr \qquad (1.1)$$

these functions allow calculating another important parameter of the porous structure—the distribution function of the pore surface S vs pore radii r.

The best known method for investigating the porous structure of different materials is the method of mercury porosimetry MMP described in [1]. This method is based on intrusion of mercury into samples of the porous material under high pressure. When an external pressure P is applied, all pores with radii $r > r_{\text{min}}$ become filled with mercury. The value of r_{min} corresponds to the condition that the

Y. M. Volfkovich et al., *Structural Properties of Porous Materials and Powders Used in Different Fields of Science and Technology*, Engineering Materials and Processes, DOI: 10.1007/978-1-4471-6377-0_1, © Springer-Verlag London 2014

mercury capillary pressure p^c in the pore is equal to the applied pressure P. The capillary pressure is determined by the thermodynamic Laplace equation:

$$p^c = 2\sigma \cos\theta / r_{\min}, \qquad (1.2)$$

where σ is the excess surface energy or "surface tension" (for mercury $\sigma = 4.67$ μJ $\times$ cm^{-2}), θ is the wetting angle of liquid mercury with the given material. Therefore, measuring the dependence of the mercury volume V_{Hg} introduced into the sample on the applied pressure P the integral distribution function of pore volumes V_{pore} versus capillary pressure (or when the value of q is known, vs. pore radii) is found. This method provides the widest range of measurable pore radii (from 2 to 10^5 nm). The accuracy of this method is very high.

For these reasons it is widely used for investigation of the geometric structure of different porous materials. Devices for automated measurements by MMP with a direct recording of the results are available on the market.

At the same time this method has some serious drawbacks. It requires the application of high pressures (up to thousands bar). This can lead to deformation or even destruction of the samples and, thus, to distortion of programs. For this reason MMP cannot be applied to investigate soft and frail materials. Also, this method cannot be applied for investigation of metals that are amalgamated by mercury. These drawbacks of MMP prevent the use of this method for most components of fuel cells such as:

(1) membranes, parts made of porous carbonaceous materials, and gas-diffusion layers which would be deformed under high pressure, or
(2) catalytic layers with metal catalysts (platinum a.o.) which would be partially amalgamated by mercury.

Several other methods for investigating the structure of porous materials are also known. Each of these methods has its own advantages and limitations. Small angle X-ray scattering [2] can be used only for pore radii from 2 to 50 nm and often leads to ambiguous results. Electronic spectroscopy is associated with difficulties in pretreatment of the samples and interpretation of the results. Centrifugal porosimetry [3], displacement of wetting liquids from the pore volume by gas pressure [4], and optical methods are practically useless for pore sizes below 10^3 nm. The method of capillary condensation [5] can be used only in the pore size range from 1 to 50 nm. Other methods that were used for the investigation of the porous structure of different materials are: hydraulic permeability [6, 7], electron microscopy [8, 9], and atomic force microscopy [10].

1.1.2 Methods for Investigation of Wetting Properties

The degree of hydrophobic properties for materials with a smooth surface can be characterized by the wetting angle of a water drop placed on this surface. However, this method cannot be used in porous materials for two reasons: (1) the surface roughness of these materials leads to a distortion of the wetting angle and (2) it reflects only the wetting properties of the external surface of the sample, but not those of inside pores.

In 1921, Washburn in [11] proposed a method for determination the internal wetting angle of water in porous media. This method is based on measuring the rate of water rise in a suspended sample of the porous material. It is not applicable for partially hydrophobic materials.

1.2 The Method of Standard Contact Porosimetry

Because of the shortcomings and limited possibilities of other methods for investigating the porous structure a new method for porosimetric measurements was developed in the 1980s at the A.N. Frumkin Institute of Electrochemistry of the Russian Academy of Sciences [2] called the Method of Standard Contact Porosimetry (MSCP). This method was described in [12–15] and discussed in detail in [13, 16–18].

1.2.1 Principles of the Method

The method is based on the laws of capillary equilibrium. If two (or more) porous bodies partially filled with a wetting liquid are in capillary equilibrium, the values of the liquid's capillary pressure p^c in these bodies are equal. In this method the amount of a wetting liquid in the test sample (V_t) is measured. Simultaneously, the amount of the same wetting liquid is measured in a standard specimen of known porous structure (V_s). The liquids in both porous samples are kept in contact. After some time a thermodynamic equilibrium is established. The measurements are performed for different overall amounts of the liquid $V_0 = V_s + V_t$. During the experiment this overall amount is changed by gradual evaporation of the liquid.

These measurements allow to establish the distribution of pore volume versus pore size of the test sample by comparing with the known PSDF of the standard sample. Figure 1.1 shows how this can be done graphically for the case when wetting of both samples by the liquid is ideal ($\theta = 0^0$, $\cos \theta = 1$). Curve 1 on the left side of this figure represents the experimental dependence of the volume V_s in the standard sample on the volume V_t in the test sample for different values of V_0. On the right side, the integral pore size distribution curves (pore volume as a function of log r) are shown. Curve 2 is the known PSDF curve for the standard sample. For a certain total volume of liquid V_0', the volumes of liquid in both bodies V_s' and V_t' are represented by the coordinates of point C. This point corresponds to point D on curve 2 and to a certain value r' of the minimal radius of filled pores. In the case of capillary equilibrium (under the assumption made), the minimal radius of filled pores in the test sample will be the same. As in this sample

Fig. 1.1 Example for
determining pore size
distribution curves by MSCP:
1 dependence of V_s on V_t, *2*
pore size distribution curve
for the standard sample, and *3*
pore size distribution curve
for the test sample

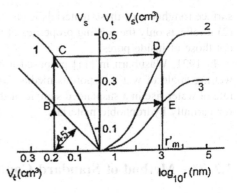

the volume of liquid is represented by point B (the line is drawn at an angle of 45°), point E corresponds to a point on the pore size distribution curve for the test sample. Thus, changing the value of the total volume V_o of the liquid, the overall distribution curve 3 for the test sample can be determined.

For these measurements mostly octane or decane are used as measuring liquid because they wet most solid materials almost ideally. Details of the experimental procedures for MSCP are described in [13, 16–18].

1.2.2 Possibilities of MSCP for the Investigation of Porous Structure

MSCP has several substantial advantages over mercury porosimetry and other porosimetric methods:

(a) *Pore size range.* This method with appropriate standard samples can be used for measurements of pore sizes in the range from 1 to 3×10^5 nm.

(b) *The accuracy* of this method depends primarily on the accuracy of measuring the structure of the standard samples by MMP, which is about 1 % of the total pore volume. The error (nonreproducibility) of MSCP is less than 1 %.

(c) *Capability to investigate all kinds of materials.* One of the main advantages of MSCP is the possibility for investigation of materials with a low mechanical strength, of frail materials and even of powders.

(d) *Possibility to measure samples at fixed levels of compression and/or temperature,* i.e., under conditions in which they are commonly used in different devices.

(e) *Possibility to use for measurements the same liquid* as that, used in real devices (i.e., leading to the same swelling degree of the sample).

(f) *Possibility of repeated measurements on the same sample.* As MSCP is a nondestructive method and measurements do not change or otherwise influence the sample's structure, this structure can be measured repeatedly at different conditions (e.g., in the case of battery electrodes at different discharge stages).

(g) *Possibility of measuring structures with corrugated pores.* As shown in [17] MSCP gives the possibility to measure the true pore size function, which is not influenced by pore corrugation. It is possible to measure the statistical distribution of the volume of the trapped (blocked) pores V_{tr} versus both their radii r_{tr} and the radii of the blocking pores (necks) r_b. For this purpose it is necessary to measure the porograms by two modes: (a) by filling of the pores with liquid, and (b) by evaporating the liquid starting from different degrees of filling.

One of the main drawbacks of MSCP is the long time required for performing measurements. Recently, an Automated Standard Porometer described in [19–21] was developed by Porotech Inc., Canada which substantially simplifies and accelerates the measurements.

1.2.3 Investigation of Wetting and Sorption Properties of Porous Materials

One of the most pronounced advantages of MSCP is the possibility to investigate the wetting (hydrophilic/hydrophobic or liophilic/liophobic) properties of porous materials. In [13, 17] it was shown that using different working liquids it is possible to measure the wetting angle of liquids with porous materials. Primarily, MSCP measures the distribution of pore volume versus the capillary pressure p^c, i.e., versus the parameter $r^* = r/(\cos \theta)$ (henceforth we call this parameter effective pore radius). For partially hydrophobic materials (for which $\theta > 0$ the porosimetric curves measured with water are shifted toward higher values of r^* in respect to the curves measured with octane that wets most materials almost ideally ($\theta = 0^0$). The value of this shift for a certain value of pore volume V_n and of the corresponding pore radius r_n allows to determine the wetting angle of water for pores with the radius i_n:

$$\cos \theta = r_n/r^*. \qquad (1.3)$$

For porous materials the wetting angles $\theta(r)$ for pores of different size can be different. In this case an average value of the wetting angle θ^* can be calculated and used.

The capillary pressure can also be represented by the thermodynamic Kelvin equation:

$$p^c = (RT/V_i)\ln(p_s/p_0) = A/V_i \qquad (1.4)$$

where V_M is the liquid's molar volume, A is the free binding energy liquid-sample, p_s, and p_0 are the values of the liquid's vapor pressure in the pore and, resp., with a plane meniscus. Therefore, measurements of porosimetric curves by MSCP allow to establish not only the dependence of the amount of liquid in the sample on the capillary pressure p^c (isotherm of capillary pressure), but also on the values of parameter p_s/p_o (sorption or desorption) isotherm) and of parameter A (energy isotherm).

For multi-component porous materials with mixed wettability that are widely used in applied electrochemistry (e.g., fuel cell electrodes containing platinum particles on carbonaceous supports along with different additives) it is possible to investigate separately the structure (pore volume distribution vs. pore size) for hydrophilic and hydrophobic (liophilic and liophobic) pores and to evaluate some important parameters such as the dependence of the fraction of the pore surface occupied by the hydrophobic (liophobic) components on the pore size.

References

1. Drake C (1949) Ind Eng Chem 41(4):780
2. Dubinin MM, Plavnik GM (1968) Carbon 6(2):183
3. Miklos S, Pohl A (1970) Bergakademie 22:97
4. Swata MJ, Jansta I (1965) Czech Chem Commun 30(7):2455
5. Gregg SJ, Sing KSW (1967) Adsorption, surface area and porosity. Academic Press, New York
6. Coalson RI, Grit WG (1972) US Pat 3,684,747
7. Bernardi DM, Verbrugge MV (1992) J Electrochem Soc 139(9):2477
8. Watkins DS (1993) Fuel cell systems. In: Blomen LMJ, Mugerva MN (eds) Plenum Press, New York, p 493
9. Schlögl R, Schuring H (1961) Z Elektrochem 10(3):863
10. Dietz P, Hansma PK, Inacker O, Lehmann HD, Herrmann KH (1992) J Membr Sci 65(1–2):101
11. Washburn EW (1921) Phys Rev 17(3):273
12. Volfkovich YM, Sosenkin VE (1978) Dokladi Akademii nauk SSSR 243(1):133 (in Russian)
13. Volfkovich YM, Shkolnikov EI (1978) Zhurnal Physicheskoy Khimii 52(1):210
14. Volfkovich YM, Shkolnikov EI (1979) Sov Electrochem 15(1):8
15. Volfkovich YM, Bagotsky VS, Sosenkin VE, Shkolnikov EI (1980) Sov Electrochem 16(11):125
16. Volfkovich YM, Bagotsky VS (1994) J Power Sources 48(3):327
17. Volfkovich YM, Bagotsky VS (2001) Colloid Surf A Physocochemical Eng Aspects 187–188:349
18. Volfkovich YM, Sosenkin VE, Bagotsky VS (2010) J Power Sources 195(17):5429
19. Volfkovich YM, Blinov IA, KulbachevskyVK, Sosenkin VE (2001) Porosimeter, US Patent 6,298,711
20. Volfkovich YM, Blinov IA, Sakar A (2006) Porosimetric device, US Patent 7,059,175
21. Web-site:http://www.Porotech.Inc

Part I
Technical Materials

Chapter 2
Paper

Yury Mironovich Volfkovich

Abstract Paper is a fibrous sheet material made from fibers based on a variety of materials: wood cellulose, synthetic polymers, mineral fibers (glass, basalt, and asbestos), and other material (wool, mica, metallic "whiskers," and graphite). The most common type of paper is writing paper. Its main feature is the capillary-porous structure that allows the absorption of inks, dyes, and graphite pencil powder. Paper (the Italian bambagia—cotton) is a fibrous sheet material. Paper with weight exceeding 250 g per m^2 is called cardboard. Distinguishing between general-purpose paper (mass and non-mass) and special, a decision was made to divide paper into a number of classes for printing (newsprint, offset, etc.), writing, typing, drawing and crayon, for paper money, for vehicles (punch-card, ticker-tape, etc.), electrical (cable, capacitor, etc.), wrapping and packaging, etc.

2.1 History

Paper is the most ancient of all humanity artificially made porous products. The Chinese Chronicle reported that paper was invented in 105 AD by Tsai Lun. However, in 1957, in the Baotsya Cave in northern China's Shansi Province, a tomb was discovered where scraps of paper sheets were found. The paper was investigated and it was found that it was made in the second century BC. Prior to Tsai Lun, paper in China was made from hemp, and before that from silk, which is made from defective silkworm cocoons. Tsai Lun pounded mulberry fibers, wood ash, cloth, and hemp. He mixed these with water and put the mass on the form (wooden frame and a screen of bamboo). After drying in the sun, he smoothed this mass with stones. The result was a solid sheet of paper. After the invention of Tsai Lun, the process of paper production began to improve rapidly. Starch, glue, natural dyes, etc., were added to enhance the strength of paper (Fig. 2.1).

At the beginning of the seventh century the papermaking method was known in Japan and Korea. And after 150 years, via prisoners of war it came to be known by the Arabs. In the sixth to eighth centuries, paper production was carried out in

Y. M. Volfkovich et al., *Structural Properties of Porous Materials and Powders Used in Different Fields of Science and Technology*, Engineering Materials and Processes, DOI: 10.1007/978-1-4471-6377-0_2, © Springer-Verlag London 2014

Fig. 2.1 Making paper by
Tsai Lun

塘 漂 竹 斬

Central Asia, Korea, Japan, and other Asian countries. In the eleventh to twelfth centuries, paper appeared in Europe, where it replaced animal parchment. Starting from the fifteenth to sixteenth centuries, after the introduction of printing, paper production grew rapidly. Paper was manufactured in a very primitive way—manual milling of mass by wooden mallets in a mortar and scooping it by the forms with a mesh bottom. Of great importance in the development of paper production was the invention of the milling machine roll during the second half of the seventeenth century. At the end of the eighteenth century rolls were allowed to produce a large number of paper pulp, but handmade casting (scooping) delayed the growth of paper production.

In 1799, N.L. Robert (France) invented the paper machine, mechanized casting of paper through the use of an infinitely moving grid. In England, the brothers G. and S. Fourdrinier, buying the Roberts patent continued to work on the mechanization of casting, and in 1806 patented a paper machine. By the middle of the nineteenth century, the paper machine had turned into a sophisticated unit that could run continuously and largely automatically. In the twentieth century, paper production became a major and highly mechanized brunch of industry with a continuous-thread processing chain having powerful thermo-electrical stations and complex chemical manufactories for the production of fibrous half-stuffs.

2.2 Production

Nowadays, paper is made from fibrous semifinished: wood cellulose, wood pulp—a product of abrasion of timber; the so-called thermo-mechanical pulp obtained by mechanical grinding (milling) of steamed wood chips; semicellulose—the product of chemical and the subsequent machining of wood; fibers of cotton, flax, hemp, and jute. Paper wastes are widely used in the manufacture of paper. Special types of paper are made from synthetic polymers, mineral fibers (glass, basalt, and asbestos), and other materials (wool, mica, metal "whiskers," and graphite).

Manufacture of paper includes a series of sequential steps: preparing pulp, paper manufacture on the paper machine, its finish processing, handing out, and packaging. Preparation of pulp is reduced to grinding, preparation of the composition, and cleaning of mass. Grinding is mechano-chemical treatment of the fibrous semifinished in water, commonly using conical and disk mills of continuous action; thus, changing the shape and size of the fibers, their swelling occurs, thin fibers—fibrils are detached from the outer surface. The composition of pulp depends on the form of paper produced. Typically, paper contains several types of semi-finished fibers, mineral fillers, sizing agents, and excipients. Thus, the composition of newsprint paper comprises 70–85 % of wood pulp and 15–30 % of wood cellulose. Then, the resulting mass is diluted and subjected to so-called screening. As a result, clusters of fibers are removed, the fibers are uniformly dispersed in water preventing the formation of fiber associates (flocculation) and chaotic sequential interlacing of fibers is provided.

Papermaking involves feeding of an aqueous suspension (dispersion) containing 0.1–1.0 % of solids in a papermaking machine, casting of the paper web in the grid part of the machine on a moving continuous grid (one or several), its pressing, drying, calendering, and coiling. In the grid part of the machine most of the water runs off and a sheet of paper is formed and sealed, passing on the grid consistently over various dewatering (suction) elements of the machine. The removed water is mainly used for diluting the pulp. In the press section of a paper machine, a paper flat is wrung out using a special cloth by several pairs of nip rolls and compacted. In the drying chamber of the machine, a paper web is pressed against the surface of the steam-heated drying cylinders by a dryer cloth. Sometimes the paper is dried on an air cushion.

The chaotic intertwining of fibers, bundles of fibrils, and individual fibrils, which are pulled together during drying to form a strong interfiber and interfibrillar links is provided in the preparation of the suspension and its dehydration. Surface finishing of the paper is going on in contact with a smooth surface of drying cylinders. The smoothness of the paper is further increased by calendering. The resulting paper is wound into a roll, sometimes gets off to further increase the smoothness (super-calendering), and then cut into sheets or rolls of predetermined size. Sometimes, during the production of paper the air is used as a dispersion medium instead of water (so-called dry method). Much of the paper is subjected to further treatment and recycling. For example, in order to improve printability, paper is subjected to so-called chalking, which is the surface coating generally containing kaolin and binder (latex, modified starch, carboxymethyl cellulose, or the like). To obtain waterproof packing, the paper surface is coated with polyethylene film, for roofing and waterproofing soft paper materials the paper is impregnated with solutions of bitumen.

2.3 The Structure, Properties, and Applications

Paper is a composite material. In addition to various fibrous reinforcing components, which create a continuous matrix, the paper may contain mineral fillers, which impart its opacity and enhancing whiteness and smoothness; as well as colorants, polymeric binders, and others. Sizing agents (rosin-based adhesive, etc.) prevent the spread of ink and China ink on the surface of the carcass paper and penetration of inks to the opposite side of the sheet. Synthetic resins, lattices, and crosslinkers provide wet strength. Common types of paper have a capillary-porous structure and are composed of fibers, bundles of fibrils, and individual fibrils, linked by hydrogen bonds, van der Waals forces, and friction. These bonds are formed during paper drying in which a vitrification of the polymer components of papermaking fibers (cellulose, hemicellulose, and lignin) goes on under considerable shrinkage stresses, which constrict fibrillar elements of the paper structure.

Hemicellulose in the production of paper can partially pass into the viscous-flow state and become vitrified during drying. This structure causes hydrophilicity of most types of paper, reduction in strength when wetting, dependence of the properties and dimensions on the relative air humidity. On the grid of the Fourdrinier machine pulp fibers are oriented predominantly in the direction of movement to a greater extent at the bottom (grid) side of the sheet and to a lesser degree on the top (front) side. Therefore, the paper is anisotropic in all directions. Anisotropy is amplified by uneven thickness distribution of fine fibers, fillers, and sizing agents. Paper and paperboard produced by multi-grid machines, as well as coated paper, such as an enameled (chalk overlay) paper have for example a multilayer structure.

A bulk weight of paper ranges from 0.40 to 1.35 g/cm3, fracture resistance—from one to tens of thousands of double folds, specific heat capacity—from 1.21 to 1.32 kJ/(kg*K). Specific volume resistivity of insulating species of an absolutely dry paper is 10–100 ohm* m, the dielectric constant—2.2–5.0. Printing and writing paper perceive printing dye, ink, Indian ink, and pencil, and possess sufficient strength and durability (the latter requirement does not apply to a newsprint paper). Wrapping paper characterized by good physical and mechanical properties: high dynamic strength (sack paper), hardness (corrugated paperboard), etc. Paper filters having predetermined capillary pore structure and high rigidity are used for purification of gases and liquids, e.g., oils and fuels in internal combustion engines. Tissue paper (toilet, sanitary napkins, diapers, paper towels, and disposable underwear) has high absorbency with sufficient mechanical and wet strength. Paper used as a carrier of information in electronic computer engineering has high mechanical strength (punched tape), flatness (punched card), and dimensional stability. Paper used as a recording object in reproduction systems for the extraction of information has "functional" coating (light-and heat-sensitive, semi-conductor paper, etc.). Sticky paper with special coatings is used for mechanized packaging and labeling, and with anti-adhesive coatings—for packing sticky materials.

2.4 Porosity

Porosity directly affects the absorbency of paper, that is, its ability to receive the ink and may well serve to characterize the structure of paper. As already mentioned, paper is a capillary-porous material with the distinction of macro and microporosity. Macropores—the space between the fibers is filled with air and moisture. The micropores or capillaries are the smallest and irregularly shaped spaces penetrating the coating layer of chulk overlay papers, and also form between the filler particles or between them and the walls of the cellulose fibers in uncoated papers. Capillaries are presented also inside cellulosic fibers. All uncoated, not too compacted papers, such as newsprint, are macroporous. The total pore volume in such papers is as high as 60 % or more and the average pore radius is about 0.16–0.18 µm. These papers absorb ink well thanks to its loose structure, which is a highly developed inner surface. Coated papers are microporous or capillary structures. They also absorb ink well but by the forces of capillary pressure. Here, the porosity is only about 30 % with a pore size less than 0.03 µm. The rest of the paper is in an intermediate position. In fact, this means that printing on offset paper leads to penetration of solvents in paints and pigments as well into the pores. Thus, the concentration of pigments on the surface is small and therefore it is impossible to achieve saturated colors. When printing on a chalk overlay paper, the pore diameter of coated layer is so small that only the solvents are absorbed into the pores, while the pigment particles remain on the paper surface. Therefore, the image is very rich.

2.5 Sorption Properties of the Paper

Absorbency is one of the most important properties of printed paper. Proper assessment of absorbency means implementation of conditions of the timely and full consolidation of paint and as a result—getting a quality print. The absorbency of paper is primarily dependent on its structure, as the interactions of paper with ink are fundamentally different processes. Before talking about the features of these interactions in certain cases, it is necessary to recall once more the main types of structures of modern printing paper. If we represent the structure of paper in the form of a scale, a macroporous paper consisting entirely of wood pulp, such as newspaper, will accommodate at one of its ends. The other end of the scale, respectively, will be occupied by a purely microporous cellulosic paper such as enameled paper. A little to the left will house a purely microporous cellulosic uncoated paper. And all the others will take the remaining gap.

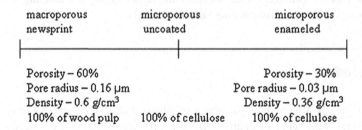

Macroporous paper is receptive to ink, absorbing it as a whole. Dyes here are low-viscous. Liquid dye quickly fills large pores, soaking in a large enough depth. Its excessive absorption can even cause "punching" of the impression, that is, the image becomes visible on the reverse side of the sheet. Improved macroporosity of a paper is undesirable, for example, in the case of an illustrative printing when excessive absorption results in a loss of saturation and gloss of paints. For microporous (capillary) papers, the mechanism of "selective absorption" is known as a characteristic one, when under the action of capillary pressure in the micropores of the surface layer of the paper, low viscous component of a dye (solvent) is absorbed predominantly and the pigment and film former remain on the paper surface. This is what is required to get a clear picture. Since the mechanism of the interaction between paper and dye in these cases is different, a variety of dyes is prepared for coated and uncoated papers.

2.6 Banknote Paper

Paper for the production of banknotes is a special, thin sheet material, consisting mainly of bonded with each other vegetable fibers. Taking into account the special characteristics of banknote paper, cotton, linen, and other fibers are used in its manufacture. Since banknotes are instruments of cash circulation, paper for them should have high resistance to wear. Banknotes must have high mechanical strength and resist repeated wrinkling, abrasion, and bending. Paper for banknotes should have good printing properties in order to receive and store the image, including the required optical properties of light color and opacity. It is also necessary to comply with certain characteristics of softness and smoothness. Very important properties are the light-fastness and durability as well as resistance to various physical and chemical influences—the necessary quality in the long process of banknotes circulation.

We applied the standard contact porosimetry method (SCPM) to study various samples of banknote paper. In Fig. 2.2 integral (a) and differential (b) porogramms which were measured using water (1) and octane (2) are presented.

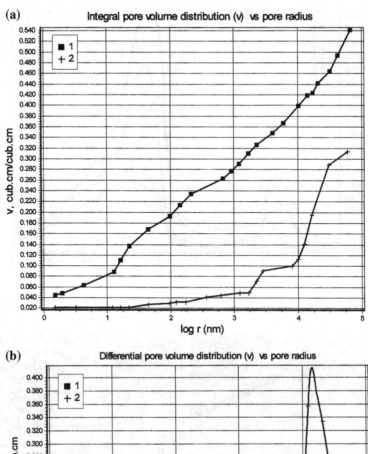

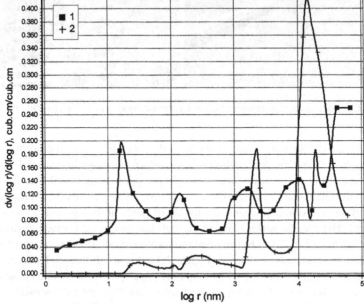

Fig. 2.2 Integral (**a**) and differential (**b**) porogramms of banknote paper measured in water *1* and octane *2*

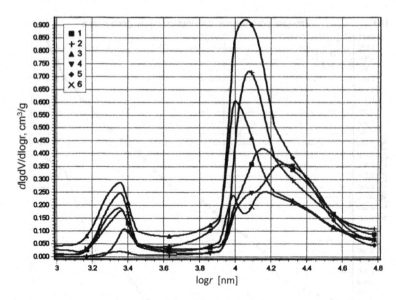

Fig. 2.3 Differential porometric curves for 6 samples of banknote paper measured in octane

Fig. 2.4 Electron microscopy images of carbon paper of Toray type—TGPH—090

As we can see, in a dry state characterized by octane curve, the porosity of a banknote is 31 % and the pore radius range extends from about 20 nm to 65 μm; the main peaks in the differential curve occur in the pores with $r \sim 2.5$ and 20 μm. When immersed in water the banknote swells strongly to porosity 54 %, the largest incremental pore volume accounts for the small pores in the range of $r < 1$ nm to $r \sim 2$ μm. The specific surface area in the wet state (46 m²/g) is much greater than in the dry state (0.7 m²/g). It follows that the sorption capacity of such paper to dyes in the wet state increases sharply. Figure 2.3 presents differential porogramms measured for six different batches of banknote paper produced by one company.

As we can see, the differences in the porous structure are very sensitive identified by SCPM. This, in particular, involves the use of SCPM in forensics.

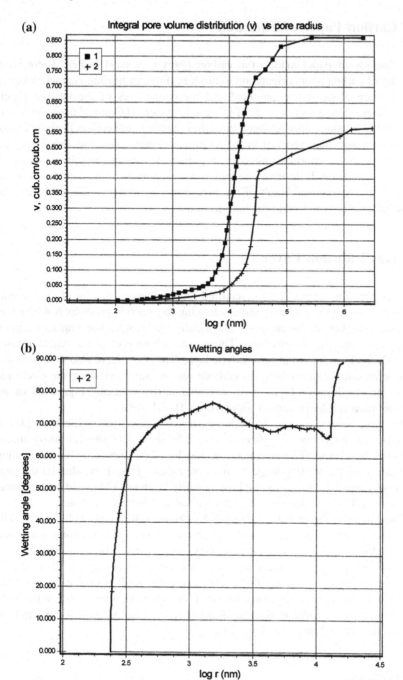

Fig. 2.5 Integral porogramms (**a**) dependence of the contact angle on pore radii (**b**) for Toray carbon paper measured in octane *1* and water *2*

2.7 Carbon Paper of Fuel Cell

One example of paper made from carbon fibers is a sheet heater. There is an ordering of carbon fibers orientation to produce different heating characteristics in the direction along the longitudinal and transverse axes of the heating paper, thereby expanding the possibility of using such paper. The aim of the invention is to provide a heating paper in which the pulp is added to the carbon fibers to obtain specific heating characteristics in the longitudinal and transverse directions, and the sheet heater comprises moreover, a polymeric coating, which serves as an electrical insulation. If desired, one may receive a variety of characteristics of the heater plate. A technical result of the invention is to improve the efficiency and reliability of the heater plate for long-term use.

2.8 Gas Diffusion Layer

The gas diffusion layer (GDL) is required for the implementation of current collection, for supplying reactants and discharging the reaction products from the fuel cell (see 2.1). Gas diffusion layers are usually made of carbon paper or carbon cloth, which are porous structures. Due to the presence of pores, reactant gases (H2, O2) freely penetrate to the catalytic layer. The pores serve also to remove the reaction products (water) from the cathode region. Since carbon is the electronic conductor, gas diffusion layer serves as both a current collector. Figure 2.4 shows electron micrographs of carbon paper Toray TGPH- 090.

We have studied porous hydrophilic structure and hydrophilic-hydrophobic properties of the Toray type paper. Figure 2.5a shows the integral porogramms, measured by octane (1) and water (2), and Fig. 2.5b shows the corresponding dependence of the wetting angle on the pore radius. This shows that (i) the total porosity is equal to 80 %, hydrophilic porosity is about 52 %, and hydrophobic porosity is 28 %; (ii) hydrophilic pores are not well wetted by water.

Consequently, the investigated paper has hydrophilic-hydrophobic pores. The gaseous components are delivered into the catalyst layer by hydrophobic pores, and water is discharged from a fuel cell by hydrophilic pores. The hydrophobic properties of Toray paper are caused by that it consists of a graphite fibers that is confirmed by spectra of X-ray fluorescence analysis. As it is known, graphite has hydrophobic properties. In many cases, if the carbon paper is not sufficiently hydrophobic, it can additionally be hydrophobized by suspension of water-repelling Teflon and carbon black (see Sect. 2.1).

References

1. http://www.bereg.net/
2. Volfkovich YM, Sosenkin VE, Bagotsky VS (2010) J Power Sources 195(17):5429

Chapter 3
Components of Power Sources/(or of Electrochemical Energetics)

Yury Mironovich Volfkovich and Vladimir Sergeevich Bagotsky

3.1 Fuel Cells and Their Components

Yury Mironovich Volfkovich and Vladimir Sergeevich Bagotsky

Abstract The operation of a fuel cell involves the flow of different gaseous and liquid components in the membrane-electrode assembly (MEA). For a better understanding of the mechanism of all processes influencing the fuel cell efficiency, for mathematical modelling of these processes, and for a possibility of their optimization, a detailed knowledge of the geometrical structure and of the wetting (hydrophobic–hydrophilic) properties of the all components of the MEA is necessary. The porous structure and wetting properties of ten different carbonaceous support materials are described as well as the modification of this structure during different preparation stages of the catalytic layer, viz. the addition of the Nafion ionomer and the deposition of the platinum catalyst. The porous structure and wetting properties of different other fuel cell components (platinum catalyst on different supports, gas-diffusion layers, homogenous and heterogeneous membranes, etc.) are also described. As a result of structure optimization, it is possible not only to improve the electrical parameters (cell voltage at a given discharge current), but also to improve the stability of these parameters under conditions of changing flooding degrees (as a result of frequent changes of discharge current, temperature, and/or conditions of water removal).

The operation of a fuel cell involves the flow of different gaseous and liquid components in the membrane-electrode assembly (MEA). Reactants must be supplied from the outside to the catalytic layer with a rate depending on the cell's discharge current and the reaction products must be removed from this layer with an analogous rate. In low temperature fuel cells operating at temperatures below $\sim 120\,°C$ proton exchange membrane fuel cells (PEMFC) and direct methanol fuel cells (DMFC) in which polymer ion conducting membranes are used as electrolyte, water is present in gaseous state as well as in liquid state. A part of the pores in these fuel cells mainly formed by metallic catalyst particles have hydrophilic properties and are easily wetted by liquid water. Another part of the pores formed by

Y. M. Volfkovich et al., *Structural Properties of Porous Materials and Powders Used in Different Fields of Science and Technology*, Engineering Materials and Processes, DOI: 10.1007/978-1-4471-6377-0_3, © Springer-Verlag London 2014

carbonaceous particles and by some additives having hydrophobic properties remain dry during fuel cell operation and thus can be used for the transport of gaseous reagents. The operational reliability and durability of this type of fuel cells depend on a proper water management connected with the choice of predetermined flow directions for vapor and liquid water in the MEA pore network [1–3].

For a better understanding of the mechanism of all processes influencing the fuel cell efficiency, for mathematical modelling of these processes, and for a possibility of their optimization, a detailed knowledge of the geometrical structure and of the wetting (hydrophobic–hydrophilic) properties of the all components of the MEA is necessary. An investigation of the porous structure of ten different carbonaceous support materials is described as well as the modification of this structure during different preparation stages of the catalytic layer, viz. the addition of the Nafion ionomer and the deposition of the platinum catalyst.

3.1.1 The Catalytic Layer

The catalytic layers of PEM fuel cell electrodes have a complex structure, which includes platinum particles deposited on carbonaceous supports, hydrophobic materials (PTFE), and an ionomer (mainly hydrated perfluorosulfonic acid introduced into the layer as a Nafion solution). The hydrophobic additive forms canals for the supply of the reacting gases (hydrogen and oxygen) to the catalyst's surface and for evacuation of the reaction product (water vapor) from this surface. The ionomer provides ionic conductivity within the catalytic layer. Hydrophilic canals (mainly in the carbonaceous support) enable an even distribution of liquid water throughout the catalytic layer. The volume ratio of carbonaceous material and ionomer in the catalytic layers must be chosen so as to attain sufficient high values both for electronic and for ionic conductivities. The structural and wetting properties of the catalyst's carbonaceous support particles influence to a great extent the properties of the catalytic layer and thus the efficiency of the fuel cell. A detailed study of these aspects began only recently.

3.1.1.1 The Porous Structure of the Catalytic Layer

The method of standard contact porosimetry (MSCP) was used for investigation of an E-TEK catalytic layer with 40 % Pt on carbon black of the Vulcan XC-72 type and 5 % ionomer (from a Nafion solution in ethanol) [4]. This investigation was not systematic enough and did not include other types of catalyst supports.

A detail investigation of the porous structure of ten different carbonaceous support materials was described as well as the modification of this structure during different preparation stages of the catalytic layer, viz. the addition of the Nafion ionomer and the deposition of the platinum catalyst [5]. The use of ten carbonaceous materials with different properties and different values of the specific surface area gave the possibility to investigate the influence of the support's pore structure on the deposition of the ionomer.

It was shown that the pore volume of these materials is formed both by intragranular (primary) mesopores with radii below 10^3 nm (mainly in the range between 0.3 and 50 nm) and by intergranular (secondary) pores with radii larger than 10^3 nm.

Upon addition of the ionomer, the total pore volume increases due to the formation of an additional intergranular porous structure. At the same time, the intragranular porosity of the granules decreases. This can be explained by the ionomer blocking the inlets of mesopores without entering into them ("gluing" the pores). This blocking diminishes the internal surface area of mesopores. For different samples of carbonaceous materials, different degrees of blocking were observed. Figure 3.1 shows integral porosimetric curves for a sample of a Siberian carbon black Sibunit 20P. Curve 1 is for the pure material and curve 2 for the same material after adding 17 % of the Nafion ionomer. It can be seen that after addition of the ionomer, the volume of pores with radii lower than 10^3 nm substantially decreases. Similar porosimetric curves were recorded for other carbon black samples, viz. for Vulcan-XC72 carbon black, a material most often used in fuel cells as catalyst support. In the case of another (nanofibrous type) carbonaceous material KVU-1, the volume decrease after adding the ionomer is not so distinctly pronounced as for Sibunit 20P and other carbon blacks. No correlation between the degree of blocking and the porous structure could be found. For this reason it was assumed that the degree of blocking depends on the surface properties of the carbonaceous material (nature and amount of surface groups).

3.1.1.2 The Wetting and Sorption Properties of the Catalytic Layer

In [6], the results of wetting properties investigations for the same carbonaceous materials as those used in [5] were reported. These properties were measured prior and after introducing the Nafion ionomer. It was found that hydrophobicity increased for most of the investigated materials after introducing the ionomer. Figure 3.2 shows for Vulcan XC-72 the dependence of the wetting angle θ on the pore radius r prior and after introducing the Nafion ionomer. It can be seen that this dependence for the pure material curve (1) is very complex (probably due to the presence of different surface groups in pores of different size). After addition of the ionomer, the wetting angle increases indicating a hydrophobizing influence of the additive. An analogous increase of hydrophobicity is also observed for some other carbon blacks. Thus, in these materials, the ionomer not only provides proton-conducting paths in the catalytic layer, but also acts as a hydrophobizing agent.

The hydrophobizing influence of the ionomer can be explained by an orientation of the sulfogroups located on the external surface of the adsorbed ionomer in the direction away from the surface of the carbonaceous material, and the fluorine-containing groups ($-CF_2-CF_2-$) pointing outwards (Fig. 3.3).

Fig. 3.1 Integral curves of pore volume distribution versus pore radius for samples of Sibunit 20P without (*1*) and with (*2*) addition of the Nafion ionomer, respectively [5]

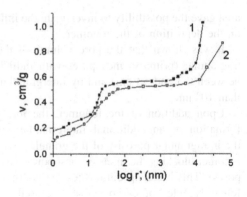

Fig. 3.2 Dependence of wetting angle θ on effective pore radius r^* for Vulcan XC-72 (*1*) prior and (*2*) after introduction of the Nafion ionomer [5]

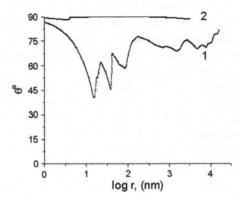

For some other carbonaceous materials, the ionomer on the contrary increases the hydrophilic properties. Thus, measurements on KVU-I show that the addition of the ionomer leads to a decrease of the wetting angle. This indicates that on these materials the sulfogroups of the adsorbed ionomer particles have an orientation in the direction opposite than mentioned above (Fig. 3.4). It seems that the orientation of the sulfogroups of adsorbed ionomer particles depends on the surface properties of the carbonaceous material. The orientation of the sulfogroups is connected with their adhesion energy to the carbon surface that is influenced by the surface properties. Thus, the surface properties of these materials influence both the degree of pore blocking and the wetting properties of the catalytic layers. A detailed knowledge of this influence is important for optimization of the properties of catalytic layers. It was shown in [7] that the nature and amount of surface groups on carbonaceous materials depend on the preparation process of these materials.

It seems that such an orientation inversion of ionogenic groups in respect to the polymer chain, which was observed in [5], is characteristic not only for the investigated system. An analogous phenomenon was observed in [7] for quite different ionogenic groups in phenol-sulfo-cationite membranes of the Polycon-type. This indicates the possibility that such inversions are quite common for ion-exchange resins. This fact is of great importance for the physical chemistry of ionites.

Fig. 3.3 Scheme of Nafion
ionomer particles on a
material of Vulcan XC-72
type [5]

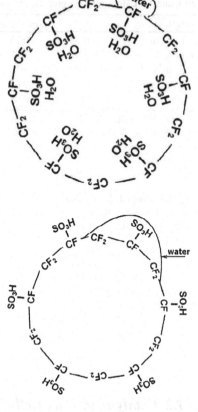

Fig. 3.4 Scheme of Nafion
ionomer particles on
carbonaceous materials of
KVU-I type [5]

This phenomenon can be regarded as a particular case of templating effects
described in [8, 9], i.e., of an influence of the support's structure on the properties of
very thin deposited layers. When making such a comparison, it must be taken into
account that such an inversion of the ionogenic groups occurs only in the surface
layer of the ionomer with a thickness of about 1 nm.

Table 3.1 shows the main integral parameters of the porous structure and
wetting properties at different preparation stages of catalytic layers containing the
carbonaceous materials Sibunit 20P and KVU-I.

Figure 3.5 shows energy isotherms (dependence of water content on the binding
energy V, log A) [calculated according to Eq. (1.4) in Chap. 1] for samples at
different production stages of the catalytic layer. These isotherms characterize the
interaction of water with the composite material. The isotherms were measured in
a range including six orders of magnitude of A. With increasing values of A, the
nature of this interaction changes: from simple filling of smaller and smaller
macropores, to capillary condensation of vapor in mesopores, then, in micropores,
to a hydration of ionogenic groups and to adsorption. Thus, these isotherms
describe in more detail the interaction of water with the catalytic layer components
than other kinds of isotherms.

Table 3.1 Main structural and wetting parameters of catalytic layers with carbon blacks Sibunit 20 P and KVU-I at different preparation stages

Samples	I	II	III	IV	V	VI
v, cm^3 cm^{-3}	0.63	0.75	0.77	0.71	0.59	0.74
S, m^2/g	228	332	324	282	95	60
S_{mea}, m^2 g^{-1}	132	99	101	73	47	38
v_{phi}, cm^3 cm^{-3}	0.47	0.65	0.74	0.63	0.56	0.76
S_{phi}/S	0.2	0.09	0.21	0.60	0.60	0.75

I Sibunit 20P, II Sibunit 20P + 17 % Nafion, III Sibunit 20P + 30 %Pt, IV Sibunit 20P + 17 % + 30 %Pt Nafion, V KVU-I, VI KVU-I + 17 % Nafion, v—total porosity, S—total specific surface area, S_{mea}—specific surface area of meso- pores, v_{phi}—hydrophilic porosity, S_{phi} —hydrophilic surface area.

Fig. 3.5 Integral distribution curves of water content versus the binding energy of water (V, log A) for the samples: (*1*) Sibunit 20P, (*2*) Sibunit 20P + 17 % Nafion, (*3*) Sibunit 20P + 30 % Pt, (*4*) Sibunit 20P + 30 % Pt + 17 % Nafion [5]

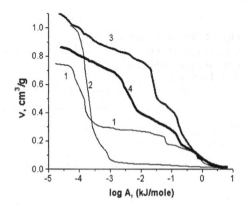

3.1.2 Catalysts in Fuel Cells

3.1.2.1 Platinum Catalysts

In [10], the structure of platinized platinum (Pt/Pt) and platinum black (Adam's platinum) was investigated. The results are shown in Fig. 3.6. The samples of platinum black were compressed under a pressure of 150 MPa. Measurements of Pt/Pt are connected with difficulties since the deposited Pt layers are very thin and the amount of working liquid in the pores of these layers is too small to be measured by analytical balances. In order to circumvent this difficulty, platinum was deposited on a platinum mesh and a stack of several discs of this mesh was used for measurements. It was found that in these Pt/Pt samples, about 50 % of the pore volume is accounted for by micropores having radii $r < 1$ nm. The radii of most of the remaining pores were not higher than 3 nm. This circumstance is very important since the size of such micropores is comparable with the thickness d of the electrical double layer and also with the size of many molecules and ions, especially organic. Both these factors influence adsorption and kinetic properties of the deposits, particularly their intrinsic catalytic activity. Such phenomena were described in [11].

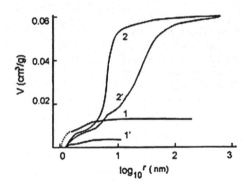

Fig. 3.6 Integral porograms for (*1, 1'*) platinized platinum and for (*2, 2'*) platinum black (*1, 2*) before and (*1', 2'*) after electrochemical sintering [10]

3.1.2.2   Platinum: PTFE Composites

In order to facilitate the transport of reacting gases to the sites on the catalyst's surface, certain amounts of hydrophobizing agents, such as PTFE, are added into the catalytic layers of fuel cells. Composites of platinum and PTFE particles are a particular case of multicomponent porous systems with mixed wettability. For a bicomponent system containing a hydrophilic component with a water wetting angle θ_1 and a hydrophobic component with a wetting angle θ_2, the average wetting angle can be found as

$$(\cos \theta)_{av} = (1 - p) \cos \theta_1 + p(\cos \theta_2) > 0, \tag{3.1}$$

where p is the fraction of the pore surface occupied by the hydrophobic component. In order to obtain information on the structural and wetting properties of such systems, it is necessary to perform porosimetric measurements with some liquids having different wetting angles with PTFE. In [12], such measurements were described in which heptane ($\theta_{PTFE} = 24°$), toluol (49°), acetophenone (73°), furfurol (80°), water (90°), and others were used as working liquids. Figure 3.7 shows differential distribution functions of the pore volume in terms of pore radii (r) and p-values, $\partial^2 V/\partial r \, \partial \log r$ for a layer containing 16 % PTFE calculated from the measured porograms. It can be seen that this function has three maxima—two for hydrophilic and one for hydrophobic pores. From these results, it can be concluded that the investigated sample contains only two types of pores—completely hydrophilic, located between platinum particles, and completely hydrophobic, located between PTFE particles. No pores with mixed wettability were recorded. No functions of such kind could be found in the literature.

In [13, 14], a macrokinetic theory was described that allows the calculation of polarization curves for hydrophobized electrodes. The main experimental parameter in equations used in this theory is the specific area S_s of the surface dividing two porous aggregates—that of platinum particles and that of PTFE particles. In [15], a method for measuring this parameter was described. For electrodes containing 4 w% platinum, the value of $S_s = 3 \times 10^4$ cm^2 cm^{-3} was found.

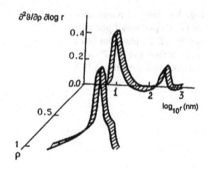

Fig. 3.7 Differential pore volume distribution versus pore radii r and versus hydrophobicity factor ρ for a platinum electrode containing 16 % PTFE [12]

Using this value of S_s and some other structural parameters measured by MSCP, the polarization curves for oxygen reduction in 30 % KOH solution were calculated. The calculated curves were in satisfactory accordance with the experimental ones. Differential pore size distribution curves of PTFE agglomerates showed that their diameter varies between 0.1 and several tens of μm, the maximal volume being connected with the range between 0.1 and 0.3 μm (the main range for non-agglomerated particles of the used PTFE suspension is between 0.05 and 0.5 μm).

3.1.2.3 Platinum Catalysts Deposited on Carbon Nanotubes

During the last decade, many investigations [16–24] revealed an increased catalytic activity of platinum deposited on carbon nanotubes CNT for the reactions of methanol oxidation and oxygen reduction.

In [25, 26], the porous structure and wetting properties of single-walled carbon nanotubes (SWCNTs) and the structure and properties of Pt and Pt/Ru deposits on them were investigated by MCSP. The SWCNTs were synthesized by an electric arc method. According to Raman spectroscopy data, they had a narrow size distribution (average diameter about 1.5 nm). Because of their high intrinsic hydrophobic properties, they were treated with a mixture of sulfuric and nitric acids in order to give them some degree of hydrophilicity (this procedure leads to the formation of hydroxyl and carbonyl surface groups). Figure 3.8 shows integral curves of pore volume distribution versus effective pore radii r^* measured with octane (1,3) and water (2,4) for the native (1,2) and (3,4) for the treated (functionalized samples denoted as $SWCNT_f$.).

In Table 3.2, the overall and the hydrophilic pore surface areas of the SWCNTs are listed, as well as the true surface area of the Pt/Ru catalyst found by measuring the amount of electric charges needed for oxidation of adsorbed carbon monoxide. The catalyst's specific surface area was found by referring the true surface area to the amount of deposited metals. For comparison, data for Vulcan XC-72 carbon black are also shown.

From the data in this table, it can be seen that the overall surface area of the treated $SWCNT_f$ measured with octane is substantially lower than that of the

Fig. 3.8 Integral porograms measured with octane (*1, 3*) and water (*2, 3*) for SWCNT (*1, 2*) and SWCNT$_f$ (*3, 4*) [26]

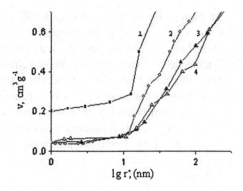

Table 3.2 Specific surface area (SPA) of carbonaceous supports and deposited Pt/Ru catalysts

I	II	III	IV	V	VI	VII	VIII	IX	X
Vulcan XC-72	205	125	80	205	125	80	100	12	49
								40	85
								76	54
								200	45
SWCNT	465	125	340	104	74	30	22	12	85
								27	85
								80	83
								230	61
SWCNT$_f$	100	52	48	121	64	57	1.21	4.1	362
								5.5	255
								6.9	166
								9.7	105
								18	138
								29.5	39
								52	37
								66	76

I support, II overall SPA measured in octane, m^2 g^{-1}, III SPA of mesopores measured in octane, m^2 g^{-1}, IV SPA of micropores measured in octane, m^2 g^{-1}, V overall SPA of hydrophilic pores, m^2 g^{-1}, VI SPA of hydrophilic mesopores, m^2 g^{-1}, VII SPA of hydrophilic micropores, m^2 g^{-1}, VIII SPA share of hydrophilic pores, %, IX amount of deposited Pt/Ru catalyst, μg cm^{-2} X SPA of deposited Pt/Ru catalyst, m^2 g^{-1}

native sample mainly as a result of a decline of the micropore volume. It can be assumed that the surface groups inside the treated SWCNT$_f$ block the access of octane into the nanotubes. BET measurements confirm the blocking effect. Furthermore, the surface groups can link adjacent tubes, which form aggregates and block parts of the external surface. The hydrophilic surface area of SWCNT$_f$ measured with water is higher than the overall surface area of the native SWCNT measured with octane. This can be explained by swelling of the nanotubes in water.

Measurements of the catalyst's specific surface area on different supports show that for very small deposit amounts (4–5 μg cm^{-2}) on SWCNT$_f$ high values (250–350 m^2 g^{-1}) of this specific area can be reached. On original SWCNT and on Vulcan XC-72 carbon black, these values are much lower.

With increasing deposition time, the crystallite size increases and the specific area decreases. For a 10 μg × cm^{-2} deposit on SWCNT$_f$, the specific surface remains 2–3 times higher than for other supports. At higher deposit amounts (70–80 μg cm^{-2}), this difference is no longer observed.

In [25, 26], it was found that the specific catalytic activity of deposited platinum catalysts (referred to a unit of the working surface) does not depend on the nature and properties of the support. But it must be noted that on SWCNT$_f$ with 5–10 μg × cm^{-2} of platinum deposits, it is possible to achieve values of the working surface up to 300 m^2 g^{-1} and thus very high overall current densities.

In [1], the investigation of a step-by-step platinum deposition on a porous titanium support was described. It was shown that during this process the main maximum on the differential porograms shifts toward lower-sized pores. This is due to platinum deposition on the surface of large pores leading to a gradual diminishing of their size.

3.1.2.4 Electron-Conducting Polymers as Catalyst Support

Some recent publications [27–30] describe the use of electron-conducting polymers (mainly polyaniline) as catalyst supports for the oxidation reactions of methanol, formaldehyde, and formic acid and mentioned the prospects of using these materials in fuel cells. The porous structure of these polymers influences the distribution of the platinum deposits and hence their properties and catalytic activity.

In [31–34], the structure of polyaniline (PAni) and polyparaphenylene was investigated. Figure 3.9 shows integral porograms for electrochemically and chemically synthesized PAni measured with water and with decane. It can be seen that in water the porosity is much higher than in decane. This increase is due to swelling of the material with the formation of a large amount of new small pores with radii from one to several tens of nm. The degree of swelling of PAni depends on the nature of the counter-ions (i.e., the nature of the emeraldine salts). According to the swelling degree, the emeraldine salts can be classified in two groups. The first group includes inorganic salts (with the anions BF_4^-, PO_4^-, Cl^-, HSO_4^-, $HCOO^-$) which are hydrophilic and have a high swelling degree. The second group includes non-swelling hydrophobic salts with $CH_3C_6H_4SO_3^-$ and CF_3COO^- anions. The porograms of the hydrophilic salts show a sharp volume increase in a narrow range of pore sizes of about 100 nm. The PAni salt structure depends to a great extent on the doping degree. Doping the emeraldine base with chloride ions increases the pore volume from 1.0 to 1.7 cm^3 g^{-1} and the specific surface area from 310 to 520 m^2 g^{-1}. It was also shown that the structure of the

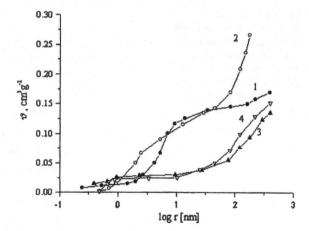

Fig. 3.9 Integral porograms for electrochemically (*1*, *3*) and chemically (*2*, *4*) synthesized polyaniline measured with water (*1*, *2*) and decane (*3*, *4*) [34]

Table 3.3 Structural parameters of emeraldine salts

Parameter	I	II	III	IV	V	VI	VII
S, m^{2}g^{-1}	590	515	490	470	430	0	0
υ, cm^3 cm^{-3}	0.05	0.45	0.55	0.60	0.40	0.06	0
R_f, nm	2.10	2.45	2.50	2.70	2.90		

I Cl$^-$, II HSO$_4^-$, III H$_2$PO$_4^-$, IV BF$_4^-$, V HCOO$^-$, VI CH$_3$C$_6$H$_4$SO$_3^-$, VII CF$_3$COO$^-$

emeraldine base prepared from different precursors (emeraldine chloride, sulfate, and phosphate) is different. This can be regarded as some sort of "memory effect."

Table 3.3 shows for emeraldine salts the values of the specific pore volumes υ, of the pore surface area S, and of the average radii R_f of single PAni fibrils. From this table, it can be seen that for hydrophilic PAni salts the specific surface area is very high—from 400 to 600 m^2 g^{-1}. It is situated mainly in the range of micro- and mesopores. These values are of the same order of magnitude as for carbonaceous catalyst supports (see Sect. 3.1.2).

Summarizing all these results, it can be said that the lability of the PAni structure and the presence of nanosized pores (from 1 to 100 nm) favor the formation of nanosized platinum particles, prevent their aggregation, and preserve a high surface area of this catalyst. However, structures with such a high lability are probably prone to changes and to degradation processes under the influence of external factors.

3.1.2.5 Catalysts for Solid Oxide Fuel Cells (SOFC)

The anodes of solid oxide fuel cells are commonly prepared from a cermet material containing highly dispersed nickel powder and a solid electrolyte ceramic material, mostly yttrium-stabilized zirconium (YSZ) that provides sufficient ionic

Fig. 3.10 Integral
porograms for: (*1*) an initial
SOFC anode, (2) the same
anode after removing all
nickel particles; curve (*3*) was
calculated by subtracting the
volumes in curve (*1*) from
those in curve [35]

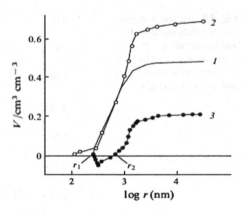

conductivity within the electrode [2, 3]. In order to enable transport of the reacting
gases to the nickel catalyst's surface sites, the anodes contain a system of interconnected pores. The electrochemical reaction of fuel oxidation proceeds at the
three-phase boundary S_s between Ni and YSZ pores and the gas phase.
For hydrophilic gas-diffusion electrodes in cells with liquid electrolytes, this
interphase can be determined from differential porosimetric curves dV/dr, r using
an equation of the type of Eq. (1.1). For SOFC anodes, this method cannot be used
since both the catalyst and the electrolyte are solid phases. In [35], a variety of
MSCP was used for measuring the value of S_s. Figure 3.10 shows porograms and
Fig. 3.11—micrographs for an anode with nickel catalyst and an electrolyte
$ZrO_2 + 8\,\%$ Y_2O_3. Figure 3.12 gives a schematic presentation of the anode's
structure.

At first the initial sample was investigated, then from this sample nickel particles
and aggregates of such particles were removed by dissolving in concentrated nitric
acid and after a thorough washing and drying the samples were investigated once
again. The curve 1 in Fig. 3.10 is a program for the initial sample and curve 2 was
measured after the mentioned procedure. In the range $r < r_2$, curve 2 shows new
pores with volume υ_1 that could not be seen in curve 1, due to dissolution of small
Ni particles formerly embedded in electrolyte aggregates (position 1 in Fig. 3.12).
These pores do not contact with the gas. Thus, their surface area is not included in
S_s. The volumes υ_2 in curve 3 were calculated by subtracting the volumes in curve 2
from those in curve 1. In the range of small pores these volumes are negative, they
correspond to pores inside Ni aggregates (position 2 in Fig. 3.2). These pores do not
contact with the electrolyte and therefore are also not included in the value of S_s.
The pores which form the surface area S_s have radii $r > r_{min}$ where r_{min} is the radius
at the minimum of curve 3. The value of S_s can be found from the program curve 3
by an expression analogous to Eq. (1.1). The values of S_s for the investigated
samples were in the range from 0.98 to 1.24 m^2 cm^{-3}.

Figure 3.11a shows micrographs for this anode. The white spots represent the
electrolyte, the grey ones—nickel particles, and the black ones—the pores.

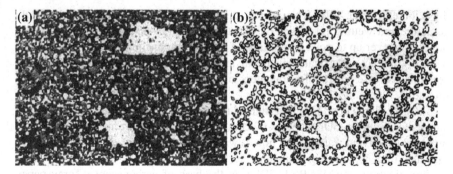

Fig. 3.11 **a** Micrograph of the initial anode; bright: electrolyte particles, *grey* nickel agglomerates, *black* pores; **b** micrograph of the same sample after nickel removal [35]

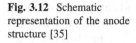

Fig. 3.12 Schematic representation of the anode structure [35]

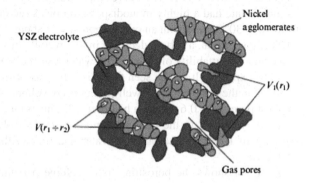

Figure 3.11b is a micrograph of this anode after removing nickel from the sample. From the micrographs, the number N of contacts on the three phase boundary was evaluated. For different samples (fresh prepared, aged, etc.), this value varied from 1.5×10^6 to 3.4×10^{10}. This value was always proportional to that of S_s and to that of the hydrogen oxidation current density. Thus, it can be stated that S_s is the electrochemically active specific surface area. From the curve 2 in Fig. 3.10, the surface area S_{Ni} of all nickel particles was found. This value is much higher than that of S_s. It follows from these data that the value of S_{Ni} has no influence on the anode performance. The values of S_s can be used for evaluating the kinetics of hydrogen oxidation.

3.1.3 The Gas-Diffusion Layer (GDL)

The porous structure and balance of wetting properties of the pores influence transport processes in GDLs and as a consequence the efficiency of fuel cells. A comprehensive review on gas-diffusion layers for PEMFCs was recently published by Cindrella et al. [36]. This review discusses in details the role of the GDL structure and wetting properties on gas and liquid transport processes within the layer. However up to now, due to lack of appropriate methods, these properties of GDLs were not investigated in detail and only few publications devoted to a quantitative assessment of parameters of these properties can be found in the literature.

In 2006, Gurau et al. [37] investigated the average values of the internal wetting angles for different GDL samples by combining the Washburn method mentioned in part 2 (involving water as test liquid) with measurements involving other liquids (hexane, toluene, acetone, methanol, etc.).

Investigations by MSCP of GDLs on the base of carbonaceous Toray paper were described in [38, 39]. Figure 3.13 shows integral distribution curves V, log r^* measured with octane and water for the pure carbonaceous paper and for the paper containing 27 w% PTFE. The curves measured with octane show that the paper in its initial state had a highly monodisperse porous structure mainly with pore radii from 10 to 30 μm and with an overall porosity of about 75 %. After treatment with a PTFE suspension, a certain decrease in the overall porosity can be observed. Due to the poorer wettability of carbon with water a shift of curves 2 to higher r^* values could be expected. It can be seen that this is the case only for the paper containing PTFE. For the pure paper, a well-pronounced volume increase in the region of smallest pores ($r < 0.6$ μm) can be observed. This is due to a swelling influence of water on the fibers in the region of small pores (with high surface area). In the presence of PTFE, the sorption of water and the swelling degree are obviously decreased.

Table 3.4 shows the porosities (pore volume per unit of sample volume) for hydrophilic pores υ_{phi}, hydrophobic pores υ_{pho}, and the overall porosity $\upsilon = \upsilon_{phi} + \upsilon_{pho}$ for Toray paper with two values of PTFE concentrations c_{PTFE}. In [38], the notion of hydrophobizing efficiency ξ was introduced which represents the ratio of hydrophobic pores (in volume percents) to the PTFE concentration (in volume percents or, as carbon and PTFE have similar density values, in weight percents). This parameter is of importance since high values of the PTFE concentration can lead to an increase of the GDL's ohmic resistance. From the table, it can be seen that the PTFE concentration does not significantly influence the volumes of both hydrophilic and hydrophobic pores. At the same time at the higher concentration, the hydrophobizing efficiency significantly declines. This is probably due to the fact that at higher concentration PTFE particles are deposited mainly at the surface of previously deposited particles, not on the carbon surface and thus do not substantially increase the hydrophobic properties.

An investigation of water wetting angles for Toray paper containing 27 % of two different hydrophobizing agents: suspensions of F-4D ($-CF_2-CF_2-)_n$ and of FEP 121A ($-CF_2-CF(CF_3-)-CF_2-)_n$ showed that the average value of the wetting angle θ^* was 65° for F-4D and 78° for FEP121A. For samples containing FEP121A, the efficiency was $\xi = 3,4$, i.e., significantly higher than that for samples containing F-4D. Thus, FEP121A containing highly hydrophobic CF_3 groups has a much stronger hydrophobizing influence on the GDL than F-4D.

Fig. 3.13 Integral
distribution curves V, log r^*
measured with octane (*curves
1, 1'*) and with water (*curves
2, 2'*) for the pure Toray
carbonaceous paper and
(*curves 1, 2,*) and for the
paper containing 27 w%
PTFE (*1', 2'*) [35]

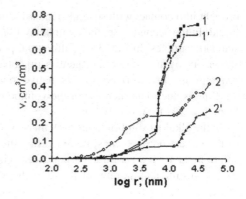

Table 3.4 Porosity of Toray paper with different amounts of PTFE

Porosity, cm³ cm⁻³	c_{PTFE}, w%		
	0	18	27
υ	0.75	0.70	0.69
υ_{hi}	0.42	0.29	0.27
υ_{pho}	0.33	0.41	0.42
ξ		2,3	1,6

3.1.4 Membranes

Different types of membranes: porous inorganic capillary membranes soaked with
electrolyte solutions or polymeric ion-exchange membranes are one of the most
important components of fuel cells [2, 3]. Membranes are also used in other fields
of applied electrochemistry such as electrolysis and electrodialysis.

MSCP is extremely well suited for investigations of membrane structures since
it is possible not only to use the same working liquid as in real fuel cells (resulting
in the same swelling degree), but also the same compression degree as that used in
fuel cell stacks.

3.1.4.1 Capillary Membranes

The structure of chrysotile asbestos, which is widely used as separator in different
electrochemical devices (alkaline fuel cells, electrolysers, etc.), was studied in [1].
As working liquids, octane, water, and a solution of 7 M KOH were used. For
alkaline (or other electrolyte) solutions the method of changing the overall amount
of liquid in the samples by its evaporation cannot be used, since during water
evaporation or condensation the solution concentration changes. A modified
method was developed in which the amount of liquid in the porous bodies is
changed by capillary soaking or drying. In the first case, the dried test sample is

brought into contact with several standard samples, filled with different amounts of liquid. In the second case, the completely filled test sample is contacted with dried standard samples. In Fig. 3.14, differential porograms obtained by this method are shown. In octane (curve 1), there is practically no swelling of asbestos and the program is characteristic for its native porous structure. In water (curve 2), swelling is due to an increase of the volume of macropores with radii in the range from 1×10^3 to 4.5×10^4 nm.

In the alkaline solution (curve 3), there is a further volume increase in the range of the largest pores. It must be noted that the formation of large pores in alkaline solutions can have serious consequences, viz. a mixing of reacting gases in the case of insufficient amount of liquid in the MEA.

3.1.4.2 Homogeneous Ion-Exchange Membranes

Ion-exchange membranes most widely used in fuel cells of the PEMFC and DMFC type are perfluorosulfonic acid (PFSA) membranes of the Nafion® type. Several versions of such membranes are known. The structure of the membranes Nafion 112, Nafion 115, Nafion117 (DuPont, USA) and the Russian-made version MF-4SK was investigated in [40] using water as working liquid. From the integral porograms shown in Fig. 3.15, it can be seen that the main part of pore volume is connected with micro- and mesopores with radii $r < 10$ nm (henceforth, this range is called nanostructure). In this range, there is no substantial difference between the curves for different versions of the membrane. It can be assumed that this nanostructure depends mainly on the chemical nature of the membrane. In the range of larger pores with $r > 500$ nm, the curves diverge. It seems that these large pores are formed due to technological factors, leading to a certain degree of surface roughness.

This assumption is corroborated by optical investigations. Measurements with octane and with water showed that these surface pores of the membranes are hydrophobic. This can be only explained by a model, according to which the sulfogroups on the membrane surface have an orientation versus the inside of the membrane. This agrees with the results of wetting angle measurements for water drops in the presence of water vapor, reported in [41] and also with the phenomenon of ionogenic group's orientation inversion with respect to the polymer chain, described in Sect. 3.1.1.1.

In [42], it was shown that any changes in the preparation method have an influence on the porous structure, for example, an increase in the ion-exchange capacity of membranes MF-4SK from 0.71 to 1.02 mg-equiv g^{-1} leads to an almost twofold porosity increase in water.

The swelling of MF-4SK membranes in different liquids was studied in [42]. It was shown that the nature of the liquid has a significant influence on the porous structure of the swollen membrane. The overall volume of the nanostructure pores in ethanol is almost twice as high as in water. This observation is of importance with respect to direct ethanol fuel cells. The swelling of ion-exchange membranes depends on the nature of the counter-ions. In membranes MF-4SK the substitution

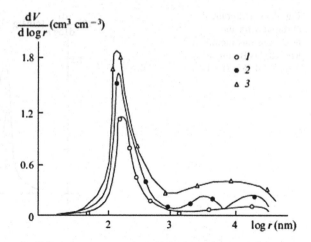

Fig. 3.14 Differential porograms for asbestos measured with different working liquids: (*1*) octane, (*2*) water, and (*3*) solution of 7 M KOH [38]

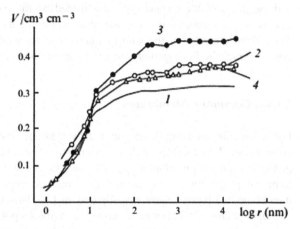

Fig. 3.15 Integral porograms for membranes: MF-4SK (*1*), Nafion 112 (*2*), Nafion 115 (*3*), and Nafion 117 (*4*) [40]

of protons by Na^+ ions lowers the porosity for about 30 %. The membranes pore volume significantly decreases upon transition from inorganic to organic counteranions.

3.1.4.3 Heterogeneous Ion-Exchange Membranes

Heterogeneous membranes contain other components in addition to the ionite. The heterogeneous cationite membrane MK-40 which is prepared by co-polymerization of styrene and divinylbenzene (DVB) was investigated in [43]. As binding agents in these membranes, polyethylene particles are used. On the differential porograms shown in Fig. 3.16, two maxima can be seen, one in the range of micro- and mesopores with radii from 1 nm up to about 100 nm, the other in the range of macropores with radii 300–3000 nm.

Fig. 3.16 Differential
porograms for the
heterogeneous membrane
MK-40. DVB content: (*1*)
2 %, (*2*) 4 %, (*3*) 8 [43]

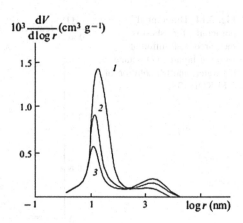

The pores of the first type are formed as a result of swelling of the ionite phase and the pores of the second type are formed by the inert polyethylene particles. The mesopores are probably formed between fragments of polyethylene and DVB. Thus, from the ratio of the sum of micro- and mesopore volumes to the overall pore volume, the homogeneity degree of membranes can be evaluated.

3.1.4.4 Composite Membranes

Polymeric fibrous composites are a new type of ion-exchange membranes. A combination of materials with a fibrous structure and those with ion-exchange properties allows to manufacture new membranes with a broad range of predetermined properties. The structural and wetting properties of such a composite membrane "Polycon" were investigated in [7]. This membrane is produced by casting a phenolsulfocationite monomer on a fibrous polyacrylonitrile matrix and a subsequent pressure application, during which exothermic heat is developed. In a swollen state, the structure of this membrane is a combination of the fibrous matrix structure and the structure of the swollen ionite. The matrix contains hydrophilic micropores, mesopores with mixed wettability, and a small amount of hydrophobic meso- and macropores. It was found that for samples of this material produced under different conditions, the surface area of the ionite pores varies from 70 to 480 m^2 g^{-1}. This, again, can be explained by a different orientation of the ionogenic groups with respect to the polymer chain.

In [44], the structure of a membrane used for phosphoric acid fuel cells (PAFC) was investigated. This membrane was prepared by pressing and sintering silicium carbide and PFTE powders. Hydrophilic pores account for about 90 % of the overall pore volume. The size of these pores varies from 0.1 to 80 μm; the main pore volume being connected with pores of size 10–50 μm.

3.1.4.5 Pore Corrugation in Membranes

In Sect. 1.2.2 of this review, it was mentioned that MSCP allows to investigate pore corrugations. In [45], such investigations were described for the homogenous membrane MA-100 and also for self-made heterogeneous membranes with carboxylic groups which were prepared by co-polymerization of methacrylic acid and DVB in a polyethylene solution acting as a binding agent. For these investigations, the integral pore size distribution was measured by MSCP for the dry membranes during octane desorption and for the swollen membranes during water adsorption and desorption. For the homogenous membranes, a considerable (up to ninefold) volume increase in the nanostructure range was observed in water. In the macrostructure range, there was practically no volume change in comparison with the dry membrane.

The most important result of these measurements is the coincidence of the adsorption and desorption curves for the swollen homogeneous membranes in the nanostructure range and a slight hysteresis in the macrostructure range. For the heterogeneous membrane, also, a complete coincidence of the curves in the nanostructure range was observed, whereas the hysteresis in the macropore range substantially increased. It follows from these results that the nanostructure pores that are formed during swelling are not corrugated whereas the macropores show a high degree of corrugation. Taking into account that pore corrugation leads to a decrease of the ionic conductivity, this result can be regarded as one of the possible explanations for the high conductivity of the homogenous Nafion membranes.

3.1.4.6 Influence of Compression and Temperature on Porous Structure

In fuel cell batteries, the membranes are compressed during compression of the whole fuel cell stack. In [45], the influence of the applied pressure on the pore structure was investigated. With increasing pressure, the volume of large pores decreases. The smallest pore size, for which this decrease can be observed, decreases with increasing pressure, i.e., small pores are less prone to deformation by pressure.

In [40], the influence of temperature on the porous structure of Nafion 117 membranes was investigated. It was shown that a temperature increase from room temperature (20 °C) to the working temperature of PEMFCs (80 °C) leads to a reduced porosity in the range of pores larger than 10 nm.

3.1.4.7 Isotherms of Capillary Pressure and Water Desorption

In [40], measurements of capillary pressure isotherms for different types of Nafion membranes were described. Figure 3.17 represents such isotherms for Nafion 117 at temperatures of 20 and 80 °C. A theoretical analysis of processes of water management in PEMFCs reported in [46] shows that capillary pressure isotherms are of great importance for optimization of these processes. In particular, high values of p^c prevent an excessive drying of the membrane near the anode at high current densities and thus improve fuel cell's discharge possibilities.

Figure 3.18 shows isotherms of water desorption, i.e., the dependence of water content V on the values of p_s/p_o for Nafion 117 membranes at temperatures of 20 and 80 °C [calculated according to Eq. (1.4) in Chap. 1]. These isotherms are of importance since, depending on mass-transport processes, the relative humidity in the gas chambers adjacent to the membranes can vary substantially. The corresponding changes in the membrane's water content influence the conductivity of the membrane and, hence, the fuel cell performance. From the figure, it can be seen that a lowering of the p_s/p_o value from 1.0 to 0.8 at 80 °C leads to a 20 % decrease of the membrane's water content that can substantially lower the performance of PEMFCs.

3.1.5 Membrane-Electrode Assemblies

In [47], the influence of flooding conditions of individual fuel cell components on efficiency and main electrical and operational parameters of fuel cells was analyzed. The membrane-electrode assemblies (MEAs) consist of several porous components (GDLs, catalytic layers of both electrodes and the membrane). All these components are soaked with liquids. In the case of PEMFCs and DMFCs, these liquids are water or aqueous methanol solutions. For PAFCs, alkaline fuel cells, and MCFCs, these liquids are phosphoric acid or alkaline solutions or, respectively, carbonate melts. Each of the porous components has its own structural, wetting, and other properties.

As shown above, MSCP allows measuring for each porous component j the isotherm of capillary pressures:

$$v_j = f(p^c), \tag{3.2}$$

where v_j is the volume of liquid in this component (flooding degree).

The overall volume of the liquid in the MEA $V = \Sigma v$ depends on the initial volume V_o and, in the case of water and aqueous solutions, on the external conditions of water removal. In the MEA, all porous components are in close contact and, thus, the liquid in all of them is in a state of thermodynamic capillary equilibrium. This implies that the values of the capillary pressure p_j^c are equalized in all porous components. The common value of p_j^c corresponds according to Eq. (3.2) to the values of v_j in the individual components. Therefore, this value depends on the overall flooding degree V:

Fig. 3.17 Isotherms of capillary pressure for the Nafion 117 membrane at temperatures: (*1*) 20 °C, (*2*) 80 °C [40]

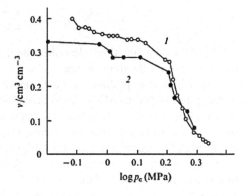

Fig. 3.18 Isotherms of water desorption for the Nafion 117 membrane at: (*1*) 20 °C, (*2*) 80 °C [40]

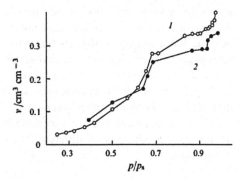

$$p_j^c = f(V). \tag{3.3}$$

During discharge of the fuel cell (current flow), the voltage losses due to each component η_j depend on its flooding degree:

$$\eta_j = f(v_j). \tag{3.4}$$

For instance, lowering of the membrane's flooding degree ("drying of the membrane") leads to a decrease in its conductivity and, thus, to an increase of ohmic voltage losses. An increase in the flooding degree of the catalytic layers hampers access of the reacting gases to the catalyst and, thus, increases losses due to concentration polarization. One of the parameters that describe the electro-chemical efficiency of a fuel cell is the cell's operational voltage U at a given discharge current density. This voltage can be represented as:

$$U = U_0 - \sum \eta_j, \tag{3.5}$$

where U_0 is the open-circuit voltage. Taking into account Eqs. (3.2–3.4), we have:

$$U = U_0 - \sum \eta_j = U_0 - \sum f(v_j) = U_0 - F(V). \qquad (3.6)$$

This "flooding equation" describes the dependence of the fuel cell voltage on the cell's overall flooding degree.

A very important problem in the design of fuel cells is that of the compatibility of the different porous MEA components. For instance, when the wetting and/or structural properties of the anode and cathode catalytic layers substantially differ there exists a possibility that one of these layers has a very low and the other layer a very high flooding degree. In this case even if the intrinsic catalytic activity of the layers is very high, the cell voltage will be low as a result of gas transportation limitations in the second of these layers. Incompatibility of the membrane and the catalytic layers exists also when in the flooding range (the overall value V of the liquid) where the polarization of the catalytic layers is minimal the flooding degree of the membrane is low. In this case not only the ohmic losses in the membrane increase, but a mixing of the reacting gases (oxygen and hydrogen) through dry pores is possible leading to a detonation danger. One of the functions of GDLs or special additional porous buffering layers is to minimize possible changes of the flooding degree of other thin components. For this reason, the thickness of GDLs is often higher than that of other layers.

In [48], a graphical method was described that allows the assessment of the flooding degree changes of different MEA components during changes of the overall value V.

3.1.6 Conclusions

It follows from the above that the distribution of the liquid's volumes among the MEA porous components is of prime importance for the fuel cell's electrical and operational parameters. For modelling processes in fuel cells and for optimizing these processes, a detailed knowledge of the structural and wetting properties of all porous MEA components is necessary. As a result of structure optimization, it is possible not only to improve the electrical parameters (cell voltage at a given discharge current), but also to improve the stability of these parameters under conditions of changing flooding degrees (as a result of frequent changes of discharge current, temperature, and/or conditions of water removal).

An essential component of most types of fuel cells is the bipolar plates that divide adjacent cells in a fuel cell stack. Problems connected with these plates are discussed in a great number of papers and reviews. One of the most important requirements for bipolar plates is a complete absence of gas permeability, i.e., the complete absence of interconnected through pores, penetrating the plates. The wettability of the plate's channels is of importance for fuel cell performance since accumulation of liquid water in them can hamper the gas supply to the GDLs.

However, in the literature no detailed investigations on the wettability of bipolar plates can be found. For these reasons, bipolar plates are not covered in this book.

3.2 Primary and Secondary Batteries

Yury Mironovich Volfkovich

Abstract The porous structure of a number of primary and secondary batteries (lithium thionyl chloride element, lead-acid batteries, alkaline cadmium–nickel and silver–zinc batteries, and lithium-ion batteries) is investigated.

Chemical power sources (CPS) are direct converters of chemical energy into electrical energy. During operation (discharge) in them, a chemical reaction of several reagents is occured. The energy of this reaction releases energy in the form of a constant electric current. CPS consists of one or more unit cells—galvanic elements. According to the principle of functionizing, galvanic cells (and CPS in a whole) are divided into groups:

(1) *Primary elements* (elements of single-use, sometimes simply called *elements*). In primary cells certain stock of reactive reagents is laid; after exhaustion of the stock (after a full charge), primary cells lose their efficiency;

(2) *Batteries* (elements of reusable, rechargeable, secondary or reversible items). After discharge the battery allows repeated charge by means of a current flow through an external circuit in the opposite direction, wherein the initial reactants are recovered from the reaction products. Most batteries authorize a large number of the charges-discharge cycles (hundreds and thousands), i.e., the total duration of their work is great, even though the process is intermittent;

(3) *Fuel cells*. In the fuel cells, new portions of all reactants are fed continuously during operation and reaction products are removed at the same time, so they can be discharged continuously for a long time.

Since a separate chapter of this book is devoted to fuel cells, in this chapter we will consider only the primary and secondary CPS.

There are a large number of not only review articles, but also monographs [49–51] devoted to CPS, so in this chapter we will describe in detail the function and patterns of multiple primary and secondary CPS only. The main attention will be paid to the specific characteristics of the porous structure and hydrophilic–hydrophobic (wetting) properties of the electrodes (and separators) of some of these CPS and their effect on the electrochemical (discharge) performance. It should be noted that there is little information on the subject in the literature. One reason for this, in our opinion, is a practical impossibility to extract valid porometric data by the most used method, namely MMP (method of mercury porometry) due to amalgamation of many metals, such as lead, cadmium, silver, etc., which are used as electrodes in CPS. In [52] to study the negative electrode of lead acid battery, MSCP was used for which the amalgamation of metals is not an obstacle.

3.2.1 Lead Acid Batteries

Since during discharge of the primary and secondary CPS, electrochemical reactions occur in electrodes with involving reagents which are not only in liquid but also in the solid phase, wherein a change in their pore structure is going on. As an example, we give the electrode reactions of charge–discharge in the lead-acid battery:

$$(+) \ PbO_2 + 3H^+ + \ HSO_4^- + 2\,e^- \underset{ch}{\overset{disch}{\rightleftarrows}} PbSO_4 + 2H_2O \qquad (3.7)$$

$$(-) \ Pb \ + \ HSO_4^- \underset{ch}{\overset{disch}{\rightleftarrows}} PbSO_4 + \ H^+ + 2\,e^- \qquad (3.8)$$

$$(cell) \ PbO_2 + Pb + \ 2HSO_4^- \underset{ch}{\overset{disch}{\rightleftarrows}} 2PbSO_4 + 2H_2O \qquad (3.9)$$

Since the density of solid reactants and products in both electrode reactions are significantly different from each other, and the volume of the electrodes is almost fixed, then the values of porosity and pore distribution curves on radii for charged and discharged conditions differ considerably from each other as well. Moreover, the porous structure not only depends on the state of charge or discharge of the electrode, but also on the mode and conditions of charge and discharge. This is clearly seen from Fig. 3.19, which shows the measured integral MSCP porograms for the same negative electrode of a lead battery after various technological and charge–discharge processes [52].

As seen, all porograms held a clear step in the $r \sim 1 \ \mu$ after formation. Further cycling saves this step, which apparently provides a stable structure and is the reason for acceptable cycling.

The effect of the formation conditions on a porous structure and on the specific capacity of the negative electrode of a lead accumulator was studied in [53]. The specific surface (by using the BET method) and porosity (by water impregnation method) were measured as characteristics of the porous structure. Conditions of the formation were changed by changing the pH. It was found that specific surface area is decreased (Fig. 3.20) and the porosity is practically not changed when increasing the pH at constant other settings.

As shown from Fig. 3.21, the specific capacitance increases with increasing the specific surface area, which takes place by lowering the pH [53].

Changes of the porous structure of the positive electrode of a lead battery during its formation and cycling were examined in [54] using mercury porosimetry and BET method. It shows a significant change of the structure during cycling due to the significant difference between the values of density of reactants and that of products of the charge–discharge process. These changes lead to a gradual decrease of electrochemical characteristics. Changing the porous structure of the electrodes of lead-acid battery, obviously, cannot be sufficiently reversible during

Fig. 3.19 Integral porograms for a lead electrode after different stages of manufacturing and cycling; (*1*) pasting, (*2*) formation, (*3*) charging, (*4*) discharge at low current density, (*5*) starter discharge at +20 °C and (*6*) −20 °C

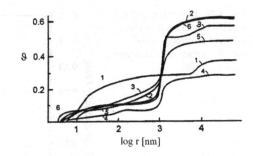

Fig. 3.20 Dependence of the specific surface of the negative electrode of a lead-acid battery on pH

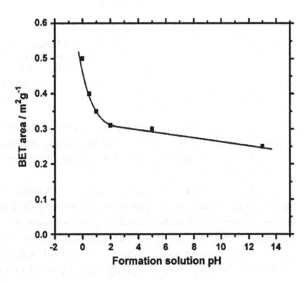

cycling. This ultimately limits a cyclability of the battery. The phenomenon of dislodging of active material of the lead-acid battery electrodes, caused by such factors, is well-known.

Among the most promising of the existing types of lead-acid batteries are sealed lead-acid batteries of VRLA Company, fitted with a valve to vent gases in the overcharging regime. These batteries contain a double-layer separator. One of these layers is a thin, finely porous polyethylene membrane, which provides sufficient strength and prevents a short circuit due to germination of active material between the electrodes. As another layer, a thicker non-woven glass separator is making use in contact with the positive electrode. Contact with the positive electrode having a high potential, does not lead to destruction of the separator, as it has anti-corrosion properties in a wide range of potential unlike polymeric membranes. Such thick separator, besides having a high porosity is essentially a buffer layer for storing a large amount of electrolyte—sulfuric acid, since, as can

Fig. 3.21 Dependence of the specific capacitance of the negative electrode on the specific surface area obtained from Fig. 3.20

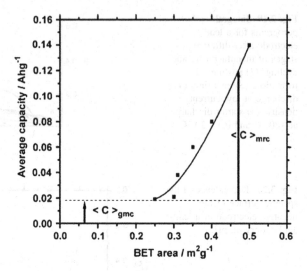

be seen from Eq. (3.21) in a charge–discharge reaction of the lead battery a large amount of sulfuric acid involves, and the battery part of the sealed type has not a free electrolyte and therefore the whole electrolyte is contained only in the pores of the electrodes and separator.

A structure of VRLA separator, comprising polyethylene layer of Daramic type and the glass separator of a non-woven glass of AGM mark, was studied in [55]. Moreover, several variants of such a separator were investigated. Specific surface area was investigated using the BET method, and the porosity was determined by measuring the weight and volume. It was found that the four types of glass separators have average pore size in the range from 0.14 to 0.22 μm, and the maximum value of the porosity ranged from 92.0 to 92.7 %. The last values are as high for separators, and meet the criteria mentioned above to their destination buffer capacity of the electrolyte. Figure 3.22 shows the differential porograms for the two versions of the AGM separator, measured by a porous diaphragm [56]. As can be seen, they have a monodisperse structure with a peak at diameter ∼6 μm.

3.2.2 Silver–Zinc Batteries

For the silver electrode of silver–zinc battery in an alkaline electrolyte, the main parameter is the efficiency of conversion of silver in AgO during the charging. This effectiveness increases with decreasing the charging current density and an increase in the discharge current density. These observations were compared with changes in the structure [52]. Figure 3.23 shows porograms of electrode made of conventional silver powder (1, 1') and the fine powder prepared by reduction of AgCl (2, 2'). Moreover, porosimetric measurements in both cases were done as

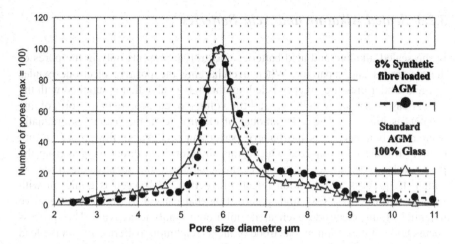

Fig. 3.22 Differential pore size distribution curves on the diameters for two versions of the separator AGM

Fig. 3.23 Integral porograms of silver electrodes prepared from (*1, 1'*) standard silver powder and (*2, 2_*) reduced AgCl measured (*1, 2*) before and (*1', 2'*) after formation)

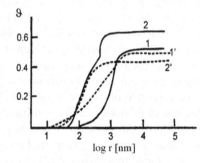

before (1, 2) and after (1', 2') formation during two cycles. For the initial electrodes, values of the specific surface area calculated from porograms were 0.20 and 2.19 m²/g, and the transformation efficiency of 77 and 47 %, respectively. After formation, the corresponding values were 0.87 and 1.14 m²/g for the specific surface area but 72 and 73 % for the efficiency of conversion. Thus, after formation, the properties of electrodes are aligned. Recrystallization during cycling leads to the loss of "memory" of the initial structure of the electrode. Increased efficiency with increasing discharge current density may be due to increased surface of silver.

3.2.3 Sealed Alkaline and Acid Batteries

In the sealed alkaline and acid batteries, electrolyte presents only in the pores of the electrodes and separator. When overcharging the positive electrode exudes oxygen. Under pressure, oxygen is led out through a valve that is supplied with the battery casing. Along with the gas, a little volume of electrolyte in the form of fine droplets is imposed. Thus, the amount of electrolyte in the battery is gradually decreasing that decreases the battery performance. Because the process is not desired, it must be minimized. Out of this situation is to direct oxygen from the positive electrode through the pores of the separator to the negative electrode, on which the process of ionization (electro-reduction) of oxygen is going on with varying rate due to a very high polarization that caused by significant negative potential of the electrode even at its minimum catalytic activity. This process occurs both on the cadmium electrode of nickel–cadmium batteries and on the lead electrode of lead-acid battery.

Problems of acceleration of the oxygen ionization process on the negative electrodes of sealed alkaline and acid batteries by using the principle of gas diffusion electrodes, which are successfully used in hydrogen–oxygen (or hydrogen–air) fuel cells are considered in [57]. To do this it is necessary to create a gas porosity in the negative electrode, whose presence forms the three-phase boundary: a liquid electrolyte–a solid electrode–gas in the pores of electrode. Electrochemical reaction of ionization of oxygen, which is delivered to the boundary through the gas pores, occurs near this boundary. In the absence of porosity, gas ionization rate of oxygen is very low ($0.01–1$ mA/cm^2), which is much less than the rate of its release at the positive electrode ($5–20$ mA/cm^2). The higher the gas porosity in the cadmium electrode of nickel–cadmium battery, the higher the ionization current, i.e., current absorption of oxygen.

It is found that even a small gas porosity (about 5 %) is sufficient to ensure the equality of ionization rates of oxygen at the negative electrode and its liberation at the positive electrode during charging hard modes (up to hour's modes). When filling with oxygen 15–20 % of the pore volume of cadmium electrodes, the ionization currents reach $150–200$ mA cm^{-2}, approaching the values of the currents on the active oxygen electrodes of fuel cells. Such high intensities of the process are determined largely by the magnitude of its polarization at the cadmium electrode (-1.2 V). Thus, under presence of the gas porosity, the negative electrode is substantially bifunctional, since when charging, except for the reaction of the cadmium deposition, reaction of the ionization of oxygen occurs also on it.

However, such an increase in the gas porosity cannot be infinite due to the increase in the ohmic losses in the separator. The combination of large pore rigid separators (such as nylon and chlorine fabrics) with small pore negative electrodes leads to the establishment of a large-jet mode of gas flow from the positive electrode to the negative electrode. In this case, the gas filling of the latter does not take place, and much of the gas is carried away to the outside in the form of "exhausts." Ionization of oxygen takes place only on the surface of cadmium electrodes, so the speed of it is small. To change this situation, it is necessary to change the flow regime in the area between the electrodes—from large-jet mode switch to small-jet or a combo regime. Such a change requires a certain choice of porous structures of separators and electrodes according to the terms of occurrence of mentioned flows. To ensure the conditions specified, two ways were used in [57]—adjusting the structure of a commercially available cadmium electrode to the structure of a suitable separator and choice of porous separator with structural parameters corresponding to the parameters of the produced negative electrode. Structures of a cadmium electrode and a separator were picked in such a way, so that their ranges of pore sizes preferably lie in the same area for both mentioned paths. At the same, high relative degrees of reduction of oxygen (95 %) have been provided.

Similar work to accelerate the ionization of oxygen during overcharging of sealed lead-acid battery by adjusting the negative lead electrode and the separator with corresponding to each other porous structures has been done in [57]. Relatively high oxygen ionization currents were achieved with small quantities of gas porosity (not greater than 10–15 %). Similar work was also done in the case of nickel–metal hydride battery. In all of these cases, for the selection of the negative electrode and the separator with porous structures corresponding to each other, MSCP was used.

3.2.4 Lithium-Ion Batteries

In the last 15–20 years, the most energy-intensive of commercially available secondary CPS—lithium-ion batteries (LIB) with capacity of 160–200 Wh/kg have been developed. These batteries are the most promising for portable equipment. In such batteries, lithium cations transferred from the positive electrode to the negative electrode during the charging. In both of the electrodes during the charging, intercalation and deintercalation of these cations occur, but during the discharging there are reversed processes. Graphite-based porous carbon electrodes are commonly used as the negative electrodes and various complex oxides are generally employed as the positive electrodes. The process of intercalation and deintercalation goes due to penetration of lithium cations into the crystal lattice of both electrodes. Examples of the electrode reactions for the system $C/LiNiCoO_2$ are shown bellow.

At the negative electrode (anode) the following reaction occurs:

$$CLi_x \leftrightarrow C + xLi^+ + xe^-, \tag{3.10}$$

where x is equal to 6 under the full capacity.

At the positive electrode (cathode) the following reaction occurs:

$$xLi^+ + xe^- + LiNiCoO_2 \leftrightarrow Li_{1+x}NiCoO_2 \qquad (3.11)$$

It is important that the processes of intercalation–deintercalation of lithium cations, having a very small size, are not accompanied by significant change in the density of reactants and products of electrode reactions, and hence significantly changing their porous structure. This is one of the main reasons for high cyclability of the batteries, unlike conventional batteries such as alkaline and acid ones (see text above).

To understand the processes and optimization of the LIB, it was necessary to create an adequate mathematical model, and the first such high-grade model was developed by Newman [58]. Newman's model was used to optimize the design of a battery [59]. Various models—from models that take into account the thermal effects and the capacity of the electrical double layer (EDL) to a model that describes a decrease of the performance of cyclic operation of the batteries—were developed afterwards on the basis of the Newman model. The approach to modeling of lithium-ion batteries is based on the theory of transport in porous media [60] developed for electrochemical systems [61] and the theory of percolation processes in the electrodes [62].

All the above models, following [58], have the geometry of the "sandwich" (Fig. 3.24). On the right side in this picture, the positive electrode—cathode (it is a source of Li^+ ions during battery charging) is shown, and on the left—a negative electrode—the anode (a source of Li^+ ions when the battery discharges) is drawn, and between them there is a neutral porous separator, which provides insulation for the electronic conductivity of the cathode and the anode.

The model shown in Fig. 3.24 is often referred to as pseudo-2D model, as in fact it has two dimensions: the dimension of the macrolevel, corresponding to the dimension of distribution of the concentration and potential along thickness of the electrodes and separator, as well as the dimension of the microlevel corresponding to the lithium intercalation in the particles of the active mass of the electrodes.

To simulate the microlevel, the porous electrodes are represented as having the structure of close-packed spherical grains of the same radius, on the surface of which there is a reaction of intercalation/deintercalation of lithium ions, controlled by the kinetics of the Butler–Volmer. The electronic conductivity of the electrodes is taken independent of the concentration of intercalated lithium, and the dependence of the equilibrium potential of the electrodes on the concentration is taken directly from the experiment at low currents.

Macro- and microlevels are compounded by using the assumption that each point on the electrode thickness corresponds to a certain microparticle of electrode and, in turn, that each microparticle of electrode is located at some point across the thickness of the electrode. This assumption implies that the electrode thickness is much greater than the diameter of the particles of which it consists. For the numerical solution of the equations describing the discharge of lithium-ion battery, a variety of approaches was used, mainly based on the method of finite differences.

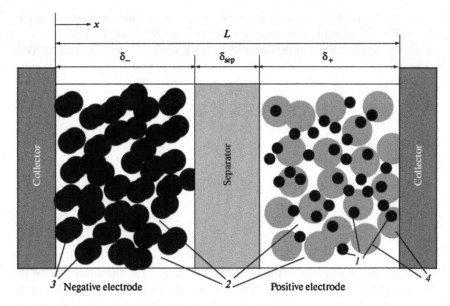

Fig. 3.24 The scheme of the lithium-ion battery

In work [63], which is discussed in more detail here, the finite element method was used for the numerical solution of the model equations. Calculated discharge–charging curves were compared with the experimental curves. By fitting the model to the experiment, kinetic parameters of reactions and coefficients of solid-state diffusion were calculated.

The fact, that in a fully discharged state (which is a result of intercalation of lithium ions in carbon) a compound of LiC_6 is formed, was taken into account in [63]. This model considered a steady state when the negative electrode has been formed a passive film. The scheme of LIB shown in Fig. 3.24 consists of three areas: composite negative electrode (LiC_6) (it includes also the particulate polymeric binder), a porous separator filled with an electrolyte, and composite positive electrode ($Li\ Ni_{0.8}\ Co_{0.2}$) (it includes also dispersed polymer binder and carbon black as a conductive additive).

During discharging the lithium cations are diffused inside the solid phase from the depth onto the surface particles of LiC_6 (deintercalation process of the negative electrode) and from there pass into the electrolyte solution holding part in the electrochemical reaction. Lithium cations by diffusion and migration mechanism then are transported from the negative electrode into the pores of the separator and then into the pores of the positive electrode. After the electrochemical reaction on the surface of the cathode material, cations of lithium are diffused into the solid phase inside the particles (the process of intercalation). The particles of anodic and cathodic materials for simplicity of calculations were considered as globules, although it is well known from the data of scanning electron microscopy that the particles of anode material have the form of scales or porous spheres, and the crystals of the cathode material have the shape of globules consist of single crystals. The following briefly describes the basic equations of the model.

It is assumed that the transport of lithium in the electrode particles is described by Fick's law:

$$\frac{\partial c}{\partial t} = \frac{D_s}{r^2} \frac{\partial}{\partial r} \left(r^2 \frac{\partial c_s}{\partial r} \right) \tag{3.12}$$

with two boundary conditions:

$$\left. \frac{\partial c_s}{\partial r} \right|_{r=0} = 0 \tag{3.13}$$

$$\left. D_s \frac{\partial c_s}{\partial r} \right|_{r=R_s} = \frac{-j^{Li}}{S_e F} \tag{3.14}$$

wherein D_s—the coefficient of the solid phase diffusion, c_s—the concentration of lithium in the solid phase, r—the current radius of the particles, $R_S = 3(1-\varepsilon_e)/S_e$—average particle radius, S_e—specific surface area, j^{Li}—volumetric flow rate of the electrochemical reaction on the porous surface ($j^{Li} > 0$ corresponds to the discharge); ε_e—porosity of the electrode, $F = 96,465$—the Faraday number.

The specific surface area is calculated from the data of the BET method, and the standard contact porosimetry method. According to the last:

$$S_e(r) = 2 \int_{r_{\min}}^{r_{\max}} \frac{\partial V}{\partial r} \frac{dr}{r} \tag{3.15}$$

where dV/dr is differential distribution function of the pore volume V on the radius r.

The transfer of lithium cations in the electrolyte is well described by the equation:

$$\frac{\partial \varepsilon_e c_e}{\partial t} = \frac{\partial}{\partial x} \left(D_e^{\text{eff}} \frac{\partial c_e}{\partial x} \right) + \frac{1 - t_+^0}{F} \left(j^{Li} + j^C \right) \tag{3.16}$$

where c_e—the effective concentration of the electrolyte, ε_e—porosity, t_+^0—lithium ion transport number in the electrolyte, D_e^{eff}—the effective diffusion coefficient of ions in the electrolyte, j^C—the current of charging of the electric double layer.

Charge transfer in the solid phase is determined by Ohm's law, which describes the potential distribution in this phase:

$$\frac{\partial}{\partial x}\left(\sigma_e^{\mathrm{eff}}\frac{\partial \varphi_s}{\partial x}\right) - (j^{\mathrm{Li}} + j^{\mathrm{C}}) = 0 \qquad (3.17)$$

where φ_s—potential of the solid phase, σ_e^{eff}—the resistance of the solid phase, I—the total current, A—the area of the battery, δ_- and и δ_+—thicknesses of the positive and negative electrodes.

The law of electrolyte transport:

$$\frac{\partial}{\partial x}\left(k^{\mathrm{eff}}\frac{\partial \varphi_e}{\partial x}\right) + \frac{\partial}{\partial x}\left(k_D^{\mathrm{eff}}\frac{\partial \ln c_e}{\partial x}\right) + (j^{\mathrm{Li}} + j^{\mathrm{C}}) = 0, \qquad (6.18)$$

(c_e—is electrolyte concentration), where the boundary conditions describes the absence of ionic conductivity on collectors:

$$\left.\frac{\partial \varphi_e}{\partial x}\right|_{x=0} = \left.\frac{\partial \varphi_e}{\partial x}\right|_{x=L} = 0 \qquad (3.19)$$

φ_e—electric potential in the electrolyte, k^{eff}—effective conductivity of the electrolyte in a porous medium, determined by the Archie relation:

$$k^{\mathrm{eff}} = k_0 \varepsilon_e^\alpha, \qquad (3.20)$$

where ε^α—attenuation coefficient of the transfer in the pores (which is due to tortuous and corrugated pores), and α—Archie exponent. Here it is assumed that $a = 1.5$.

The k_0^{eff} coefficient may be obtained from the theory of diluted electrolytes.

The kinetic equation of the electrochemical reaction at the interface is in the form of the Butler–Volmer equation (the theory of slow discharge):

$$j^{\mathrm{Li}} = S_e i_0 \left\{ \exp\left[\frac{\alpha_a F}{RT}\eta\right] - \exp\left[-\frac{\alpha_c F}{RT}\eta\right] \right\}, \qquad (3.21)$$

where α_a and α_c—the transfer coefficients of reaction, i_0—the exchange current density, η—overvoltage as the difference in electrical potential in the electrolyte and solid phase minus the thermodynamic equilibrium potential:

$$\eta = \varphi_s - \varphi_e - U(c_{s,e}), \qquad (3.22)$$

where $U(c_{s,e})$ is defined as a table function of the concentration of lithium on the surface of particles (usually it extrapolated by some polynomial—own for each material).

Fig. 3.25 Comparison of the experimental (*points*) and calculated (*lines*) battery discharge curves at different discharge currents

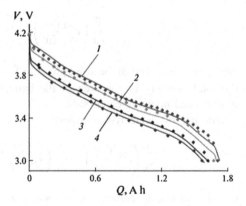

The equations for the charging of the EDL are as follows:

$$j^C = S_e C \frac{\partial(\varphi_e - \varphi_s)}{\partial t},$$
(3.23)

where C is the EDL capacity.

The simulation results and comparison of the battery discharge curves (i.e., the voltage dependence on the amount of electricity produced) with the experimental curves is shown in Fig. 3.25 [63]. The basis of the negative electrode was graphite from Taiginskoye fields in Russia, and the basis of the positive electrode was a mixture of oxides of lithium, nickel, and cobalt (Baltiiskaya Manufactura). As the electrolyte, 1 M $LiPF_6$ solution in the equimolar mixture of ethylene carbonate, dimethyl carbonate, and methylethyl carbonate was used.

As shown from Fig. 3.25, the agreement between the calculated and experimental curves is quite satisfactory, indicating about sufficient correctness of an accepted model.

One of the important phenomena that influence the LIB characteristics is an irreversible process that occurs mainly in the first cycle and leads to the formation of a passive film on the negative porous carbon electrode. The most common name of this film is SEI (solid electrolyte interface). The MSCP was applied in [64] to study the SEI structural characteristics. The formation of SEI in a carbon electrode leads to changing its porous structure. The graphite oxide was used as working electrode in this study. This material has a particular globular structure, the globules consist of a relatively close-packed graphene sheets. The material has a specific surface area of about 20 m^2/g.

Electrodes of the particulate material (graphite oxide modified) for porometric and electrochemical studies were prepared by pasting the active material (90 % of graphite and 10 % polyvinylidene fluoride as the binder, and N-methylpyrrolidone served as the solvent) on a nickel grid, followed by drying in vacuum at temperature of 120 °C. Each electrochemical cell contains a working electrode (graphite oxide), and the lithium both the auxiliary electrode and reference electrode.

Fig. 3.26 Integral
porograms for electrode made
from modified graphite oxide:
1 source electrode, *2* after the
first cycle, *3* after the fourth
cycle, *4* after the eleventh
cycle of the charge–discharge

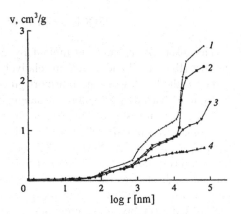

The electrolyte was 1 M of $LiClO_4$ in a mixture with propylene carbonate (PC)–
dimethoxyethane (DME) (7:3). Cells were cycled in the potential range from 0.75
to 0.01 V (with respect to the lithium electrode) at the current density of 20 mA/g.
Following cycling, the electrodes were washed in DME and dried, and afterwards
the porometric study was conducted.

Figure 3.26 shows the integral porograms of source electrode and the electrode
after the first, the fourth, and eleventh charge–discharge cycles. Analysis of the
source electrode has shown that it has the pores with characteristic size of ~ 100,
1,000 and 20,000 nm (20 μm). During the cycling, the volume of pores with
dimensions of 100 and 1,000 nm decreases due to formation of a passive film on the
surface of the graphene layers. Volume of macropores with dimensions of 20 μm
was also decreased very much and these pores are almost completely disappeared
after of 5–7 cycles. This could be due to the accumulation of large quantities of
material of passive films and the destruction of the secondary structure (destruction
of globules) without changing of the primary mesoporous structure.

According to simple calculations using porograms shown in Fig. 3.26, taking
into account the source electrode surface area and mesopore volume reduction, it
was found that the average thickness of the passive film for investigated electrodes
is 2.9–4.4 nm, which is consistent with literature data obtained by other methods.

3.2.5 Lithium Thionyl Chloride Primary Elements

Lithium thionyl chloride element has the highest energy density among the primary CPS. In some samples, it reached 600 Wh/kg. A lithium anode and cathode
from a porous carbonaceous material, usually on the base of carbon black, are used
in this CPS. As the oxidant (cathode reactant) and simultaneously solvent thionyl
chloride is exploited. Current-generating reaction during discharge can be represented by the equation.

$$2SOCl_2 + 4Li \rightarrow 4LiCl + SO_2 + S \tag{3.24}$$

The discharge capacity is limited by the precipitation of insoluble reaction product (IRP)–LiCl and sulfur (in relatively small amounts) into the pores of the carbon cathode, slowing the transport processes in the pores, and leading eventually to a sharp drop in voltage. Changing of the porous structure of particulate cathodes was investigated in [65] using the MSCP during the discharge of a lithium thionyl chloride cell. The electrolyte solutions with concentrations of salts $LiALC_{14}$ $C = 1$, and 1.35 M in $SOCl_2$ (TX) were operated. Octane was applied as the measuring liquid when making use of the MSCP. Before porosimetric measurements, discharged cathodes were thoroughly rinsed with thionyl chloride (TX) until removal of $LiALCl_4$ from the pores and then dried under vacuum. Disassembly and assembly of clamping device with the test and reference samples were carried out in a dry box, allowing aqueous vapor to avoid contact with the hygroscopic salt LiCl. It was found that one of the main features of the structure changes in the particulate cathodes during discharging is that they swelled, resulting in an increase of its thickness and area. For the different studied electrodes, the maximum degree (at the end of full charge) of swelling λ, expressed by the ratio of the volumes of the discharging and the source electrodes was strongly dependent on the elasticity of the original electrodes and their mode of discharge, and ranged from 1.06 to 2.6. The λ value increases when the current density j decreases and the discharge time increases. Changes in the structure of the cathode during discharge is a complex process of superposition of two phenomena—clogging pores with IRP and swelling of the electrode as a result of pressure on the elastic walls of the pores from the side of IRP.

Figure 3.27 shows the integral porograms for electrodes in the initial state (curve 1) and after discharge at $j = 0.5$ mA/cm^2, and $C = 1$ M of $LiALCl_4$ up to 50 % (curve 3), 75 % (curve 2), and 100 % (curve 4) of the total capacitance (110 mA h/cm^2). As can be seen, the IRP initially are formed in the largest pores and localized within the finer pores during the discharge process. This pattern is consistent with the thermodynamics of surface phenomena in disperse systems in the case of the absence of specific adsorption (or chemisorption) of IRP in respect to carbon. From the Kelvin equation the following expression for the degree of supersaturation of LiCl in the pores with different radii can be written:

$$\ln\left(\frac{c_r}{c_\infty}\right) = \frac{2\sigma \cos\theta}{rRT} \tag{3.25}$$

where C_r and C_∞ are the equilibrium concentrations of Cl$^-$ ions in the pores of radii r and $r = \infty$, σ—interfacial tension of solid crystals of LiCl in a standard electrolyte solution of TCh saturated with lithium chloride; θ—effective three-phase angle of "wetting" in the system LiCl—walls of carbon pores—the solution, $r/\cos\theta$—average radius of curvature of the sediment.

Fig. 3.27 Porograms for
particulate cathodes: *1* source
electrode, *2*, *3*, *4* cathode
after discharge to 50 % (*3*)
75 % (*2*) and 100 % of total
discharge capacitance

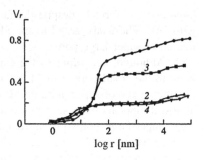

Fig. 3.28 Schematic
representation of porograms
for: *1* original and *2*
discharged electrodes

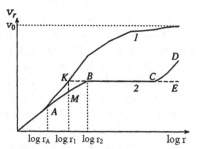

From this equation, it follows that the degree of supersaturation increases with
decreasing r, and therefore, the probability of precipitation of IRP in large pores
(wherein the degree of supersaturation is small) is higher than in the fine pores. In
the case of ideal distribution of IRP in a pore volume (i.e., according to the
thermodynamics), all large pores with the pore radii r, that exceed a certain limit
value r_1, must be filled with a sediment, while pores with $r < r_1$ must be kept free
of it. So, the program for the last pores should almost coincide with the corresponding part of the program for a source cathode, at the same time porosity V_r
should remain constant for the region of $r > r_1$, i.e., this part of the porogram
should be a straight horizontal line (segment BC of the curve 2 in Fig. 3.28).
Experimental data show that the horizontal portions in the large pore region occur
on most program for electrodes, especially discharged at low currents. For
example, in Fig. 3.27, such portion extends from $r = 40$–5,000 nm. At the same
time, the outside of this portion the IRP distribution deviates from an ideal thermodynamic one (see the segment AB of the curve 2 in Fig. 3.28). When the
current increases and the temperature drops, these deflections increase, probably
due to difficulties in kinetic diffusion processes, crystallization, etc. As a result, a
significant filling of small pores with IRP begins before the moment when a part of
the larger pores are filled with the IRP.

All integral porograms for discharged electrodes (as opposed to initial electrodes) have a "step" in the field of super-large pores with $r \sim 10^4$–10^5 nm
(section CD in Fig. 3.28 and respective portions shown in Fig. 3.27). This indicates that during the discharge and swelling of electrodes, the volume V_e for super-

large pores is formed, despite the fact that smaller pores are completely clogged with IRP. The analysis led to the identification of the following reasons of formation of super-large pores.

1. Although, according to the Eq. (3.25) the sequence of blocking from the large to the small pores should be kept, dependence of the factor of order itself, i.e., the degree of supersaturation (C_r/C_∞), on r in the region of super-large pores almost disappears because here we have $(C_r/C_\infty) \sim 1$.

2. Super-large pores play a major role in the swelling of the cathode. Their absence would make the process virtually impossible. These pores separate the largest agglomerates each from other. Clogging of the smaller pores with IRP leads to mechanical pressure of IRP on the walls of the larger pores. The result is a slight increase in the size of the agglomerates and their repulsion each from other, which leads to swelling of the cathode due to formation of super-large pores. A continuous process of swelling of the elastic cathode prevents clogging of super-large pores by the IRP sediment.

From Fig. 3.27, it is shown that in the area of small pores and low porosity, porograms for partially and fully discharged electrodes cross the source electrode program. This indicates the formation of a relatively small volume of secondary porosity in the larger pores, which are clogged by the IRP. Consequently, these pores are located between crystals of LiCl in large pores. Secondary pore formation under not very stringent conditions of discharge (not very large currents and low temperatures) is probably specified due to steric effects (a crystal form, etc.).

3.2.6 Lithium-Air Batteries

In the most recent years, researches on the development of lithium-air batteries, which are the most high-energy of all existing batteries, have been appeared [67]. Its maximum theoretical energetic density reaches up to 11,500 Wh/kg which is several orders of magnitude greater than the corresponding values for other CPS. On the positive electrode of the battery based on the fine-dispersed carbon, the reaction of ionization—release of oxygen is going on, and on lithium electrode—ionization—allocation of lithium. The overall reaction is as follows:

$$2Li_{(s)} + O_{2(g)} \underset{\text{charge}}{\overset{\text{discharge}}{\longleftrightarrow}} Li_2O_{2(s)} \qquad (3.26)$$

Oxygen in the air falls on the carbon electrode with the back of its hand through a special porous diaphragm. To prevent the entry of oxygen to the lithium electrode, the latter is covered with special protective ion-exchange film which is ion permeable to lithium cations, but impermeable to oxygen. This battery is most commonly used non-aqueous electrolyte, although recently the aqueous electrolyte

Fig. 3.29 Integral
porosimetric curves for the
initial particulate cathode (1)
and this cathode after 50 %
and 100 % of the discharge

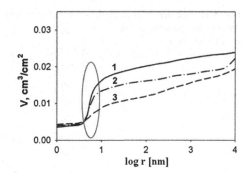

Fig. 3.30 Schematic
representation of the
localization of the IRP in
pores of different sizes

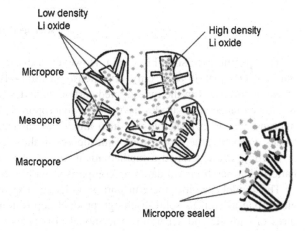

started to be used as well. In the latter case, the protective film should not be
allowed to enter the water at the lithium electrode. The product of reaction (3.26),
Li_2O_2, is poorly soluble in non-aqueous electrolyte. The insoluble residue clogs
the pores of carbon cathode, slowing the transport processes in the pores, that
ultimately leads to a drop in voltage and limits the amount of storage capacity.
Thus, the mechanism of the functioning of the cathode of lithium-air batteries is
very complex. On the one hand, it is similar to the cathode of hydrogen-air fuel
cell under the discharge, and on the other hand, (with its pores clogged by
insoluble reaction products—IRP), it is analogous to the cathode of carbon thionyl
chloride–lithium cell (see above). Therefore, the problem of optimization of the
porous structure of the cathode of Li/O_2 battery is very difficult. In [66], it was
found using MSCP that the precipitate Li_2O_2 is initially localized within the small
mesopores with $r \sim 4$ nm, and under discharge the clogging front moves into
larger pores. It follows from Fig. 3.29, which shows the integral porosimetric
curves for the initial particulate cathode (1) and for this cathode after 50 % and
100 % of the discharge.

   Schematically, the process of blocking by the IRP is illustrated in Fig. 3.30 (see
Fig. 3.29).

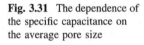

Fig. 3.31 The dependence of the specific capacitance on the average pore size

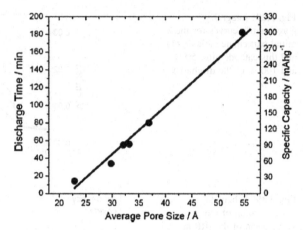

The specific pore size (Fig. 3.31). However, a disadvantage of this work is, in our opinion, that it used only porometric method—BET which is capable to measure pore sizes in the investigated carbon cathodes only up to 50 nm, while it is known that such electrodes have a large macropore volume to the size of tens of microns, which in principle can also affect the capacitance of the carbon electrodes. Probably, the proportion of mesopores in these electrodes was very small.

In [67], we developed a mathematical model of a discharge of a lithium-air battery that has been validated by comparison with experimental data.

This model describes the transport of molecular oxygen and Li cations in liquid electrolyte as well as solid discharge product deposition within the porous structures of cathode and separator. The model also takes into consideration the diffusion of reaction products, their deposition governed by the Kelvin equation, changes in porosity as the solid discharge products precipitate, and the influence of these porosities changes on transport processes.

The modeled region spans the space between the left edge of the cathode compartment, which is in direct contact with the protected anode, and the right edge of the cathode compartment, where molecular oxygen enters the cathode. The origin of the x coordinate is located at the left edge of the separator. L_S is the separator thickness, L_C is the cathode thickness, $L = L_S + L_C$ is the total thickness of the modeled region. The schematic drawing of the Li–O_2 cell is presented on Fig. 3.32. It is assumed that the electrode is fully flooded.

The mass-transport equations for the battery's porous layers can be written as follows:

$$\frac{\partial \varepsilon c_{Li}}{\partial t} = \frac{\partial}{\partial x}\left(D^{eff}_{M,Li}\frac{\partial c_{Li}}{\partial x}\right) - \frac{1 - t^0_+}{F}j^r, \tag{3.27}$$

$$\frac{\partial \varepsilon c_{O_2}}{\partial t} = \frac{\partial}{\partial x}\left(D^{eff}_{M,O_2}\frac{\partial c_{O_2}}{\partial x}\right) - \frac{j^r}{2F}, \tag{3.28}$$

Fig. 3.32 Schematic
representation of the lithium-
air battery

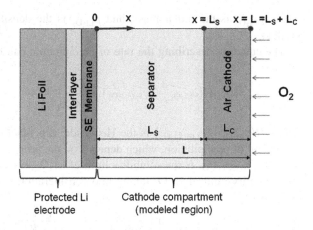

$$\frac{\partial \varepsilon c_{Li_2O_2}}{\partial t} = \frac{\partial}{\partial x}\left(D^{eff}_{M,Li_2O_2}\frac{\partial c_{Li_2O_2}}{\partial x}\right) + \frac{j^r}{2F} - J_{dep}. \tag{3.29}$$

Equations (3.27–3.29) are the effective diffusion equations for Li cations with concentration c_{Li}, dissolved molecular oxygen with concentration c_{O_2}, and the reaction product with concentration $c_{Li_2O_2}$, which take into account the volumetric reaction current j^r and volumetric rate of precipitate formation J_{dep}. Here, ε and $D^{eff}_{M,s}$ are the porosity of the medium M and the effective diffusion coefficient of species s in the porous medium M, respectively, t^0_+ is the transference number for Li cations, F is Faraday's constant.

The effective diffusion coefficient of species s in the porous medium M is defined by the Bruggeman equation through diffusion in a free solution: $D^{eff}_{M,s} = D_s \varepsilon^{\alpha_M}$, where α_M is the Archie exponent of the medium M.

The volumetric reaction current j^r is zero at the separator and is defined by the specific surface density S and Butler–Volmer kinetics at the porous electrode:

$$j^r = Si_0\left(\frac{c_{O_2}}{c_{0,O_2}}\exp\left(\frac{\alpha\eta F}{RT}\right) - \exp\left(-\frac{(1-\alpha)\eta F}{RT}\right)\right), \tag{3.30}$$

where i_0 and α are the exchange current density and the transfer coefficient, respectively, c_0, O_2 is the solubility limit of oxygen in electrolyte, $\eta = \varphi_e - \varphi_c$ is the overpotential of the reaction equal to the difference of electrolyte potential φ_e and cathode potential φ_c. Specific surface density S and porosity ε are functions of the coordinate along the thickness of the modeled region as well as of time:

$$\frac{\partial \varepsilon}{\partial t} = -\frac{\mu_{Li_2O_2}}{\rho_{Li_2O_2}}J_{dep}, \tag{3.31}$$

where $\mu_{Li_2O_2}$ is the molar mass, and $\rho_{Li_2O_2}$ is the density of the solid discharge product.

The equation describing the rate of precipitation can be written as follows:

$$J_{dep} = K_M H(c_{Li_2O_2} - c_{Lim,Li_2O_2}) * \left(\frac{c_{Li_2O_2}}{c_{0,Lim,Li_2O_2}} - \frac{c_{Lim,Li_2O_2}}{c_{0,Lim,Li_2O_2}} \right)^2, \qquad (3.32)$$

where $H(c_{Li_2O_2} - c_{Lim,Li_2O_2})$ is the Heaviside step function and K_M is the kinetic coefficient of precipitation, which depends on the properties of the porous medium M. Here, c_{0,Lim,Li_2O_2} is the solubility limit of Li_2O_2 in free electrolyte, c_{Lim,Li_2O_2} is the solubility limit of Li_2O_2 taking into consideration modified Kelvin equation.

$$\ln \left(\frac{c_{Lim,Li_2O_2}}{c_{0,Lim,Li_2O_2}} \right) = \frac{2\gamma_M V_{Li_2O_2}}{r_{min} RT}. \qquad (3.33)$$

In this equation, the effective surface tension $\gamma_M = \gamma_{E,W} - \gamma_{Li_2O_2,W}$, where $\gamma_{E,W}$ is the surface tension between the liquid electrolyte and the pore walls and $\gamma_{Li_2O_2,W}$ is the surface tension between the discharge product and the pore walls. Also $V_{Li_2O_2}$ is the molar volume of the reaction product, and r_{min} is the radius of the smallest pores filled with electrolyte; all pores with radii smaller than r_{min} are filled with the solid discharge product. For $\gamma_M < 0$, the surface tension between the discharge product and the pore walls is greater than the surface tension between the liquid electrolyte and the pore walls, and thus the precipitation of discharge product occurs in the small pores, which contribute most of the electrode surface area. In our opinion, the modified Kelvin Eq. (3.33) can be applied to the general case of precipitation as the result of various processes of the new solid phase in the pores filled with the liquid.

The charge transport equation in the electrolyte is:

$$\frac{\partial}{\partial x} \left(k_M^{eff} \frac{\partial \varphi_e}{\partial x} \right) + j''(x) = 0. \qquad (3.34)$$

As above, the properties of the porous structure are described by the model through the use of the Bruggeman equation for effective conductivity of electrolyte in the porous medium M : $k_M^{eff} = k\varepsilon^{\alpha_M}$ where the conductivity of the free solution k generally depends on concentration in a complex way.

Experimental Li–O_2 cells were built with a PLE, a porous air cathode and a zirconia ZYF-50 separator. The air cathode was prepared from an active mass containing Ketjenblack carbon black and PTFE dispersion. The air cathode and the separator were completely filled with the non-aqueous electrolyte containing 0.5 M of $LiN(CF_3SO_2)_2$ salt dissolved in DMF. The cells were assembled and disassembled in the atmosphere of dry argon and discharged in the atmosphere of dry oxygen.

Fig. 3.33 Comparison of experimental (*solid lines*) and calculated (*dashed lines*) discharge curves at two current densities (*curves 1* and *2* at 0.25 mA/cm², *curves 3* and *4* at 0.1 mA/cm²)

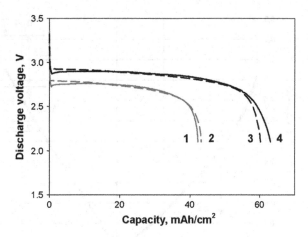

In Fig. 3.33, it was compared the experimental (curves 1 and 4) and calculated (curves 2 and 3) discharge curves at two different discharge current densities. As shown, the proposed model yields discharge curves that fit the experimental discharge data quite well. Good agreement between the discharge curves derived from the model of volumetric pore filling and the experimental discharge curves lends further support to this model.

3.3 Electrochemical Supercapacitors

Yury Mironovich Volfkovich

Abstract Electrochemical devices, in which the quasi-reversible electrochemical charge–discharge processes proceed, and the shape of the galvanostatic charge and discharge curves are close to linear, i.e., close to that of the corresponding relationships for conventional electrostatic capacitors are called electrochemical supercapacitors (ECSC). ECSC used in devices of pulse performance, as storage of electrical energy, to run their starters and regeneration of brake energy in cars, locomotives, and ships. ECSC are also employed in portable information devices (radio telephones, cameras, music players, etc.). Recently, ECSC are applied as drive power for peak smoothing of electrical loads networks.

3.3.1 Main Properties of Electric Double-Layer Capacitors

In recent years, along with the primary and secondary chemical power sources (CPS) as well as fuel cells, a new object of electrochemical energetic–electrochemical supercapacitors (ECSC) appeared. According to Conway [68, 69],

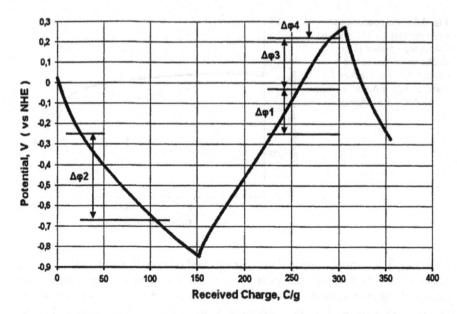

Fig. 3.34 Galvanostatic charge–discharge curves for the real electrode of the DLC

electrochemical supercapacitors are electrochemical devices in which quasi-reversible electrochemical charge–discharge processes are going on, and the shape of the galvanostatic (under direct current) charge and discharge curves (dependences of the voltage or potential on time) which is close to linear, i.e., close to that of the corresponding relationships for conventional electrostatic capacitors. ECSC are divided to double-layer capacitors (DLC), which use the energy of recharging an electric double layer (EDL) at the electrode/electrolyte interface; pseudo-capacitors (PC), which use the pseudo-capacity of fast Faraday's quasi-reversible reactions; and also hybrid capacitors (HC). Inside HC various fast processes occur at different electrodes (positive and negative). For example, recharging the EDL is available at one electrode, while some fast Faraday's reaction takes place at the other electrode.

Figure 3.34 shows the charge–discharge galvanostatic curves (i.e., dependences of a potential on the capacity of the charge or the time of charge and discharge under direct current) for one of the real electrodes of DLC.

For a more detailed electrochemical analysis of the ECSC electrodes, as well as other electrochemical systems cyclic voltammetric (CVA) or potentiodynamic curves are measured, i.e., a current versus a potential at a given speed of the potential sweep (V/s). Figure 3.35 shows the schematic CVA dependences. The top picture exhibits such dependences for an electrode of ideal DLC, that there is only charging the EDL in which. These dependencies are straight lines parallel to the x-axis. In the lower diagram, the schematic CVA are shown for the PC electrode with the pseudo and EDL capacitances.

Fig. 3.35 Schematic CVA dependences for an electrode of the ideal DLC, in which the charging of the EDL (*upper panel*) occurs and for electrode of the PC with a pseudo-capacity (*bottom*)

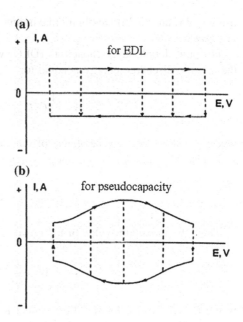

ECSC are used in pulsed devices and sinusoidal currents techniques—as a reactive element having a variable resistance to current practically without losses of energy, as a drive power.

For DLC, as well as for all the capacitors, the capacitance is inversely proportional to the thickness of the electrode:

$$C_s = \frac{\varepsilon}{4\pi d},$$
(3.35)

where ε—the dielectric constant, d—the thickness of the capacitor plates. For example, the classic paper capacitors have the paper plates of capacitors disposed between the electrodes. Its thickness is several tens of microns, and therefore, such capacitors have a low value of the specific capacitance. In DSC electrode, the thickness of the sheath is the thickness of EDL, which is of the order of tenths of nanometers. As a result, the specific capacitance C_s per unit of the true surface of the "electrode/liquid electrolyte" interface for many orders of magnitude is greater than that of conventional capacitors—$C_s \sim 10^{-5}$ F/cm^2. So, the specific capacitance per gram of electrode is:

$$C_g = S \times C_s$$
(3.36)

where S—the specific surface area (cm^2/g). To achieve a high capacity in DLC, electrodes having a high specific surface area $S = 1{,}000–3{,}000$ m^2/g based on highly dispersed carbon materials (HDCM), activated carbons (AC), aerogels, carbon black, carbon nanotubes, nanofibers, graphenes, etc., are applied As a result using (3.35) and (3.36) we obtain maximum values of $C_g = 100/300$ F/g for DLC electrodes.

For pure double-layer capacitors (DLC) with perfectly polarizable electrodes, the specific discharge energy is equal to:

$$A = \frac{1}{2} C \left[U_{max}^2 - U_{min}^2 \right] \tag{3.37}$$

where C—the average capacitance of the electrodes, U_{max} and U_{min}—beginning and finishing values of the discharge voltage. If $U_{min} = 0$, then:

$$A = A_{max} = \frac{1}{2} C U_{max}^2. \tag{3.38}$$

Thus, the maximum value of the power density is:

$$W = \frac{U_{max}^2}{4 R_s} \tag{3.39}$$

where R_s—resistance for serial equivalent RC—circuit.

The difference $(U_{max} - U_{min})$ is called "a window" of potentials. The wider the window, the higher the value of specific energy and specific power of DLC.

Using of non-aqueous electrolytes in ECSC with electrodes on the basis of fine carbonaceous materials ensures high (3–3.5 V) values of the window of potentials that significantly increases the energy, but limits the power of capacitors decrease due to the low conductivity of electrolytes. Aqueous solutions of KOH and H_2SO_4 with concentrations from 30 to 40 wt.% allow to reach sufficiently high capacity due to the high conductivity, while a low range of operating voltages (~ 1 V) reduces energy characteristics of ECSC.

Unlike batteries, ECSC can function in a very wide range of charge–discharge times: from fractions of seconds to hours. Accordingly, ECSC are classified into two main types—power with high specific power and energetic having a high density of energy. Power ECS capacitors include double-layer capacitors. Power supercapacitors allow the charge and discharge processes in a very short time (from a fraction of a second to minutes) and getting the high power characteristics from one to tens of kW/kg in concentrated aqueous electrolytes. Measurements for superfine carbon electrodes under modes of energetic capacitors typically provide the specific capacitance values ranging from 50 to 200 F/g of [70]. For carbon materials, the limit capacity of 320 F/g is achievable due to the substantial contribution of pseudo-capacity of quasi—reversible redox—reactions of surface groups of coals [71]. However, this is not a DLC in a pure form.

A very wide range of characteristic times of charge–discharge is illustrated in Fig. 3.36, which contains the Regone diagrams for any rechargeable electrochemical devices. The diagrams constitute zones of operation of these devices in the following coordinates: specific power–specific energy. This illustration shows

Fig. 3.36 Regone diagrams for various electrochemical rechargeable devices

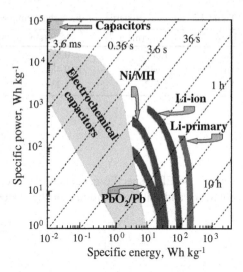

the different types of batteries (lead-acid, nickel–metal hydride and lithium-ion) as well as electrochemical supercapacitors. As one can seen, the range of ECSC operation extends by 7 orders of the characteristic time magnitude that many orders of magnitude greater than any type of battery. ECSC have own "ecological niche." As compared with the batteries ECSC have lower values of specific energy, but much larger quantities of specific power. Furthermore, they have a magnitude of cyclability more of many orders than that of batteries. The cyclability of some types of ECSC (especially DLC) reaches hundreds of thousands or even millions of cycles, whereas the battery—only from a few hundred to a few thousand cycles.

Macrokinetics of DLC was considered in [72]. If we assume that the porous electrode of DLC is ideally polarizable, the basic processes occurring during operation DLC, in general, will: non-stationary process of the EDL charging at the developed electrode/electrolyte interface in the pores, the diffusion–migration transport of ions in the electrolyte contained in the pores, and intraohmic energy losses. The equation for the potential distribution across the thickness of the electrode E, i.e., coordinate x, is followed from the theory of porous electrodes:

$$\kappa \frac{\partial^2 E}{\partial x^2} = Si(E), \tag{3.40}$$

wherein κ is effective conductivity of the electrolyte in the pores, S—specific surface area of the electrode [cm^2/cm^3], $i(E)$—the dependence of the local current density i at the electrode/electrolyte interface of the potential. The current density i is determined by the charging of EDL:

Table 3.5 Physical properties of the carbon samples

Carbon	S_{BET} (m^2 g^{-1})	V_{mic}^b (cm^2 g^{-1})	d^c (nm)	Capacitance (F g^{-1})
T01	1,005	0.51	0.70	126
T61	935	0.48	0.71	142
T63	2,312	1.23	0.63	113

$$i = C_s \frac{\partial E}{\partial t} \qquad (3.41)$$

where C_s—EDL specific capacitance per unit of the surface area, t—time. The boundary and initial conditions for the system of Eqs. (3.40) and (3.41) are:

$$I = \kappa \frac{\partial E}{\partial x}\bigg|_{x=L} = Si(E)_{x=L}; \quad \frac{\partial E}{\partial x}\bigg|_{x=L} = 0, E\big|_{(x,t=0)} = E_0 \qquad (3.42)$$

where I—the current density per unit of apparent surface of an electrode.

The boundary conditions correspond to the galvanostatic mode of operation of a porous electrode, having a flat shape, with known value of the potential $E = E_0$ at the initial time $= 0$. Here, L—thickness of the electrode; $x = L$—front side of an electrode facing to the counter-electrode; $x = 0$—its rear ("sealed") side. The system of equations and boundary conditions (3.40)–(3.42) has an analytical solution.

In [72], we compared the calculated and experimental discharging and charging curves for a symmetric DLC with electrodes based on activated carbon fabric ACF–600 (activated carbon fabric) with specific surface area $S = 600$, and got their coincidence that says about the correctness of the model adopted.

It was found in [73] for a number of activated fibers based on carbonized polyacrylonitrile (PAN), that if the specific surface area increases then specific capacity decreases. This occurs only if there is a reduction in the average size of the micropores and relatively large molecules and ions are not placed in the micropores of small size. Table 3.5, where the volume (V_{mic}^b) of micropores, average diameter (d^c) of micropores and specific capacitance are shown, demonstrates this.

In that work, measurements of a capacitance were made in an aqueous solution of KOH, but said conclusion mentioned above, to an even greater extent, relates to an organic electrolyte with large molecular size and ions.

Formation of EDL on the surface of the carbon electrode in the EDLC is the main mechanism. It has been found in a large number of studies that a big surface area is preferred for high-capacity EDLC, but a definite link between the observed capacities and S_{BET} never been received. It was noted that the actual determination of the real surface is very important for understanding the performance of the carbon electrode EDLC and it should not be limited with S_{BET}. So, it is recommended to use different techniques: analysis of an adsorption/desorption isotherms

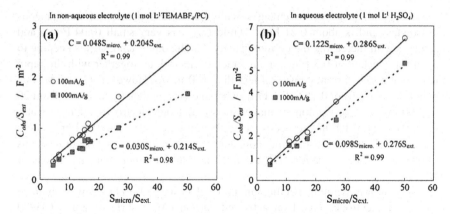

Fig. 3.37 Dependences of the ratio C_{obs}/C_{ext} on the fraction S_{micro}/S_{ext} for any AC for non-aqueous (**a**) and aqueous (**b**) electrolyte and the current densities of 100 and 1,000 mA/g [7]

for different procedures, as well as measurements of the enthalpy of immersion into liquid.

Observed capacity C_{obs} (F/g) was analyzed and divided into two parts: the capacity associated with the surface S_{micro} of micropores and the capacitance associated with the surface of larger pores (macropores and mesopores), which was called the outer surface S_{ext} [74]:

$$C_{obs} = C_{ext} \times S_{ext} + C_{micro} \times S_{micro} \tag{3.43}$$

where C_{micro} and C_{ext} are the capacities given by 1 m^2 of micropores and large pores, respectively. It is important to note that S_{ext} is determined primarily by mesopores and this quantity may be replaced by S_{meso}. Equation (3.43) can be rewritten as:

$$\frac{C_{obs}}{S_{ext}} = C_{ext} + C_{micro}\left(\frac{S_{micro}}{S_{ext}}\right) \tag{3.44}$$

This implies a linear relationship between the parameters (C_{obs}/S_{ext}) and (S_{micro}/S_{ext}), both of which are determined experimentally. Figure 3.37 shows appropriate parts for various activated carbons, covering a wide range of S_{BET}, S_{micro}, and S_{ext}. The capacity was measured at the same carbons in aqueous and non-aqueous electrolytes (1 mol/l of H$_2$SO$_4$ and 1 mol/l TEMABF$_4$/PC, respectively) at different current densities between 100 and 1,000 mA/g.

Relations between C_{obs}/S_{ext} and S_{micro}/S_{ext} are well approximated by using the least squares method to a linear dependence in both aqueous and non-aqueous electrolytes with different current densities, as it is shown for two current densities in Fig. 3.37. The results show that contribution of the micropores and the larger pores to the measured capacitances, i.e., C_{micro} and C_{ext}, is different in aqueous and non-aqueous electrolytes.

Value of C_{ext} shows no changes with the current density in non-aqueous electrolytes, and is about 0.21 F/m^2, while C_{micro} is very small (0.04 F/m^2) and decreases with increasing the current density. On the other hand, C_{ext} is keeping to be almost constant at 0.3 F/m^2, in an aqueous electrolyte, together with the rapid increase at a low current density of less than 100 mA/g. However, C_{micro} shows a gradual decrease from 0.15 to 0.1 F/m^2 when the current density increases from 20 to 1,000 mA/g. C_{micro} is much smaller in non-aqueous electrolytes than in water (about 0.04 and 0.1 F/m^2, respectively), although C_{ext} ($\sim$0.2 and 0.3 F/m^2) is not so different. These results become more apparent upon the fact that the dimensions of aqueous electrolyte cations H$^+$ are much less than the non-aqueous cations of TEMA$^+$.

A very high maximum specific amount of electricity (1,560 C/g) was obtained in [75] for electrodes based on activated carbon (AC) fabric of grade CH900 having the specific surface area of 1,500 m^2/g. This value has been achieved in the process of deep cathodic charging of AC to high negative potentials from -0.3 to 0.8 r.h.e. (relative hydrogen electrode). It was assumed on the basis of these data and other experiments that such a high amount of electricity is obtained by implementing the process of intercalation of hydrogen to carbon of AC, which is limited by solid phase diffusion. It was concluded in the current work that the most probable is the maximum amount of electricity corresponding to the formula C$_6$H, which is similar to the formula C$_6$Li for negative carbon electrodes of lithium-ion batteries.

3.3.2 Nanoporous Carbons Obtained by Different Methods

3.3.2.1 Activated Carbon

Activated carbons (AC) are one of the most common electrode materials for ECSC. Typically activated carbons are obtained during carbonization and subsequent activation of a variety of natural and synthetic carbonaceous materials. Vegetable and animal raw materials (wood, sugar, coconut, nut shells, fruit pits, coffee, bone, etc.), mineral materials (peat, coal, pitch, tar, and chark), synthetic resins, and polymers may be used to prepare the AC. Decomposition of starting materials (precursors) and the removal of non-carbon elements are going on during carbonization. Available carbon atoms form elementary crystallites, chains or amorphous carbon. Oxidation, burning of unorganized carbons and elementary crystallites, and formation of developed porous structure of AC particles hold in the process of gas activation by aqueous steam, carbon dioxide, and oxygen at temperatures of 500–900C. Depending on technology of production, specific surface area of activated carbons measured by BET is in a very wide range of about 500–3,000 m^2/g. Activated carbons comprise three types of pores: micropores, mesopores, and macropores. The largest contribution to the specific surface area gives firstly micropores, and secondary—mesopores. Simultaneously

Fig. 3.38 Voltage-capacitance cyclic curves ACF-900. Sweep rates of potentials are 0.5 (*curve 1*), 1.0 (*2*), and 2.0 mV/s (*3, 4*) [8]

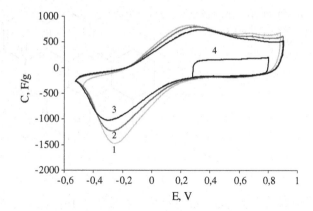

with the change of the porous structure in the carbon materials during their processing, the formation of chemical structure of the surface (two-dimensional) and changing of electrical characteristics is occurred. The chemical composition of the surface of the AC, as well as their electrical properties and pore structure, has an important influence on the electrophysical and capacitive properties of ECSC. The values of specific capacity due to charging of the EDL are usually in the range of 50–200 F/g.

Porous carbons with high capacity are obtained simply by pyrolysis without activation from some samples of raw. The AC which was originated from poly(vinylidene) chloride (PVDC) at 700 °C showed a high specific capacitance (C_g) under different current densities of 1–100 mA/cm^2. Carbons made from seaweed (sodium alginate) with carbonization at 600 °C showed a high value of C_g and excellent stability during cycling. Five porous carbon materials were obtained from various precursors (peach pits, furfural resin, saran, etc.) with or without activation. The AC with $S_{BET} = 2,100$ and $2,700$ m^2/g almost stored the capacity (reduction in capacity was only 10 and 25 %, respectively) at low temperature to −40 °C that significantly less than reduction in capacity of other coals with a lower S_{BET}.

Electrochemical properties of electrodes on the base of activated carbon ADH and activated carbonized fabrics CH900 and TSA with specific surface area of 1,000–1,550 m^2/g in concentrated H$_2$SO$_4$ solutions in a wide potential range between −1 and +1 r.h.e. were investigated in [75]. Figure 3.38 compares the cyclic voltammetric curves measured in two segments of potential: in the field of reversibility (from 0.2 to 0.8 V— curve 4), and in-depth charging field (from −0.8 to 1—curves 1, 2, 3).

Charging of the EDL and fast redox reactions of surface groups occur within the reversibility segment (curve 4). The Faraday processes with the very large pseudo-capacity are observed in the region of negative potentials (<-0.1 V). It is assumed that the hydrogen chemisorption processes and electrochemical intercalation of hydrogen to carbon are going on. Intercalation process is controlled by slow solid state diffusion of hydrogen that may explain the very high maximum

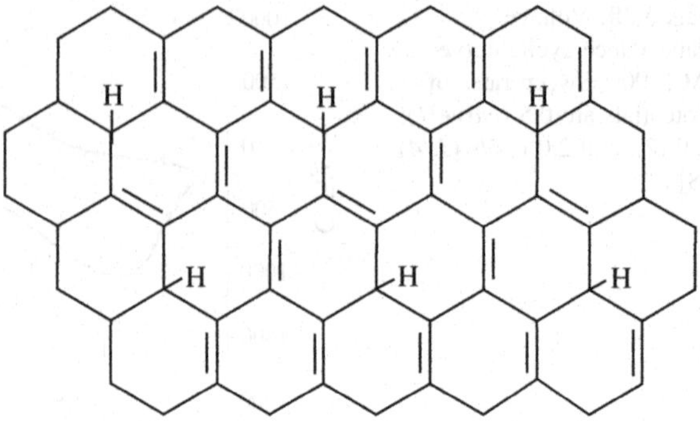

Fig. 3.39 The structural
formula of C_6H composition
[75]

time of charging in dozens of hours. The maximum total specific charge $Q_{max} = 1,560$ C/g, which corresponds to the maximum specific capacitance of 1,110 F/g, was obtained in the literature at the first time. Based on the obtained value of Q_{max} and Faraday's law, it was suggested that in the limiting case of deep cathodic charging of the AC, the combination C_6H is formed (Fig. 3.39), which is similar to C_6Li for carbon negative electrodes of lithium batteries.

It was also shown that increasing the AC charging time, a substantially increase (order of magnitude) of the electrical resistivity of the electrode takes place. These results can be explained by changes in the chemical composition of the bulk phase during the charging, since electro-conductivity of C_6H is substantially lower than that of C_6.

Two-dimensional mathematical model for the charge–discharge of highly carbon electrode was developed in [75] on the basis of experimental data, taking into account the diffusion–migration ion transport in the pores, charging of the EDL, kinetics of the electrochemical reaction, kinetics of adsorption, intercalation of hydrogen to carbon, solid state diffusion, distribution of polarization on the electrode thickness and characteristics of its porous structure. Comparison between the calculated and experimental discharging curves showed their satisfactory agreement, indicating the correctness of the model adopted.

3.3.2.2 Activated Carbon Fibers and Fabrics

Carbonaceous fibers and fabrics, for example, polyacrylonitrile (PAN), rayon, nylon, silk are used as the raw materials for producing of activated carbon fibers (ACF). Despite the ACF have many properties in common with the AC, they also have the clear differences. Namely, micropores are directed in a straight line to the surface of the fibers in the ACF meanwhile they are formed on the walls of macropores and mesopores in the AC. Thus, one can expect the rapid processes of adsorption and desorption of the adsorbent molecules onto the ACF. Figure 3.40 shows the dependence of specific capacitance for the various activated carbons (AC) and the activated carbon fibers on the specific surface area measured by the BET method for the non-aqueous solution of $LiClO_4$ (Fig. 3.40a) and aqueous

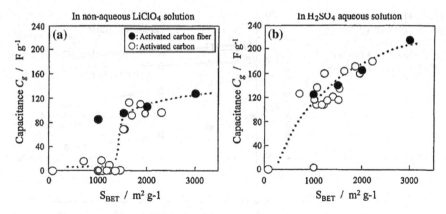

Fig. 3.40 Dependences of the specific capacity for various AC and ACF on the specific surface for a non-aqueous solution of $LiClO_4$ in propylene carbonate (Fig. 3.40a) and an aqueous solution of H_2SO_4 (Fig. 3.40b) [3]

solution of H_2SO_4 (Fig. 3.40b). These materials were obtained from various precursors (peach pits, furfural resin, saran, etc.) with or without activation.

As can be seen, although there was some variation of characteristics, increasing the specific surface area increases the specific capacitance, but the average level of the specific surface area achieved in an aqueous solution is significantly higher than in non-aqueous solution. It has been shown that the fine-dispersed carbon materials (FDCM) with larger pores should be better suited for high power capacitors due to its ability to discharge at high speeds. Reactivation of commercially available AC has been also undertaken to improve the performance of EDLC, and the positive effects of this procedure was reported in [70].

It is shown from Fig. 3.40 that ACF can provide relatively high capacitance for both electrolytes, assuming that there are differences in porous structures between granulated AC and fibrous ACF. EDLC capacity was analyzed according to the ratio between the size of solvated ions and the pore size in carbon electrodes of the AC and ACF types. Mesoporosity has been formed in the microporous ACF by simultaneous carbonization and activation of a pair consisting of phenolic resin of novolac-type and Ni-complex [76]. Adding of Ni markedly increased V_{meso} from 0.16 ml/g (without Ni) to 0.86 ml/g, which led to a large C_g.

Besides the usual ACF, so-called nanofibers have been developed in recent years, and the properties of the carbon nanofibers were investigated in relation to the ECSC. Carbon nanofibers, most of which have been synthesized by the catalytic vapor diffusion (CVD) method, often called as multiwalled carbon nanotubes (MWCNT). Nevertheless, they are different in structure and should be differentiated from carbon nanotubes, including one-, two-, and multiwalled. Carbon nanofibers can be activated by conventional methods using water vapor and air in order to increase their surface area.

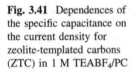

Fig. 3.41 Dependences of the specific capacitance on the current density for zeolite-templated carbons (ZTC) in 1 M TEABF$_4$/PC

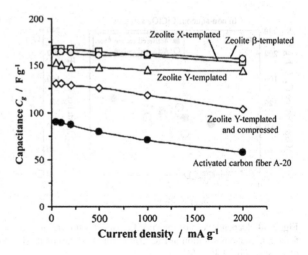

3.3.2.3 Templated Porous Carbons

The method of templates is used for producing microporous and mesoporous carbons in dependence on the templates themselves and the raw materials [76]. The method suggests impregnating the fine-porous precursors with organic solvent (e.g., furfuryl alcohol), carbonization of the solvent inside fine pores, and dissolution of the precursors.

By carbonizing precursors inside nanochannels of zeolites of various types, microporous carbons (zeolite-templated carbons—ZTC) were prepared, including carbon with the highest S_{BET} of 4,000 m^2/g and total pore volume of 1.8 cm^3/g, respectively [77, 78]. Since the pores formed in the ZTC were originated from the zeolite channels, they consist of the micropores with homogeneous size and morphology and a small fraction of mesopores. The CVA curves, built for ZTC in non-aqueous solution, had a non-rectangular form, which indicates the presence of pseudo-capacity mainly due to the presence of oxygen functional groups on the carbon surface. However, the value of C_g is not correlated with the oxygen concentration. Figure 3.41 shows the capacitance versus current density for various ZTC together with the analogous dependence for commercial ACF.

The results showed a very good speed of discharge for the most of ZTC, i.e., almost constant capacity even at a high current density of 2 A/g in addition to high values of C_g. Such discharge rate is associated with a three-dimensionally ordered system of micropores, which gives a low resistance in the micropores. Other microporous ZTC were prepared using two-step molding process using 13X-zeolite as the template and subsequent activation with KOH. High capacities persist until 94–100 % at a current density of 2 A/g for non-activated and activated ZTC.

Mesoporous carbons were prepared using the mesoporous silica-gel: a two-dimensional hexagonal mesoporous silica-gel (for example, SBA-15) produces carbons with two-dimensionally ordered mesopores, and three-dimensional cubic mesoporous silica-gel (e.g., MCM-48) produces carbons with three-dimensionally ordered mesopores. By using mesoporous silica-gel, the first impregnation with furfuryl alcohol resulted in mesoporous bimodal carbons with peaks at pore size of 2.9 and 16 nm, and carbons obtained by further impregnation have the unimodal pore size of 2.8 nm. S_{BET} was 1,540–1,810 m^2/g, and more than a half of the total pore volume was charged to the mesopores, while the capacitance C_g was 200 F/g at the current density of 1 mA/cm^2, but markedly decreased with increasing current density. Three-dimensional network of mesopores in carbons gave better performance than two-dimensional network. Combination of sugar and MCM-48 showed the best performance among the many combinations of carbon sources (precursors) with templates [79]. A study conducted with 25 silica-gel/template carbons, having different pore structure and surface area of 200–1,560 m^2/g, led to the conclusion that these carbons do not show obvious advantages in comparison with ACs in aqueous and non-aqueous solutions [80].

3.3.2.4 Carbide Derivatives of Carbon

As it turns out, the various metal carbides (TiC, B$_4$C, SiC, etc.) are capable to produce highly microporous carbons by heat treatment at a temperature of 400–1,200 °C in a steam of Cl$_2$ [81]. Thus, S_{BET} is 1,000–2,000 m^2/g. The porous structure of these carbons is strongly dependent on the carbide source and the heat treatment temperature (HTT). In these carbide-derived carbons, mainly micropores are formed to 800 °C, but when heated above 800 °C mesopores begin to predominate, and consequently, S_{BET} has the maximum at 800 °C for most carbons [82, 83]. The EDLC based on carbide microporous carbons, mentioned above, were created and their behavior was studied in various non-aqueous solutions [84, 85] and solutions of H$_2$SO$_4$ [86]. The effects of influence of the electrolyte ions size and use of organic solvents were also discussed. Derived carbons in the case of TiC were produced at 500–1,000 °C, whereby the average size of the micropores and S_{BET} varied from 0.7 to 1.1 nm and from 1,000 to 1,600 m^2/g, respectively. So, as it was shown, the capacity normalized to S_{BET} increases with decreasing the average pore size [84]. The B$_4$C derivatives of carbon showed good results in the work velocity in the solution of KOH, and 86 % retention of the capacity under the scan rates from 2 to 50 mV/s [87].

3.3.2.5 Carbon Aerogels and Xerogels

Aerogels is a class of materials that constitute the gel, whose liquid phase is substituted completely with gaseous phase. Such materials have extremely low density and exhibit a number of unique properties: hardness, transparency, heat resistance, very low thermal conductivity, etc. The first samples of carbon airgel were produced in the early 1990s. Aerogels are a class of mesoporous materials, whose cavities occupy at least 50 % by volume. Typically, this percentage reaches to 90–99 % and the density varies from 1 to 150 kg/m^3. The structure of aerogels is a tree network of nanoparticles with sizes of 2–5 nm and pores with sizes up to 100 nm, which are combined into clusters.

Carbon aerogels having a large number of mesopores were produced mainly by pyrolysis of resorcinol–formaldehyde aerogels [88, 89]. The primary particles of carbon aerogels, as a rule, have a size of about 4–9 nm and are interconnected with each other, forming a network of interparticle mesopores.

Changing conditions for obtaining of carbon aerogels lead to a strong influence on the capacity of EDLC. The use of carbon aerogels in the EDLC electrodes requires activation of the primary particles [90]. Activation using CO_2 leads to the transformation of a relatively large number of micropores in the mesopores, and approximately doubles the capacitance in a non-aqueous solution. Surface modification of the carbon airgel can give a large capacity at a high current density, which may be associated with improved wettability of the surface of carbon in organic electrolytes [90].

The *xerogel* is a gel, whose liquid medium has been removed from it, so that the structure is compressed, and the amount of porosity is a reduced one to some extent due to surface tension forces acting in the course of removal of the liquid. Xerogels are ensembles of contiguous spherical particles, whose size and packing density depend on the method of preparation. The value of C_g for the carbon xerogel increased from 112 to 171 F/g upon activation of CO_2. This change in C_g is associated with a significant increase of S_{micro} from 530 to 1,290 m^2/g together with increasing S_{meso} from 170 to 530 m^2/g [91].

3.3.2.6 Carbon Nanotubes

Advantages of Carbon Nanotubes

Carbon nanotubes (CNT) are also promising for use as electrode materials in the ECSP. Their distinctive features are not only a large area of opened surface and storage spaces for different electrolyte ions, but also high electrical conductivity. Cavities in carbon nanotubes are shown in Fig. 3.42.

The outer surface of the CNT walls (1 in Fig. 3.42) consists essentially of a basal plane of graphite. Large capacity of EDL takes place on a perfect surface like CNTs due to the high polarization with a large window of potentials. Most of the CNT, as it is known, are connected with each other by the van der Waals forces. Here, only the outer tubes in the bundle are exposed to the action of an electrolyte and to the stratification of the spaces between the bundle of tubes (3 in Fig. 3.42), which is hardly used for forming the EDL. The inner surface of the nanotube (2 in Fig. 3.42) is also suitable, in principle, for penetration of the electrolyte ions. However, there are certain limitations: first, most of the ions can not penetrate inside nanotubes because of the small internal diameter (1.3–1.6 nm), and secondly, aqueous electrolytes not wet the interior space because of the hydrophobicity of the inner surface (see below). The interlayer space in the walls of the

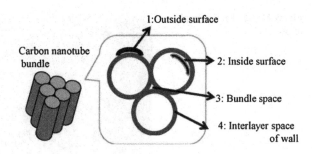

Fig. 3.42 Schematic representation of spaces in the CNT for the storage of electrolyte ions

multiwalled nanotubes (4 in Fig. 3.42) may be accessible for intercalation of the electrolyte ions, such as Li^+. Thus, the intercalation as well as the Faraday reaction is possible, while giving pseudo-capacitance.

Single-Walled Carbon Nanotubes (SWCNT)

The most common method to produce SWCNT is the voltaic arc synthesis using different catalysts, for example, Y/Ni catalyst [92]. Often afterwards, nanotubes have a narrow diameter distribution 1.4–1.6 nm and the average length of ~ 0.7 μm. This is followed by a purification step of the initial materials of the arc synthesis using oxidation in air and washing, for example, in hydrochloric acid. Due to the high surface energy of SWCNT, nanotubes in the SWCNT powder are aggregated into strands (bundles) of flattened structures up to 5–10 μm in length and a broad distribution of strands in the transverse size from 6 nm to about 100 nm [93]. Another method of obtaining SWCNT is ethylene pyrolysis.

Capacitive properties of single-walled carbon nanotubes (SWCNT) have been investigated on the basis of various electrolytes from organic electrolytes to aqueous solutions in a variety of different literature. Contributors of most articles usually come to the conclusion that SWCNT have excellent properties as the electrode material at high charge and discharge rates as compared to AC.

High power characteristics (excess of 20 kW/kg) under high currents up to several hundred of A/g [92] were obtained for the electrodes on the basis of SWCNT in sulfuric acid electrolyte. Such high power characteristics are explained by regular structure of the pores, which are located between the individual nanotubes and their strands (see Fig. 3.43). The regularity of the porous structure means the virtual absence of twisted and crimped pores, and hence, the maximum conductivity of the electrolyte in the pores.

Figure 3.44 shows the capacitance-voltage potentiodynamic curves measured for the mesoporous activated carbon FAS and SWCNT, at a very high scan rate equaled to 1,000 mV/s. The ordinate axis is scaled in the values of capacitance calculated from the experimental CVA curve dividing the current by the rate of potential sweeping. Figure 3.44 shows that the curve for SWCNT has an almost perfect form, which is a characteristic of the EDL charging (compare with the upper illustration in Fig. 3.35). While the curve for AC FAS has a distorted view

Fig. 3.43 SEM micrograph
of SWCNT

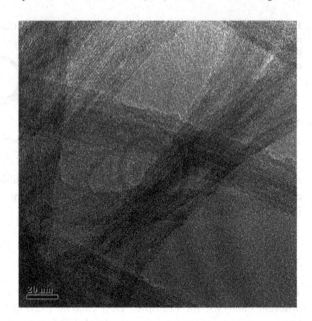

Fig. 3.44 Voltage-
capacitance cyclic curves at
the scan rate of 1,000 mV/s
computed in the range from
−400 mV to 200 mV: *1* FAS,
2 SWCNT. The electrolyte is
dilute sulfuric acid with a
concentration of 35 wt.%
[25]

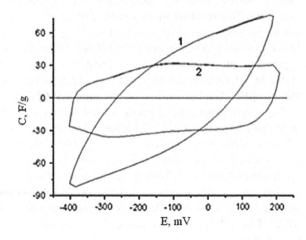

of "fish" due to the fact that the equilibrium charge–discharge processes do not
have time to occur at high scan rate.

Unlike carbons, SWCNT possess significant hydrophobicity [92]. Quantitatively, this property can be estimated by comparing the integral porosimetric
curves obtained by the MSCP using octane and water (Fig. 3.45).

Porosimetric curves for all pores are measured by octane while by water—only
hydrophilic pores. Figure 3.45 shows that the volume of SWCNT micro- and
mesopores wetted with octane, almost 5 times more than the pore volume, wetted
with water. Taking into account that the SWCNT surface wetted with electrolyte is

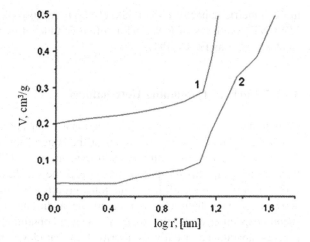

Fig. 3.45 Integral porosimetric curves for SWCNT obtained by the MSCP: *1* by octane, *2* by water [25]

about 5 times smaller than the surface of the carbon FAS (specific capacitancies differ three times only), it was shown that the capacitance value of the specific surface is close to 30 SWCNT mcF/cm^2 of true hydrophilic surface and the specific surface capacity of FAS approaches 20 mcF/cm^2.

In order to use the inner surface of SWCNT, the opening of some holes at the ends of the tubes was conducted by oxidation. If the inner surface of the opened SWCNT can be fully utilized for the formation of the EDL, then the capacity is doubled. During the optimization process of the oxidation conditions of super-grown SWCNT, values of S_{BET} over 2000 m^2/g were obtained, and the capacitor, which used $TEABF_4/PC$, provided high specific energy of 24.7 Wh/kg and also high power density of 98.9 kW/kg [93]. Nevertheless, the increase in capacity is not proportional to the increase of S_{BET}. The main disadvantage of SWCNT is their high cost.

Double-Walled and Multilevel Carbon Nanotubes

As compared with SWCNT, properties of capacitors based on double-walled carbon nanotubes (DWCNT) are not often published because clear DWCNT are very difficult to obtain. Basically, DWCNT and multiwalled carbon nanotubes (MWCNT) have a smaller surface area for the EDL formation. On the other hand, a lot of papers devoted to the capacitive performance of MWCNT was published, since MWCNT are relatively easy to synthesize and significantly cheaper compared to SWCNT. Depending upon the methods of synthesis and modification, various types of MWCNT with different values of specific surface area have been prepared. Their specific capacities, obtained in aqueous and non-aqueous electrolytes, comprise from 10 to 200 F/g. However, specific capacities are not so surprising in comparison with the AC. On the other hand, it should be noted that the volumetric capacity C_v is relatively high because of the high bulk density of MWCNT. Synthesis of MWCNT is often conducted by ethylene pyrolysis using catalysts such as Fe–Co [94].

3.3.2.7 Carbons Containing Heteroatoms

Carbon materials can have various functional groups at the surface; most of them should be linked to the carbon atoms at the edge of the hexagonal carbon layers. These functional groups often contain oxygen, e.g., –COOH, = CO, and others, that strongly depends on the precursors and the conditions of preparation of the carbon material. Some of these functional groups are acidic and electrochemically active and make their contribution to the additional EDLC capacity, called a pseudo-capacitance. In the case of oxygen-containing functional groups, the pseudo-capacitance usually is ascribed to Faraday's redox-reactions of these groups with ions of the electrolyte. Oxygen-containing functional groups are formed by conventional activation and independently appear on the surface in the oxidation of carbon with oxygen, HNO_3, and the electrochemical oxidation.

Besides oxygen, functional groups of carbons may contain nitrogen, boron, sulfur, and various metals. In many cases, these groups contribute to the pseudo-capacitance. The electrodes, with almost reversible Faraday processes occur on their surface, are often used as electrodes of the ECSC of the energy type. Such electrodes are electrodes based on electron-conducting polymers—ECP (polyaniline, polythiophene, polypyrrole, and similar) as well as the electrodes on the basis of some oxides of transition metals (oxides of ruthenium, iridium, tungsten, molybdenum, zirconium, manganese, etc.) [68, 69]. The effect of the porous structure of the ECP on capacitive properties of the ECSC electrodes was studied in [95].

References

1. Volfkovich YM, Bagotsky VS (2001) Colloid and surfaces. A: Physocochem Eng Asp 187–188:349
2. Vielstich W, Lamm A, Gasteiger HA (eds) (2003) Handbook of fuel cells. Fundamentals technology and applications, vol 1–4. Wiley, Chichester
3. Bagotsky VS (2009) Fuel cells: problems and solutions. Wiley, Hoboken
4. Baturina OA, Volfkovich YM, Sakars AV, Wynne KJ, Wnek GE (2005) In: 207th meeting of the electrochemical society, Quebec, Canada. 15–20 May 2005 (Meeting Abstracts)
5. Volfkovich YM, Sosenkin VE, Nikolskaya NF (2010) Russ J Electrochem 46(4):438
6. Tarasevich MR (1984) Electrochemistry of carbonaceous materials. Nauka, Moscow [in Russian]
7. Volfkovich YM, Kononenko NA, Cherniaeva MA, Kardash MM, Shkabara AI, Pavlov AV (2008) Critical technologies. Membranes 39(3):8 [in Russian]
8. Napolskii KS, Barczuk PJ, Vassiliev SY, Veresov AG, Tsirlina GA, Kulesza PJ (2007) Electrochim Acta 52(28):7910

9. Kyotani T, Xu W, Yokoyama Y, Inahara J, Touhara H, Tomita A (2002) J Membr Sci 196(2):231

10. Gladysheva TD, Shkolnikov EI, Volfkovich YM, Podlovchenko BI (1982) Sov Electrochem 18(4):337

11. Podlovchenko BI, Gladysheva TD, Vyaznikovskaya OV, Volfkovich YM (1983) Sov Electrochem 19(3):424

12. Volfkovich YM, Shkolnikov EI (1983) Sov Electrochem 19(6):586

13. Volfkovich YM, Shkolnikov EI (1979) Sov Electrochem 15(1):5

14. Volfkovich YM, Shkolnikov EI (1983) Sov Electrochem 19(9):1177

15. Vol'fkovich YM, Shkolnikov EI, Dubasova VS, Ponomarev VA (1983) Sov Electrochem 19(6):765

16. Wu G, Chen Y-S, Xu B-Q (2005) Electrochem Commun 7(12):1237

17. Frackowiak E, Lot G, Cacciaguerra T, Beguin F (2006) Electrochem Commun 8(1):129

18. Li X, Hsing I-M (2006) Electrochim Acta 51(25):5250

19. Guo D-J, Li H-L (2006) J Power Sources 160(1):44

20. Wang C-H, Shih H-C, Tsai Y-T, Du H-Y, Chen L-C, Chen K-H (2006) Electrochim Acta 52(4):1612

21. Prabhuram J, Zhao TS, Liang ZX, Chen R (2007) Electrochim Acta 52(7):2649

22. Chen C-C, Chen C-F, Chen C-M, Chuang F-T (2007) Electrochem Commun 9(1):159

23. Tsai M-C, Yeh T-K, Tsai C-H (2006) Electrochem Commun 8(9):1445

24. Wang HJ, Yu H, Peng F, Lv P (2006) Electrochem Commun 8(3):499

25. Tusseeva EK, Mayorova NA, Sosenkin VE, Nikol'skaya NF, Volfkovich YM, Krestinin AV, Zvereva GI, Grinberg VA, Khazova OA (2009) Russ J Electrochem 44(8):884

26. Mayorova NA, Tusseeva EK, Sosenkin VE, Rychagov AY, Volfkovich YM, Krestinin AV, Zvereva GI, Zhigalina OM, Khazova OA (2009) Russ J Electrochem 45(9):1089

27. Wu G, Li L, Li J-H, Xu B-Q (2006) J Power Sources 155(2):118

28. Santosh P, Gopalan A, Lee K-P (2006) J Catal 238(1):177

29. Jiang C, Lin X (2007) J Power Sources 164(1):49

30. Hu ZA, Ren LJ, Feng XJ, Wang YP, Yang YY, Shi J, Mo LP, Lei ZQ (2007) Electrochem Commun 9(1):97

31. Volfkovich YM, Bagosky VS, Zolotova TK, Pisarevskaya EY (1996) Electrochim Acta 48(13):1905

32. Volfkovich YM, Levi MD, Zolotova TK, Pisarevskaya EYu (1993) Polymer 34(11):2443

33. Volfkovich YM, Zolotova TK, Levi MD, Letuchy YA (1993) Adv Mater 5(4):274

34. Volfkovich YM, Sergeev AG, Zolotova TK, Afanasiev SD, Efimov ON, Krinichnaya EP (1999) Electrochim Acta 44(10):1543

35. Divisek J, Wilkenhöner R, Volfkovich YM (1999) J Appl Electrochem 29(2):153

36. Cindrella L, Kannan AM, Lin JF, Saminathan K, Ho Y, Lin CW, Wertz J (2009) J Power Sources 194(1):146

37. Gurau V, Bluemle MJ, De Castro ES, Tsou YM (2006) J Adin Mann Jr, Zawodzinski Jr TA (2006) J Power Sources 160(2):1156

38. Volfkovich YM, Sosenkin VE, Nikolskaya NF, Kulova TL (2008) Russ J Electrochem 44(3):300

39. Gostik T, Fowler MW, Ioannidis MA, Pritzker MD, Volfkovich YM, Sakars AV (2006) J Power Sources 156(2):375

40. Divisek J, Eikerling M, Mazin VM, Schmitz H, Stimming U, Volfkovich YM (1998) J Electrochem Soc 145(8):2677

41. Gottesfeld S, Zawodzinski TA (1997) In: Alkire RC, Gerischer H, Kolb DM, Tobias CW (eds) Advances in electrochemical science and engineering, vol 5. Wiley-VCH, Weinheim, p 195

42. Volfkovich YM, Dreiman NA, Belyaeva ON, Blinov IA (1998) Sov Electrochem 24(7):624

43. Berezina NP, Volfkovich YM, Kononenko NA, Blinov IA (1987) Sov Electrochem 23(7):912

44. Khrizolitova MA, Volfkovich YM, Mikhaleva GM, Tabakman LS (1988) Sov Electrochem 24(6):709

45. Volfkovich YM, Luzhin VK, Vanyulin AN, Shkolnikov EI, Blinov IA (1984) Sov Electrochem 20(5):613
46. Eikerling M, Kharkats YI, Kornyshev AA, Volfkovich YM (1998) J Electrochem Soc 145(8):2684
47. Volfkovich YM (1978) Sov Electrochem 14(3):460
48. Volfkovich YM (1978) Sov Electrochem 14(10):1282
49. Bode H (1977) Lead-acid batteries. Wiley, New York
50. Dasoyan MA, Aguf IA (1975) Modern Theory of lead: acid battery. Energiya Publishers, Leningrad
51. Bagotzky VS, Skundin AM (1980) Chemical power sources. Academic Press, London [in Russian]
52. Volfkovich YM, Bagotsky VS (2001) Colloid Surf A: Physocochem Eng Asp 187–188:349
53. D'Alkaine CV, de O Brito GA (2009) J Power Sources 191(1):159
54. Ferg EE, Geyer L, Poorun A (2003) J Power Sources 116(1–2):211
55. Soria ML, Valenciano J, Ojeda A, Raybaut G, Ihmels K, Deiters J, Clement N (2003) J Power Sources 116(1–2):61
56. Clement N, Kurian R (2003) J Power Sources 116(1–2):40
57. Khomskaya EA, Kazarinov IA, Semykin AV, Gorbacheva NF (2008) Macrokinetics of gas cycles in hermetic batteries. Saratov University publishers, Saratov [in Russian]
58. Doyle M, Fuller T, Newman J (1993) J Electrochem Soc 140(6):1526
59. Fuller TF, Doyle M, Newman J (1994) J Electrochem Soc 141(1):1
60. Chizmadzhev YA, Markin VS, Tarasevich VR, Chirkov YG (1971) Macrokinetics of processes in porous media. Nauka, Moscow [in Russian]
61. Newman J, Tiedemann W (1975) AIChE J 21(1):25
62. Chirkov YG, Rostokin VI, Skundin AM (2011) Russ J Electrochem 47(3):288
63. Bograchev DA, Volfkovich YM, Dubasova VS, Nikolenko AF, Ponomareva TA, Sosenkin VE (2013) Russ J Electrochem 49(2):115
64. Kulova TL, Nikolskaya NF, Skundin AM (2008) Russ J Electrochem 44(5):558
65. Bagozky VS, Volfkovich YM, Kanevsky LS, Skundin AM, Broussely M, Chenebault P, Caillaud T (1995) Power sources 15. In: Attewel A, Keily, Crowborough T (eds) International power sources Symposium p 359
66. Nimon VY, Visco SJ, De Jonghe LC, Volfkovich YM, Bograchev DA (2013) ECS Electrochem Lett 2(4):A33
67. Tran C, Yang X, Qu D (2010) J Power Sources 195(7):2057
68. Conway BE (1999) Electrochemical super capacitors. Kluwer Academic/Plenum Publishers, New York
69. Volfkovich YM, Serdyuk TM (2002) Russ J Electrochem 38(9):935
70. Inagaki M, Konno H, Tanaike O (2010) J Power Sources 195(24):7880
71. Bleda-Martinez MJ, Agull JA, Lozano-Castell D, Morall E, Cazorla-Amors D, Linares-Solano A (2005) Carbon 43(13):2677
72. Volfkovich YM, Mazin VM, Urisson NA (1998) Russ J Electrochem 34(8):740
73. Lee J, Kim J, Kim S (2006) J Power Sources 160(2):1495
74. Wang L, Toyoda M, Inagaki M (2008) New Carbon Mater 23(2):111
75. Volfkovich YM, Mikhailin AA, Bograchev DA, Sosenkin VE, Bagotsky VS (2012) Chapter 7 in book: recent trend in electrochemical science and technology. INTECH open access publisher. www.intechopen.com, p 159
76. Inagaki M (2009) New Carbon Mater 24(3):193
77. Hou PX, Yamazaki T, Orikasa H, Kyotani T (2005) Carbon 43(12):2624
78. Nishihara H, Itoi H, Kogure T, Hou P, Touhara H, Okino F, Kyotani T (2009) Chem Eur J 15(21):5355
79. Vix-Guterl C, Frackowiak E, Jurewicz K, Friebe M, Parmentier J, Beguin F (2005) Carbon 43(6):1293
80. Sevilla M, Alvarez S, Centeno TA, Fuertes AB, Stoeckli F (2007) Electrochim Acta 52(9):3207

81. Morishita T, Tsumura T, Toyoda M, Przepiyrski J, Morawski AW, Konno H, Inagaki M (2010) Carbon 48(10):2690
82. Gogotsi Y, Nikitin A, Ye H, Zou W, Fischer JE, Yi B, Foley HC, Barsoum MW (2003) Nature Mater 2(9):591
83. Dash RK, Yushin G, Gogotsi Y (2005) Micropor Mesopor Mater 86(1–3):50
84. Chmiola J, Yushin G, Dash R, Gogotsi Y (2006) J Power Sources 158(1):765
85. Chmiola J, Yushin G, Gogotsi Y, Porter C, Simon P, Taberna PL (2006) Science 313(5794):1760
86. Lin R, Taberna PI, Chmioda J, Guay D, Gogotsi Y, Simon P (2009) J Electrochem Soc 156(1):A7
87. Chmiola J, Yushin G, Dash RK, Hoffman EN, Fischer JE, Barsoum MW, Gogotsi Y (2005) Electrochem Solid-State Lett 8(7):A357
88. Wang H, Gao Q (2009) Carbon 47(3):820
89. Tao Y, Endo M, Kaneko K (2009) Recent pat Chem Eng 1(3):192
90. Fang B, Wei Y-Z, Maruyama K, Kumagai M (2005) J Appl Electrochem 35(3):229
91. Lin C, Ritter JA, Popov BN (1999) J Electrochem Soc 146(10):3639
92. Volfkovich YM, Rychagov AYu, Sosenkin VE, Krestinin AV (2008) Elektrokhimicheskaya Energetika 8(2):106
93. Hiraoka T, Najafabadi AI, Yamada T, Futaba DN, Yasuda S, Tanaike O, Hatori H, Yumura M, Iijima S, Hata K (2009) Adv Funct Mater 20(3):422
94. Usoltseva A, Kuznetsov V, Rudina N, Moroz E, Haluska M, Roth S (2007) Phys Status Solidi B 244(11):3920
95. Volfkovich YM, Sergeev AG, Zolotova TK, Afanasiev SD, Efimov ON, Krinichnaya EP (1999) Electrochim Acta 44(10):1543

Chapter 4
Powder Metallurgy

Mikhail Mikhailovich Serov

Abstract Classification of porous materials (PM), their properties and character-istics is studied, a comparative analysis is done. An overview of a number of modeling-analytical descriptions of the physical properties of porous bodies is provided. The general issues for producing porous materials by powder metallurgy techniques including various methods of molding and sintering processes are under consideration. The potential applications of porous materials are demonstrated.

In dependence of the chemical composition, the porous materials can be divided into two types: metallic materials and nonmetallic materials. Porous powder materials (PPM), porous fibrous materials of inorganic fibers (PFM), porous mesh materials (PMM), highly porous cellular materials (CM) are produced by powder metallurgy of powdered metals, alloys and refractory compounds.

Porous materials, depending on the type of the pores, can be classified as materials with perforating, closed, and dead-end porosity. The total porosity is the sum of these types of porosity. If the total porosity is less than 7–10 %, all pores are closed, when total porosity is 20–30 % then closed porosity is no greater than 2–3 %. Due to porous structure, the PM can be divided into isotropic and anisotropic ones. Regular alternation of homogeneous structural elements in a space is specific property of isotropic structures, which is typical for materials consisted of spherical particles of the same size as well as for meshes of a regular structure. Anisotropic pore structures have some distribution in numbers, sizes, etc., in one or more directions.

4.1 Comparative Analysis of the Properties of the Porous Body

Properties of PM are dependent on the chemical composition and shape of the feedstock, processing methods and modes of obtaining the material, its porosity, and other factors [1–3].

Y. M. Volfkovich et al., *Structural Properties of Porous Materials and Powders Used in Different Fields of Science and Technology*, Engineering Materials and Processes, DOI: 10.1007/978-1-4471-6377-0_4, © Springer-Verlag London 2014

The pore sizes depend mainly on the size of the particles (fibers, wires, etc.) of the source materials and porosity. Comparative assessment of different kinds of materials is convenient to carry out in the coordinates: relative pore size—porosity (Fig. 4.1). The highest average pore sizes are typical for cellular materials and for the materials made of the wires and fibers. Maximum porosity of PMM from grid fabrics is limited by the porosity of grids and minimal degree of their compression required to produce the material. The potential of PM can be estimated for practical use basing on the ranges of porosity. From Fig. 4.1 follows that it is possible to use materials of powder particles and fiber as the materials of the filter elements of fine purification, which are characterized by a small pore size, while grid materials—for the more coarse filter. It is known that materials with the porosity of 0.6–0.7 and more are necessary to use as sound-absorbing materials and therefore fiber fabrics, mesh knitted and cellular materials are among them. In the cases where the permeable materials of low porosity with high hydraulic resistance are required, the powder materials consisted of lnonspherical particles and woven meshes are applied.

An important property of porous materials is their permeability. Minimum values of the coefficient of hydraulic resistance in all modes of filtration have been observed in the case of porous materials made of smooth fibers, wires, spheres. The increase of hydraulic resistance of materials compared to similar materials made from smooth particles and fibers is observed with increasing roughness of the precursor particles and fibers, as well as with increasing the tortuosity of porous channels.

Filtration properties of PM are mainly determined by pore size distributions, and mechanisms of filtration of liquid or gas through the pores. Porous materials made of rough particles of nonspherical shapes have the best retention properties during filtration of liquids and gases. Uneven distribution of porosity on the surface of PM significantly reduces quality of the filtration. Permeable materials produced of meshes as well as materials of the fibers have the best strength properties.

Cost of PM is composed of material costs up to 40–60 %. A fraction of overheads increases under conditions of pilot production.

Comparative analysis of the properties of PM shows that several options can be recommended for a specific application. The final choice of material should be carried out taking into account the additional requirements arising from the conditions of PM exploration. The most effective cases of application of materials are those that use multiple properties of PM.

The properties of all materials, including porous ones, can be classified into two groups. Properties that do not depend on the structure, in particular, the coefficients of thermal expansion, magnetic susceptibility and specific heat capacity, which are the same for the porous and nonporous states of materials. Other properties are structurally sensitive. Here we can distinguish properties determined by porosity characteristics of the object: magnitude of porosity, pore sizes, pore shapes and distributions, and tortuosity. These properties include hydraulic and capillary, filter characteristics, etc. Studies of bodies, formed from ultrafine and nanoparticles,

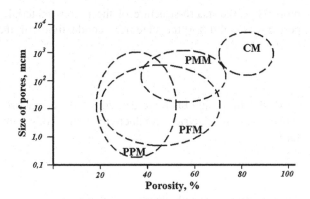

Fig. 4.1 The ranges of porosity of permeable materials [3]

have shown the existence of dependence of their properties, including physical ones, on particle size and, as a consequence, on the pore size. These effects occur when the average crystal grain size not greater than 100 nm, and most clearly seen when their size is less than 10 nm.

Very important and wide area of durable and successful application of small particles in the PM production is catalysis of chemical reactions. Catalyzed reactions are typically run at lower temperature than uncatalyzed reactions and are more selective. The number of atoms in an isolated metal particle is small, so the distance between the energy levels $\sim \delta \sim E_F/N$ (E_F—Fermi energy, N—number of atoms in the particle) is comparable with the thermal energy kT. In the limit when $\delta > kT$, the levels are discrete and the particle loses metallic properties.

With decreasing particle size the surface contribution to the free energy is increased. To reduce the total energy of the system such deformation of the crystal, which will decrease the surface energy, can be more profitable. Surface energy has a minimum for the close-packed structures, so face-centered cubic (FCC) or hexagonal close-packed (HCP) structure are most preferred for nanocrystalline particles, that is observed experimentally. Electron diffraction study of niobium, tantalum, molybdenum and tungsten nanocrystals of 5–10 nm showed that they have the FCC or HCP structure, whereas in a normal state, these metals have a body-centered cubic (BCC) lattice.

Dependence of surface energy on the particle size determines relationship between the melting point and the size of the nanoparticle. Experimental lowering the melting temperature of small particles was observed in many studies. Maximum reduction of the melting temperature of the clusters of Sn, Ga and Hg was 152, 106, and 95 K, respectively.

Among the properties of porous objects most research attention are paid to the properties of the conductivity, in particular, electrical and thermal conductivity, as well as certain mechanical characteristics, mainly tensile strength and modulus of elasticity. In the descriptions of the influence of porosity on the conductivity of the body, assumptions of different authors vary considerably, however, lead to approximately identical results, which are in satisfactory agreement with the experimental data that generally correlates with the degree of sensitivity of this property to the macro-structure of the porous material. A linear dependence on porosity is valid for the electrical conductivity of the porous material (with $\varepsilon < 0.667$):

$$\lambda = \lambda_0(1 - 1.5\varepsilon) \tag{4.1}$$

M. Balshin, using his representation of the contact section of the porous body, showed that the electrical conductivity of the body with connected pores has the form:

$$\lambda = \lambda_0(1 - \varepsilon)^2 \tag{4.2}$$

while for the body with isolated pores we have:

$$\lambda = \lambda_0(1 - \varepsilon)^{1.5} \tag{4.3}$$

With the amount of porosity of more than 70 ... 80 % very significant differences in the conduction of bodies with matrix and random structure can be observed.

Accounting for formation of random clusters and network of physically continuous phase in the statistical mixture of quasi-spherical particles are proposed by V. Skorokhod as early as 1959 by introducing also randomly oriented infinite cylinders along with the spherical particles, as components of the continuous phase, into a model of the porous body. The following dependence of conductivity on the porosity for such an ideal model of the body is obtained:

$$\frac{3\lambda_0}{2\lambda + \lambda_0}\varepsilon_s + \frac{(3\lambda + \lambda_0)\lambda_0}{2(\lambda + \lambda_0)\lambda}\varepsilon_c = 1 \tag{4.4}$$

where ε_S and ε_C—the proportion of spherical and cylindrical phases of matter, respectively, and $\varepsilon_s = 1 - \varepsilon(1 + \log\varepsilon)$, $\varepsilon_c = \varepsilon \log \varepsilon$. Equation (4.4) is valid for any porous body with perfect inter-particle contacts: powder, fiber or with a mixed structure in the whole range of porosity values.

Significant dependence of structure-sensitive properties of the powder body on quality of interparticle contacts is reflected in the ambiguity of relation of such properties with porosity of the material.

The most common dependence of tensile strength on the porosity is the power-law dependence based on the concept of the contact section of porous bodies which was introduced by M. Balshin:

$$\sigma = \sigma_B(1 - \varepsilon)^m \tag{4.5}$$

wherein σ_B—tensile strength of nonporous material; $m = 2 ... 10$. Minimum values of the exponent m, corresponding to the maximum value of the contact section and strength are 2 and 3, respectively, for porous, fiber bodies, and powder compacts.

B. Pines, A. Syrenko, and N. Sukhinin suggested the dependence of strength on the porosity, which takes into account the weakening of inter-particle contacts by pores:

$$\sigma_B = \sigma_{B_0}\left[1 - \frac{3}{2}\varepsilon(1 + 2S) + \frac{9}{2}S\varepsilon^2\right] \tag{4.6}$$

where S is the weakening coefficient.

V. Troshchenko, assuming spherical shape for pores, the independence of their size and shape on the porosity as well as changes of porosity being proportional to the initial value during loading, obtained the following relationship:

$$\sigma_B = \sigma_{B_0}\left\{1 - A\left[1.5\varepsilon + B\sqrt{\left(\frac{6\varepsilon}{\pi A} - \frac{36\varepsilon^2}{\pi^2}\right)}\right]\right\} \tag{4.7}$$

where A—coefficient of the porosity variation (tension—$A > 1$, compression—$A < 1$), B—coefficient of inhomogeneity of the pore distribution on sections of a prototype. It should be noted that the relation (4.7) satisfactorily approximates the experimental data to the porosities in the range of 45 ... 50 %. For large values of porosity, the relation (4.7) loses its meaning, because the spherical shape, adopted in the model, does not correspond to the actually observed shape in highly porous objects.

Mention should be made concerning dependence of strength on the porosity, obtained by other authors also, although their practical significance is significantly limited by the complexity of decoding various coefficients. These dependencies are the following expressions:

F. Knudsen's:

$$\sigma_B = kL^{-a}e^{-b\varepsilon} \tag{4.8}$$

where L—size of the grain; κ, a, b—constants;

D. Harvey's:

$$\sigma_B = \sigma_{B_0}\left[\frac{(1-k)^3}{k^3 + (1-k)^3}\right] \tag{4.9}$$

where $k = \frac{\sqrt[3]{3\varepsilon/\pi}}{2}$;

D. Hasselman's:

$$\sigma_B = \sigma_{B_0}\left[1 - \frac{A\varepsilon}{1 + (A-1)\varepsilon}\right] \tag{4.10}$$

where A—parameter;

A. Vail's:

$$\sigma_B = \sigma_{B_0} \frac{1 - \varepsilon}{1 + A\varepsilon} \tag{4.11}$$

where A—parameter;
R. Herman's:

$$\sigma_B = k \frac{\sigma_{B_0}}{D} (1 - \varepsilon)^m \tag{4.12}$$

where $\varepsilon = 10 \ldots 40\%$; $m \sim 4 \ldots 7$; $D = 4 \ldots 60$ μm—diameter of spherical particles of the initial powder.

Phenomenological approach of V. Skorokhod, based on the concept of rms stresses and strains in the material of the porous body, deserves special attention when accounting for the strength of the structure factor. V. Skorokhod proposed the following expression for the dependence of tensile strength on the porosity of the ductile-hardening material with taking into consideration the reduction of the cross-section during deformation process:

$$\sigma_B = \sigma_{B_0} \frac{2(1 - \varepsilon_{cr})^2}{(4 - \varepsilon_{cr})^{0.5}} \tag{4.13}$$

where the fracture porosity ε_{cr} is connected with the initial porosity ε (before a test) using the expression of true strain by relative necking of nonporous material ψ_0:

$$\ln \frac{1}{1 - \psi_0} \approx 4/3 \ln \frac{\varepsilon_{cr}}{\varepsilon} - 1/2(\varepsilon_{cr} - \varepsilon) \tag{4.14}$$

Some dependencies for the yield strength look like the same as for the tensile strength. As an example, the binomial power dependence proposed by M. Balshin in the form of Eq. (4.3), which also describes the relative yield strength of the material in compression as a function of porosity, may be noted. There are known the following basic expressions suggested by M. Nakamura, it is based on the analysis of the elastic stress distribution around the pores (A and B—coefficients):

$$\sigma_T = A + B\varepsilon^{2/3} \tag{4.15}$$

N. A. Fleck and R. A. Smith:

$$\sigma_T = \sigma_{T_0} \left(1 - \varepsilon^{2/3}\right)^2 \tag{4.16}$$

Shalak:

$$\sigma_T = \sigma_{T_0} e^{-b\varepsilon} \tag{4.17}$$

where b—factor expressing the intensity of the influence of porosity on the yield stress.

Porous materials made of the compacted metal powders and sintered are usually anisotropic objects due to the anisotropy of the size and shape of the initial dispersed particles. Rather simple estimate of the expected anisotropy of properties can be carried out according to the normal elastic modulus. Under the assumption of independence of the integrated elastic rigidity of a plastically deformable porous body on strain and under absence of formation of the new contacting surfaces, the elastic modulus of pressurized body has the following expressions:

in a direction perpendicular to the pressing direction,

$$E_\perp = E_0(1 - \varepsilon_0)(1 - \varepsilon) \tag{4.18}$$

in a direction parallel to the pressing direction,

$$E_\parallel = E_0 \frac{(1 - \varepsilon_0)^4}{1 - \varepsilon} \tag{4.19}$$

where ε_0 and ε—initial and current porosity.

The most prominent feature in this case is to reduce the modulus of elasticity in the direction of compressing with decreasing porosity of the material.

Further development of theoretical approaches to the description of the elastic properties of porous bodies is using the method of "self-consistency" to the statistical mixture of quasi-spherical elements, with the formation of random clusters and networks of physically continuous phase, which was applied by V. Skorokhod. Corresponding equation, describing the dependence of the shear modulus of the porosity in the whole range of change, has the form:

$$\frac{5\mu_0}{3\mu + 2\mu_0}\varepsilon_S + \frac{(7\mu + \mu_0)\mu_0}{2(3\mu + \mu_0)\mu}\varepsilon_C = 1 \tag{4.20}$$

The above equation is valid for any type of structure—powder, fiber and mixed one.

4.2 Processes of Powder Metallurgy for Porous Materials

The process of creating the porous material by powder metallurgy method typically includes three basic steps: preparation of the dispersed particles; imparting a predetermined shape to a conglomerate of particles—molding, providing a predetermined set of physical and mechanical properties resulting from high temperature processing—sintering.

4.2.1 Methods for Forming Porous Materials

Pressing in closed molds are widely used for making articles of simple shape (disk, cone, hub, etc.). Distinguish between unilateral and bilateral pressing. A unilateral compression is used to mold products with a ratio of height to diameter not greater than 1. In all other cases, a bilateral compression is applied. The disadvantages of the method of compression in closed molds should include the limited shapes and sizes of manufactured products, as well as uneven distribution of porosity in the compacts as a result of frictional forces arising between the powder and the walls of the mold, which leads to uneven distribution of permeability and pore sizes in finished products. Advantages of the method are high-dimensional accuracy and higher performance. Increased dispersion of powders, especially of nanoparticles is accompanied by a marked decrease in their compressibility.

Isostatic pressing is a method of compressing the powder in the elastic shell under hydrostatic compression. A variation of this process is hydrostatic and gasostatic pressing and also pressing in thick-walled flexible shells placed into the steel mold.

The method of the hydrostatic compression is based on transmission of the pressure by fluid in the high pressure vessel to metal powder enclosed in an elastic shell. Pressing pressure is usually not more than 15–20 MPa. This method is mainly used to manufacture pipes and sleeves with bottomed. Magnitude of the hydrostatic pressure and pressing power to produce uniform density briquettes is smaller during the hydrostatic pressing than when pressing in a closed mold. This is due to the lack of pressure losses for external friction, due to that under the uniform compression of the powder its slipping relative to the shell does not occur. The disadvantages of hydrostat pressing include difficulty of obtaining products with exact geometric dimensions, the relatively high cost of manufactured products, low productivity. The advantages of the method include the possibility of obtaining products of large size and complicated shape with a uniform distribution of porosity.

By pressing in the resilient thick-walled shells, an elastic matrix inserted into a steel mold plays the role of a pressure-transmitting medium. The shell material should provide a uniform and thorough transfer of pressure on a compressible powder, and have the Poisson coefficient of 0.5. Rubber, a resin, a wax, paraffin, polyurethane, etc., are used as the shell materials. The compression scheme in the elastic shell is shown in Fig. 4.2. Disadvantages of this method of formation are the difficulty of obtaining accurate product shape and size, and the advantage is the ability to manufacture complex shapes.

Rolling of the powders is a continuous molding of pre-forms from powders by rolls. The essence of the method consists in feeding the powder from the hopper into a nip between two rotating rolls toward one another. Powder is entrained in the gap by friction between it and the surface of the rotating rollers and pressed into a strip that is strong enough to carry in a sintering furnace. The process is realized in rolling mills in dissimilar ways, different with arrangement of the axes plane of rolls (vertical, horizontal and inclined rolling) and powder feeding

Fig. 4.2 Pressing scheme in
the elastic shell: *1*—matrix,
2—punch, *3*—elastic shell,
4—powder

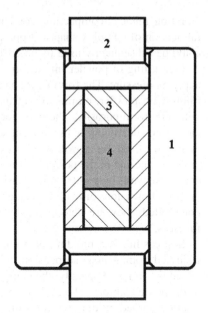

(gravity, forced). The main advantage of the rolling is the possibility of PPM manufacturing in thin ribbons and strips of width up to 550 and of thickness up to 5 mm with the porosity of more than 20 %. The disadvantages include the simplicity of the form and low strength of PPM.

Extruded pressing is a molding of performs from the mixture of powder with a plasticizer by forcing it through the opening. Bakelite alcoholic solutions, a starch paste, a wax, polystyrene solutions, etc., are used as the plasticizers. Its amount in the mixture is usually 6–10 % (by weight). Porous powder materials in the form of rods or tubes of great length (1 m) with uniform distribution of porosity in length are obtained by extruded pressing. The porosity of produced materials is 60–70 %. The disadvantage is the need to introduce suitable plasticizers into a powder, that complicates the subsequent sintering process and contaminate the sintered body with undesirable impurities. This method can produce PM without the use of plasticizers, but with addition of the pore-forming agents.

Vibratory compaction is a compression of the powders in the molds with vibration. The method is based on the drastic reduction of frictional forces between the particles of powder as well as between the powder and the walls of the mold. The static pressure applied simultaneously to the vibration is typically 0.5–5 MPa for various metal powders. The absence of such pressure can lead not to compaction, and to loosen the powder. Compaction of powders under the action of vibration occurs in a few seconds. Vibratory compaction mode is characterized with frequency that varies from 5 to 100 Hz and the amplitude of 5–30 microns. A high frequency of vibration, an amplitude increase, and application of a greater load to the powder, which should increase with decreasing surface roughness and accuracy of particle shape, must be chosen with decreasing particle size. Application of vibration reduces compaction pressure tenfold. This significantly facilitates the fabrication of complex shapes. Appropriate to apply the graphite matrixes, which could be subsequently used to sinter the product.

Free filling of powder into a form is the easiest way to forming PPM without applying pressure. The form under backfilling is subjected to vibration for the best filling with powder, elimination of the "arch effect" and production of the PPM with a uniform porosity. The main requirement for the material of the form is the lack of powder sintering to the form at the sintering temperature. Steel or graphite molds are used for powders of nonferrous metals, and ceramic—for iron. To prevent sintering of the powder particles to the walls of the mold, it is necessary to coat those walls by a suitable suspension (e.g., alumina with water) and subsequent drying. Dimensions of a billet, molded by free backfill of powder into the form, are not limited; they are determined by the size of the working space of sintering furnaces.

Slip casting is a molding of preforms by pouring the slurry, which is a concentrated homogenous suspension of powder in a liquid, into a porous form followed by drying. Typically, the slurry fills the plaster mold, which is the negative of the porous preform of a desired configuration. After filling a large portion of the liquid absorbed by the gypsum form; solid particles dry out and adhere firmly to one another. The mold is then opened and the casting is recovered dried up, and subjected to final drying and sintering. Fine powders (1–2 microns, at most 10 microns) are used to produce the slip, because when using a coarse powder a stable and uniform suspension can not be obtained.

Water is used as the coupling fluid for nonoxidizing metal powders and alcohols—for oxidizing metal powders. To avoid the formation of coagulant in a solution, the additions of hydrochloric and acetic acids, alkalis, etc., are used. The amount of solids introduced into the slurry is 40–70 % by volume. Slip casting process can be intensified by vacuuming a form, applying overpressure to the slip or simply heating the slurry. Slip casting is used for manufacturing articles of complex shape and large sizes, which are difficult to obtain by conventional powder metallurgy techniques.

Table 4.1 shows the advantages, disadvantages, and recommended fields of the application of molding techniques that have found the greatest use in the manufacture of PPM.

In the manufacture of porous fiber materials (PFM), the following methods of molding are used: various types of compression, rolling, hot extrusion, and sintering under pressure. Pre-fabricated fiber felts are made using several embodiments of felting: liquid, air, gravity, under the action of electric or magnetic fields, vibration.

When the *wet felting* is applied, slurry of fibers in a viscous fluid (e.g., glycerol) is prepared, and it is poured into a porous mold. A shape corresponding to the configuration of the finished product is subjected to a vacuum, whereby when the liquid is sucked off, the fibers are deposited on the bottom and walls of the mold, forming a billet in the form of felt with entangled fibers. During the deposition, density is self-regulating. Permeability decreases in a high-density layer, and the slurry supply slows, so the deposition of fibers is inhibited and in areas with a

Table 4.1 Advantages, disadvantages, and recommended fields of the application of molding techniques in the manufacture of PPM

Dignities of the method	Disadvantages of the method	Recommended fields of application
Static compression		
The high dimensional accuracy, High performance	Inhomogeneous permeability, Limitations of shapes and sizes	Manufacturing of elements of small size like bushings and disks
Hydrostatic and hydrodynamic pressing		
Preparation of larger articles, Uniform distribution of porosity	Low productivity	Fabrication of highly porous and long pipes of different diameters
Extruding		
Getting larger pipe sizes with high uniform permeability, High performance	The necessity of introducing a plasticizer, Limited form	Fabrication of highly porous and long pipes
Rolling		
Ability to produce thin sheets and strips, High performance	Getting products with only simple forms, Products with low strength	Fabrication of highly porous sheets and strips of different thickness
Loose filling of the powder into the mold		
The possibility of complex shapes producing, Ease of fabrication	Low productivity	Manufacture of complex shapes from spherical powders

lower density (more permeable) the fibers are deposited until the density of the felt is aligned.

When the *air felting* is applied a porous material is obtained by precipitating the fibers from the air stream. However, consistency and uniformity of felt a little bit worse than in the wet felting.

A method of forming a porous material in the production of metal fibers from a hanging drop of melt by a heat-receiver, which is rotating, is called the pendant drop melt extraction (PDME). When using the PDME the lower end of a vertical rod is melted to form a hanging drop of the melt. Vertex of the working edge of the rotating cooled heat-receiver is in contact with the drop, which is tuned in the form of an isosceles triangle. The melt is cooled in the zone of contact at a speed up to 10^6 K/s, which leads to the formation of a microcrystalline or amorphous structure. Due to the rotation of the heat-receiver, the solidified material is taken out of the melt in the form of fibers, and is discharged from the top of the working edge with the speed of about 30 m/s under the influence of centrifugal forces. A fiber with equivalent diameter of 30–80 microns falls on the receiving surface, which moves relative to the place of falling of the fiber. As a result, a porous material from fibers of steel and nickel-based alloys, copper, etc., having the porosity up to 98 % (Fig. 4.3), is formed.

Fig. 4.3 Schematic of the
formation of PFM using the
PDME. *1*—Transported
substrate; *2*—Formed porous
material; *3*—Dispersible
preform; *4*—Rotating heat-
receiver

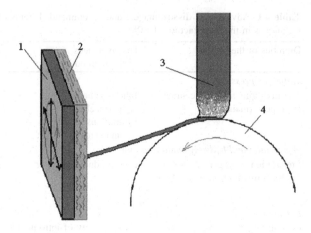

Molding due to vibration produces felt with oriented location of fibers. The intensity of this method of felting depends on the mode of vibro-processing and on sizes of fibers. For example, formation of cylindrical and sleeve billets most effectively occurs at frequencies of 50–100 Hz, the amplitude of 40–50 microns and a peak pulse shape fluctuations; fibers of 5–7 mm in length fit almost two times faster than the same fibers with length of 10–12 mm.

Application of the *rolling* enables to form continuous sheet of a single- or a multilayer material of fibers with controlled porosity. The thickness of sheets is from several tens of micrometers to hundreds of millimeters. The degree of compression of the order of 75 % does not cause a noticeable closing of the pores at the surface of products. During compaction of metal fibers, a contact, as well as reversible elastic and irreversible plastic bending deformation occurs. After removing the compaction pressure, instantaneous purely elastic deformation should be realized, then further prolonged (tens of hours) relaxation of residual stresses with decreasing rate, and the total strain reaches tens of percent, is followed. Large deformation effects related to the design of the fiber body, close to the structure of the spring. Relatively small elastic deformations of the individual fibers, overlapping, eventually give greater total deformation. The maximum strain in the compacts of the fibers is observed in the direction of application of pressing force, i.e., along the height and 4.5 times superior to that of the width. In carrying out of the rolling, the suspension of metal fibers in the viscous liquid is fed onto the grid coated filter paper. Grid is moved continuously with the help of rollers and passes over chambers for suction of liquids. The molded felt is then passed through the nip rolls and introduced into a sintering furnace.

When forming PFM by extrusion, a significant difference in the processes of structure formation compared with extrusion powders was found. Samples of the extruded fibers have three deformation zones. The first (peripheral) zone differs due to defectiveness and has a thickness of 0.5 mm. The second (middle) zone is widest whose fibers are oriented along the extrusion axis. In the third (central) zone the fibers are arranged randomly. With increasing draw ratio, the middle zone is expanding at the expense of the peripheral and central zones. Preliminary impregnation of the felt with wax, resin or a mixture of salts (44.5 % KNO_3 and 55.5 % $NaNO_3$) at the temperature of 190 °C allows using the method of extrusion in the manufacture of porous products in the form of tubes and cylinders of different sizes.

Porous mesh materials (PMM) are produced on the basis of knitted and woven nets. Production of crosslinked materials is to prepare nets (washing, degreasing), cutting, a package stacking, pressing and sintering the package. The form of a grid is the criterion for selecting the type of manufacturing operations and their modes.

Highly porous cellular metallic materials (CM) with a porosity of 80–98 % are formed by several groups of ways:

1. Casting methods,
2. Methods of powder metallurgy,
3. Metal deposition from the gaseous state,
4. Electroplating methods.

Detailed consideration of the ways is given in the review [4].

Methods of foaming molten metal (usually aluminum and its alloys, zinc, etc.) are realized in two basic ways: a gas is introduced into the molten metal from an external source, or gas bubbles are formed in the melt by dissociation of the particles of chemical compounds. In the first case, for deceleration of the floating rate of gas bubbles in order to increase its viscosity, from 10 to 20 % (by volume) of SiC, Al_2O_3, or MgO particles of 10–20 microns are previously introduced into the melt. The advantages of this technique are high-performance and low-density ($\varepsilon = 80$–98 %). In the second case, the molten metal is pre-injected by calcium of 1.5–3.0 %, which forms dispersed particles of CaO, $CaAl_2O_4$ that increases the melt viscosity. After reaching a predetermined value of viscosity, from 0.5 to 1.6 % of ZrH_2 or TiH_2 particles are introduced into the melt. As a result of the decomposition of hydrides, hydrogen is generated, and foams the melt. The method allows to achieve high uniformity of the pore size distribution in the material.

PM can be obtained by the casting method without foaming of the metal. According to that process, the porous preform of polymer with interconnected porosity, e.g., polyurethane is filled with a liquid solution of sufficiently refractory material, for example a mixture of mullite, phenolic resin, and calcium carbonate. The polymer is removed then by heat treatment, and the melt is poured into the open interconnecting voids which replicate the original polymer foam structure. After removal of the ceramic filler material (e.g., by pressurized water), metal structure which accurately replicates the original polymer structure is obtained. Difficulties in this process include the achievement of complete filling of the structure of the polymer with metal and the filler material removal without damaging the structure. The method is used to produce porous products from aluminum alloys, copper, and magnesium.

Metal powders can be use to duplicate highly porous structure of reticulate-cellular polymer such as polyurethane foam. Polymer with open through porosity is used for that. Polymer billets are cut into the samples of specified sizes, and are impregnated with prepared suspension of a metal powder. The average particle diameter of the powder used is determined by the fact that the powder must form the slurry layers at the reticulate-cellular porous polymer surface containing minimally 3–5 rows of metal particles along its thickness. The average particle diameter should be no greater than 5 microns to produce high quality preforms. Impregnation is carried out in immersion baths and by deformation of billets. After impregnation, the excess of slurry is removed and its distribution throughout the volume of the preform is aligned by the compression operation. The degree of compression is adjusted so as to provide a desired preform density. The preform is a composite material, since the polymer surface is uniformly coated with the dried slurry. With increasing density of preforms their structure changes somewhat and films of suspension, spanning part of the pores between the cells, appear. Removal of the polymer from the workpieces is carried out at the temperature of 200–600 °C and the heating rate is not more than 100 deg/h. A content of no more than 1–1.5 % of the starting polymer remains in blanks at the temperature of 550 °C. Spatial structure of the workpiece is not violated.

Metal ions from the electrolyte are deposited onto the open cells polymeric foam, which is later removed when using methods of electroplating. Application of electrochemical methods requires the electrical conductivity of the polymer. This is achieved by lowering the polymer in a liquid conductive solution, based on graphite. The polymer can be removed by heat treatment after electrodeposition. This method provides a porous preform made of nickel, copper or nichrome. Porous nichrome is prepared making layered coating of nickel and chromium. The material is then annealed at high temperatures. As a result of diffusion, the alloy is formed.

Technology of nickel carbonyl production can be applied when using methods of CVD. The porous preform of the polymer can be maintained at the temperature of nickel carbonyl decomposition (~ 120 °C). The resulting nickel is deposited onto the polymer, forming a metallic coating. After cooling, the polymer can be removed by thermal or chemical treatment.

4.3 Sintering of Porous Materials

Sintering is a thermally activated process, spontaneous or initiated by the external influence of the transition of contacting solids or porous medium in a thermodynamically equilibrium state by reducing the free surface.

Conventionally, shrinking process (compaction) of the powder body in an isothermal sintering can be subdivided into three sequential steps.

The early stage. The density of the powder body is small and the rate of densification is determined by processes occurring in the contact areas, the structure and geometry of which play a significant role. The rate of displacement relative to each other and the volume deformation of the particles, leading to shrinkage of the porous structure, are high.

The intermediate stage. The density of the powder body is large enough and reduction of the volume of each of the pores, which in the aggregate constitute a single ensemble, can occur almost independently. Porous matrix of the particles behaves like a viscous medium and its densification is evenly throughout the volume (with a uniform distribution of pores).

The late stage. Powder body contains some isolated pores that are healed (overgrown) by the diffusion dissolution in the matrix with outlet of vacancies on the external (overall) surface of the sintered preform.

There is no sharp and distinct boundaries between these three stages, the first of which is typical for nonisothermal sintering, and densification of powder body at an intermediate stage can be determined by the processes, which are characteristic for the early and (or) late stage.

One of the characteristic features of the shrinkage of the heated powder body is slowing its rate under isothermal exposure. With an increase in the sintering temperature the rate of density enlarge also increases, but this rate decreases more intense with an increase in the sintering temperature. General behavior of the analytic curves of isothermal volumetric shrinkage $V/\Delta V$ can be expressed as power functions in the following form:

$$\frac{\Delta V}{V} = K\tau^{0.5} \tag{4.21}$$

where V and ΔV—the current pore volume and its change in the reporting time τ of isothermal exposure, respectively; K—constant. Equation (4.21) is valid in a relatively small interval of time, usually no more than a few hours.

V. Ivensen showed that the speed ratio of the pore volume reduction during isothermal sintering of any metal powders at the two arbitrarily moments of time τ_1 and τ_2, for selected isothermal sintering, is equal to:

$$\frac{\left(\partial V/\partial \tau\right)_1}{\left(\partial V/\partial \tau\right)_2} = \left(\frac{V_1}{V_2}\right) \tag{4.22}$$

where V_1 and V_2, are the total pore volumes at time τ_1 and τ_2, respectively. This regularity under constant value of the exponent n (which characterizes the intensity of braking shrinkage) is maintained throughout the sintering process, including the very rapid reduction of the pore volume at the beginning of the sintering and the slow reduction—at the end. The ratio of (4.22) was proved valid for isothermal sintering duration >50 h, as well as for sintering powders of different metals with any methods for their preparation.

The dependence of shrinkage on the sintering duration, described by the Eq. (4.22), is explained by the fact that the supply of the free energy per unit mass of powder is determined not only by its dispersivity, but also by the surface state and the degree of defectiveness of the particles to a considerable extent. Powders with a high surface area and high density of defects are compacted with the highest rate, as having a large supply of free energy, in the process of isothermal exposure at sintering. Attenuation of the shrinkage is due to occurring decrease in free energy of the powder body during isothermal exposure.

An essential feature of the shrinkage of compacts is that in the case of further temperature increase after prolonged isothermal sintering shrinkage, when shrinkage is almost ceased, the shrinkage rate increases again. Figure 4.4 shows the effect of stepwise heating on shrinkage of silver powder compacts at three temperatures, and each new rise in temperature leads to the intensification of the shrinking process, proceeding at different stages about the same.

V. Ivensen proposed the dependence, taking into account the change of the pore volume in the sintering and the concentration of defects:

$$\ln\frac{V}{V_0} = -\frac{a}{b}\exp\left(-\frac{\Delta E}{RT}\right)\ln\left[aN_0\exp\left(-\frac{E_a}{RT}\right)\tau_0\right] \qquad (4.23)$$

where $\Delta E = E_b - E_a$, E_b, E_a—the activation energies of flow processes of the crystalline solid and the elimination of defects, respectively; V_0, N_0—the pore volume and concentration of defects, respectively, at the beginning of isothermal sintering; τ_0—time. Equation (4.23) allows to describe simultaneously both the timing and the temperature dependence of shrinkage.

Sintering of fine powders has a number of features. With their sintering, the densification processes go so intense that a significant portion of shrinkage occurs already during heating to the sintering temperature. The densification rate sharply is attenuated with isothermal exposure. Initial densification rate for highly active superfine powders during sintering can be six to eight orders of magnitude larger than predicted by the theory of diffusion of vacancies for diffusion- viscous flow, even taking into account the role of grain boundary diffusion. The observed effects can be explained by the following reasons.

Systems of nanopowders are subjected to increased capillary pressure, since the radius of curvature of the quasi-spherical nanoparticle is equal to half of its linear dimension. Laplace's pressure on the particle surface at $r = 10$ nm and surface tension s ~ 1.5–2 kJ/m^2 is 150 ... 200 MPa.

Size effect has two aspects in relation to diffusion. First, the reduction of grain size increases the concentration of the boundaries along which the diffusion coefficients significantly larger than those in the bulk. In addition, it is found experimentally that self-diffusion coefficients, in nanocrystalline materials at relatively low temperatures, are higher by many orders of magnitude than expected ones which are extrapolated from high temperatures for grain -boundary diffusion. Thus, diffusion processes take place in nanostructured systems with much higher speeds than in conventional systems. A difference between diffusion parameters gradually disappears when the temperature rises due to relaxation of the nonequilibrium structure.

A key feature of sintering nanocrystalline PM is a competition between two major processes: shrinkage and enlargement of the microstructure. Both processes are controlled by the same mechanisms of mass transfer: surface and grain-boundary diffusion. Parameters of the thermal activation of these processes are

Fig. 4.4 Shrinkage during
isothermal sintering with
stepped temperature rise
(by V. Ivensen): *a*—600 °C,
b—740 °C, *c*—880 °C

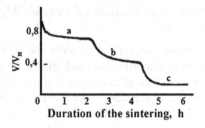

similar to each other and, as a consequence, shrinkage and evolution of the microstructure occur simultaneously.

During sintering of nanoparticles their size remains approximately constant over a wide range of porosity and their intensive growth begins at the porosity below 10 % and depends on the porosity type as well as the pore size distribution and the pore volume. During the sintering, it is necessary to control the pore size distribution while maintaining the channel open porosity. The grain size distribution is controlled at the final stage. During sintering of nanopowders, the special tendency to agglomerate and isolation of local shrinkage resulting in inhibition of densification, especially observed in the later stages of sintering, should also be considered. A good example of a fully localized sintering, not accompanied with densification of pressing in a whole, is sintering of tin oxide nanopowder. Compacts from SnO_2 powder with the particle size of 100 nm, having the porosity of 45 %, were sintered in air for 1 h at 500–1100 °C. Total shrinkage of briquettes was absent during sintering. The pore size distribution was measured and specific surface area was determined by mercury porosimetry for the porous samples. As a result of high-temperature sintering of active nanopowder, the pore size was increased fourfold, indicating the intensive zonal isolation process which takes place in the absence of overall shrinkage. The only way to combat the negative impact of localization shrinkage is to ensure maximum uniformity of particle size distribution of the powders and to use special (e.g., slurry) molding methods, excluding the formation of large pores while still forming.

Sintering the porous fibrous material is carried out under sintering conditions of powders of similar composition. Enlargement of sizes of fiber pellets during sintering occur most strongly in the direction of the previous extrusion (along a height). In the plane perpendicular to the direction of the forming force in the felting and pressing, a sample size is increased much smaller. Growth of samples is enhanced with increasing porosity along the height and width.

Progressive increase in the volume of samples with increasing porosity is a phenomenon inherent only for materials of a fiber structure, in which changes of the number and strength of contacts between the individual fibers under the influence of residual stresses, generated during the pressing, occur. Relaxation of residual stresses in the fiber compacts ends in the early stages of sintering (at heating to the temperature of exposure). Although further shrinkage will occur, a cumulative effect of manifestation of reversible and irreversible deformations in the form of a "growth" of samples is fixed more often at this stage.

4.4 Application of Porous Materials

Porous materials are widely used for the purification of liquids and gases, environmental protection, and human health protection. PM are used for purification of potable, waste and process water, alkali, acid and salt solutions, fuels and lubricants, food products (milk, wine, juices, vegetable oil), varnishes and paints, melts of polymers and metals, liquefied and compressed gases. They are used for deferrization of water by a method based on the oxidation of iron ions, followed by filtration of oxidized water during electro-filtration decontamination of drinking water and food liquids, disinfection of sewage water, desalination of water in an electric field using porous electrodes. Filtration is also proposed to use in an electric field through the highly porous cellular materials for purification of the air from fumes having the particle size less than 1 micron.

One of the promising areas of application of porous materials, determined by their high capillary and filtering properties, is cleaning and drying of compressed gases from mechanical impurities and moisture.

They are used for separation processes in the chemical, petrochemical and biological production.

Important properties of the filter are the degree of purification, high permeability, good reconditioning, high service life, and good corrosion resistance.

Materials with a high interconnected porosity could be used as a heat exchanger between a gas and a liquid or solid. Another application of the PM—is the heat pipes.

The efficiency of catalysis is critically dependent on the high magnitude of surface contact between the catalyst and the liquids or gases which are involved in the reaction. Therefore, the catalyst is a highly porous structure or, if not, is applied to a porous ceramic material. Porous metals may substitute ceramics, even if they cannot compete with it in the specific surface area, because they have high thermal conductivity and ductility. New area of catalysis at the nanoparticles is photocatalysis using semiconductor particles and nanostructured semiconductor films, which is promising, for example, for photochemical purification of wastewater from a variety of organic pollutants through their photocatalytic oxidation and mineralization.

One of the oldest applications of porous powder metal materials is self-lubricating materials in which the oil is stocked in the pores between the particles and is released in the operation process during the heating, replacing the oil used in this manner. Furthermore, the porous materials can reduce unwanted liquid flow in partially filled containers.

The most common sound-absorbing materials are porous materials. Losses of sound energy in the PM occur due to viscous friction of air in the narrow pores of the material, as well as sequential compression and rarefaction of air in contact with the sound wave, which is accompanied by the air heating, and due to the reflection of sound waves from the gas, disposed in the pores of the material back to the source. To reduce noise, dual mufflers are used at both low and high frequencies. Porous materials may also restrain sudden pressure changes occurring in the compressors, or pneumatic devices.

Saturation of liquids with gas by passing it through a porous elements—an effective way to intensify the processes in chemical technology, biological treatment of wastewater, fine cleaning and disinfection of drinking water and process water, as well as food, pharmaceutical and microbiological industry, municipal and industrial water treatment systems.

Porous metals may be used to stop spread of flame in the fuel gases.

Porous fiber material from alloys of Fe–Cr–Al is used in the neutralization of exhaust gases of diesel automotive engines. Highly porous aluminum can ideally perform three functions: absorption of energy in emergency situations, sound isolation, and heat extraction.

Porous titanium is used for the manufacture of prosthetic or dental implants because of their biocompatibility with living tissues.

References

1. Kostornov AG (2002) Material science of dispersible and porous metals and alloys. Naukova Dumka, Kiev (in Russian)
2. Belov SV (ed) (1987) Porous permeable materials Handbook, Metallurgiya, Moscow (in Russian)
3. Tumanovich MV et al (2010) Porous powder materials and items on their base for protection of human health and environment: production, properties and applications. Byeloruss, Navuka, Minsk (in Russian)
4. Liu PS, Liang KM (2001) J Mat Sci 36:5059
5. Banhart J (2001) Prog Mater Sci 46:559
6. Colombo P, Degischer HP (2010) J Mater Sci Technol 26(10):1145
7. Luyten J, Mullens S, Thijs I (2010) Kona Powder Particle J 28:131

Chapter 5
Thermal Insulating Materials

Yuliya Sergeevna Dzyazko and Boris Yakovlevich Konstantinovsky

Abstract In this section, some classifications of porous thermal insulation materials are given and different techniques for porosity investigation are described. Special attention is focused on the standard contact porosimetry, which provides no destruction of the samples and gives a possibility to determine pores in a wide diapason of sizes. Owing to these advantages, the technique allows us to research evolution of porous structure at different stages of the product preparation and identify the synthesis phase, when functional properties of the material are transformed to diametrically opposite ones. Effect of porosity on such properties as thermal conductivity and compression strength is estimated, the appropriated correlations are represented. The information dealt to research of thermal conductivity is given, the heat transfer through porous media is considered. It is noted, that the main way to reduce thermal conductivity is to increase porosity of the material, the contribution of solid phase can be diminished by this manner. This principle is used for manufacture of most of thermal insulators such as widespread polymer and inorganic foams. They are characterized by extremely low thermal conductivity, the order of magnitude of which is 10^{-2} and 10^{-1} W m^{-1} K^{-1}, respectively. Modern approaches to development of new thermal insulating materials and structures are considered. These approaches are based on a decrease in heat conductivity of gaseous phase. In order to minimize the fluid contribution, inert gases, which are characterized by lower conductivity in a comparison with air, can be encapsulated in closed pores. Other ways are degassing of thermal insulating materials and decrease of their pore sizes, simultaneously high porosity has to be provided.

5.1 Functions and Types of Thermal Insulation

Thermal insulation plays an important role in modern industry and building technology. It allows one to solve problems of life support, organization of technological processes, and energy saving. Thermal insulating structures are essential constituents of protective elements of industrial equipment, pipelines, and

Y. M. Volfkovich et al., *Structural Properties of Porous Materials and Powders Used in Different Fields of Science and Technology*, Engineering Materials and Processes, DOI: 10.1007/978-1-4471-6377-0_5,
© Springer-Verlag London 2014

buildings. Insulation increases safety, durability, and efficiency of exploitation of buildings, structures, and equipment.

Thermal insulation performs the following functions [1, 2]:

- maintains comfortable living conditions for residential houses;
- reduces heat loss to environment from the buildings, structures, equipment, pipelines, etc.;
- provides optimal temperature conditions of technological processes in apparatus;
- gives a possibility for long-range transport of vapor, gas, hot water, and different industrial products without sufficient change in their initial temperature;
- keeps predetermined temperatures of technological processes;
- ensures normal temperature conditions for operating personnel;
- decreases temperature stresses in metallic structures, refractory lining, etc.;
- allows one to decrease thickness and weight of cladding, as a result, the consumption of building materials, such as bricks, wood, cement etc., can be diminished;
- maintains predetermined temperatures in refrigerators and cooling devices;
- protects liquified gases and light oils against evaporation, when they are stored in insulated tanks;
- performs complex function, which involves also soundproofing, fire-resistance, protection against corrosion, fungi, bacteria, etc.

In order to perform these functions, the thermal insulators have to possess insignificant thermal conductivity (sometimes called "K-factor"), low density, slight hygroscopicity and vapor permeability, significant mechanical strength, elasticity, fire- and frost-resistance, uniform porous structure, absence of odors, and insensitivity to them. Moreover, these materials must not cause corrosion of the insulated surface or promote it. They must be harmless to human health and resistant against biogenic factors, cheap, convenient for transportation, installation, and repair. The lifetime of these materials should be long, no special care is to be be required. Known thermal insulators cannot satisfy all the requirements. Owing to this, the materials have to be selected rationally to provide the necessary properties of the insulating structure.

Thus, the main function of thermal insulating materials is to keep temperature of buildings, equipment, pipelines, etc., at the predetermined level. According to this, the thermal insulation can be subdivided into two types. First, heat insulation prevents heat losses to the environment. Alternately, cool insulation precludes heating of objects being isolated.

From the point of view of structure and manner of attachment to the surface, the classification involves the following types of porous insulating materials:

- tiled (foam plastic, foam glass, cork tiles, etc.);
- fibrous (mineral wadding in semi-rigid and rigid tails bounded with synthetic binder, glass fibers in tiles, etc.);
- loose (cork crumbs, fumed silica);
- sprayed and flooded (polymer foams).

The thermal insulators can be also divided into two types from a geometrical point of view. In accordance with this, the first type includes materials, in which the solid phase is a continuous medium (foams, cork, etc.). The other type is fibrous materials with discontinuous solid phase. The fibers can be held together with a bonding agent (mineral wadding in tiles), the space between them can be also free from a binder (cottonwool, wool).

Other classification includes inorganic (mineral and glass wadding, foam glass), organic (expanded and extruded polystyrene, polyurethane, melamine and phenol foams, cork), combined (gypsum foam, wood wool) as well as advanced materials (transparent and dynamic ones) [1].

Glass and stone wool insulators are fibrous products—this structure provides their outstanding thermal performance. They are produced using cheap and renewable natural resources, such as sand and basalt rocks, which are available elsewhere. Glass wool is made of sand, the manufacture process involves adding of recycled glass (cullet) and fluxing agents. Stone wool is made of slag and basalt, which are melted at 1,600–1,800 K. The fibers are formed by centrifugation through drilled disks.

Some wastes and disposals produced by agriculture can be also used for production of thermal insulating materials, for example, coconut husk, bagasse, corn cob. Jute, flax, and hemp are manufactured from renewable agricultural resources. Among the insulators of biogenic origin, the cork, whose structure is comb-like, probably is the most known material.

The new technology materials are intensively developed only in this century. Dynamic insulators (vacuum panels) are the core-shell constructions, the thermal conductivity of which can be controlled within a desirable range [3, 4]. The panels (15–25 mm of a thickness), which are made of a covering plastic foil containing a filling material (silica, polyurethane foam, etc.), are evacuated in order to achieve low pressure (10–100 Pa). The nonporous transparent materials for windows production, combine thermal insulating function, and lighting transmittance [5].

The classifications, which are developed in the countries of the former USSR, involve also light concrete as an insulator [6], though its thermal conductivity is much higher in a comparison with above-mentioned materials [7, 8]. In owing to this, the light concrete occupies weak positions in the world market of thermal insulators. However, this material performs other engineering functions, it has an intrinsic economies benefit over other building materials. Thus, the function of thermal insulation should be considered as additional for this type of concrete.

5.2 Porosity and Its Measurements

A number of functional properties, such as thermal conductivity, ability to absorb water, mechanical strength, heat capacity are affected by porosity, the first three ones depend on geometry and size of pores. Taking into consideration a shape, the pores, some types of them are given in Fig. 5.1, can be subdivided to four groups:

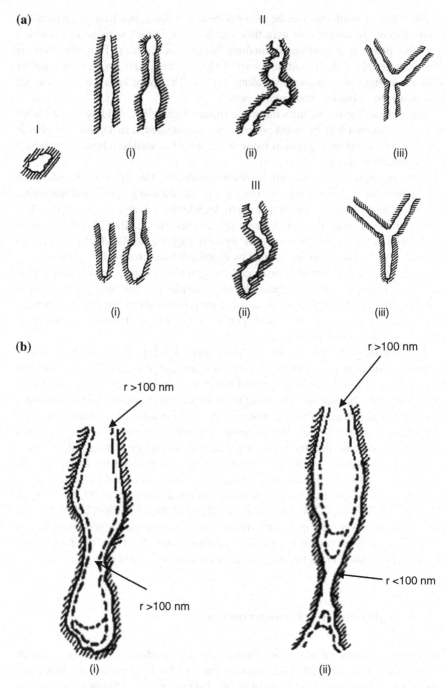

Fig. 5.1 Types of pores (**a**) and their filling with water (**b**). Pores: **a**, *I*—closed; **a**, *II*; **b**, *ii*—opened ones, which form channels; **a**, *III*; **b**, *i*—dead-end, **a**, *II*, *i*; **a**, *III*, *i*; **b**—straight; **a**, *II*, *ii*; **a**, *III*, *ii*—tortuous corrugated; **a**, *II*, *iii*; **a**, *III*, *iii*—branched

- according to shape of cross-linking section: tubular, bottle-like, wedge-shaped, slit-shaped, and their combinations;
- according to extension: straight, tortuous;
- according to continuity: opened (which form channels), dead-end (opened from one side), closed (pores, which cannot be found with any porosimetry method except pycnometer (Archimedes) technique).

Among the rangings, which are based on pore sizes, the classification proposed by M. M. Dubinin is the widespread [9]. The main criterion is a possibility of water condensation in various voids. The pores, a radius (r) of which is less than 1 nm, are related to micropores. In these voids the capillary condensation occurs according to the mechanism of volume filling (Fig. 5.1b). As a result of capillary condensation in larger pores ($r = 1–100$ nm), their walls are covered with a thin water film. These voids are considered as mesopores. At $r > 100$ nm, no capillary condensation occurs, these voids (macropores) can be filled with a liquid only under a direct contact. No water absorption occurs in macropores. Moreover, the macropores lose water, which was located there initially. It is exactly the classification, which will be used further.

The porosity of thermal insulating materials is represented by randomly distributed voids of variable cross-linking section with a radius from 1–2 nm to 50 μm, for some pores the r value can reach 50 μm–5 mm. Total porosity is usually 0.75–0.98 %. Regarding foam materials, the terms like "cell", "gas-filled cell" are usually used in a literature. The morphological unit of foams is the primary spatial structure consisting of the cell, its walls and edges (strands). This volume element, which includes gaseous and solid phases, is repeated at regular intervals and high degree of order, it forms a whole structure of the foam (Fig. 5.2 [10]). The primary structure is characterized by shape and size, packing type, and certain pattern of distribution in a bulk of the material.

In opposite to a number of polymer foams, closed pores make a great contribution into total porosity of inorganic materials, such as light concrete,

A very important characteristic, which is related to porosity, is the bulk density (ρ_b): this parameter involves the volume of solid phase and voids. The methods of its determination have been standardized for building materials and described elsewhere. It should be noted, that the fibers have to be compacted before the measurements of geometric sizes of the sample. The ρ_b values are summarized in Table 5.1 [1, 7] for several insulating materials, these characteristics are seen to be extremely low. The light concrete shows the highest ρ_b values, however, they are much lower in a comparison with compact concrete (2,200 kg m^{-3}).

The particle density (ρ) is attributed only to solid phase. Further we will use this term for all the materials, though it is incompletely exact for polymer foams. The pycnometer (Archimedes) method for measurement of this parameter, which allows one to determine both opened and closed pores, has been also standardized. In the case of polymer foams, this value is obtained using the formula [11]:

Fig. 5.2 Honeycomb
structure of polystyrene
foam [10]

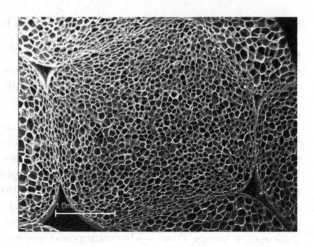

Table 5.1 Density and
thermal conductivity of some
insulating materials [1, 4, 7]

Material	Bulk density, kg m^{-3}	Thermal conductivity, W m^{-1} K^{-1} (at ≈ 298 K)
Glass wool	13–100	0.030–0.045
Stone wool	30–180	0.033–0.045
Extruded polystyrene	20–80	0.025–0.035
Expanded polystyrene	18–50	0.029–0.051
Polyurethane foam	30–80	0.020–0.027
Binderless coconut	250–350	0.046–0.068
Binderless bagasse	250–350	0.049–0.055
Flax	20–100	0.035–0.045
Flax and hemp	39	0.033
Fumed silica	70–220	0.020
Foam concrete	300–1,600	0.100–0.660

$$\rho = \frac{\rho_b}{1 - 1.33\pi r^3 N}, \tag{5.1}$$

where N is the number of pores per volume unit, which is determined using microscopy technique. Porosity of foam polymers can reach even 0.98.

The next important parameter is the porosity (ε), which can be found from a simple relation of $\varepsilon = 1 - \frac{\rho_b}{\rho}$. The space occupied by closed pores and solid phase is measured by comparing the external volume of the sample with the volume of gas, which is displaced from the hermetic chamber of special device after the sample insertion. The difference of these volumes is attributed to the opened pores. This technique can be applied only to macroporous materials.

Since the thermal insulating materials contain pores, a size of which is in a wide diapason, different methods are necessary for their diagnostics.

The use of high resolution optical or scanning electron microscopy (SEM) followed by automatic image analysis is very useful to determine the pore size distribution for cellular materials (polymer or inorganic foams) [7, 11]. In the case of glass fibers, the ways of sample preparation, microscope setting, and algorithms of statistical analysis of the SEM images have been developed [12]. The images give information about the solid phase, porosity, anisotropy, and shape of the pores. An X-ray tomography method allows one to obtain 3D images of polymer foams, based on which the program can be plotted [13]. Transmission electron microscopy (TEM) is attractive for nanomaterials like fumed silica.

All the porosimetry methods allow one to recognize only opened pores. Mercury intrusion is generally carried out to investigate inorganic materials. However, no voids, which radius is higher than 100 μm, can be determined using this technique. These cavities can be identified only with optical microscopy. Moreover, the mercury porosimetry is unsuccessful for polymer foams due to their rupture under high pressure. Nevertheless, sometimes this method is used for analysis of porous polymers [14].

The measurement of gas permeability indirectly characterizes the pore structure [7]. Only continuous pores, which permit the gases through the entire thickness of the material, can be identified. Nitrogen adsorption technique is used to determine micro- and mesopores.

The method of standard contact porosimetry (MSCP) (see Chap. 1) was applied to the carbonic fibers [15]. After the additional fluffing, this material can be used as a thermal insulator. However, the materials of this type can be transformed into refractories, which must be characterized by significant bulk density and high conductivity in opposite to thermal insulators. The function transformation can be achieved by modification of fibers with silicon carbide. The MSCP allows one to investigate the evolution of porous structure after each manufacture stage. This is an important task, since the maintenance of porosity during chemical vapor deposition provides uniform filler precipitation on the fibers. Let us describe this application of the MSCP in more detail.

The procedure of the composite synthesis involved: (I) obtaining fibrous matrix by calcination of carbon plastic in inert media at 1,273 K, (II) chemical vapor deposition from methylsilane at 973 K, (III) annealing in inert media at 1,873 K, (IV) second SiC insertion.

As seen from Fig. 5.3, the highest volume of the opened pores is provided by macropores ($r > 100$ nm), however, micropores ($r < 1$ nm) make the greatest contribution to the specific surface (except the annealed sample after the stage III). Increase of microporosity after the modifier deposition is caused by formation of secondary porosity attributed to SiC. Decrease of micropore volume after the annealing is due to a change of structure of the substrate and modifier.

SiC deposition inside the carbon matrix causes a decrease of volume of macro- and mesopores. According to thermodynamics of capillary processes, first of all the largest pores must be filled with a precipitate [16]. This is due to minimization of energy, which is necessary for formation of a new phase, since the specific surface area is inversely proportional to the pore size. During the deposition, the front of filling moves to finer pores, especially for relatively low deposition rates.

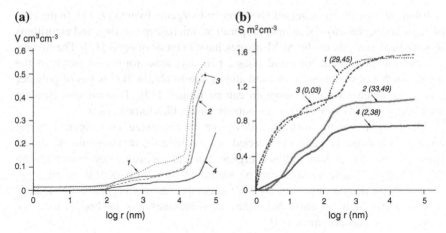

Fig. 5.3 Integral pore volume (**a**) and specific surface area (**b**) distributions for carbon fibrous matrix: unmodified (*1*), once modified with SiC (*2*), once modified and annealed (*3*), modified repeatedly after annealing (*4*). Numbers in *brackets* correspond to specific surface area ($m^2 \ cm^{-3}$) caused by micropores (**b**) (adapted from [15])

Fig. 5.4 SEM image of carbon fibers modified with SiC (adapted from [15])

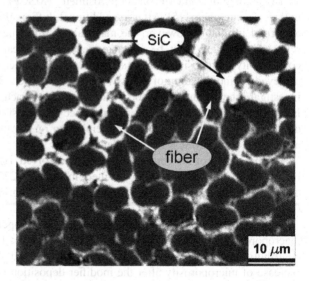

However, if the reaction rate in pores of the substrate is limited by diffusion of reagents, the finer pores are also filled partially.

Disappearance of micropores and increase of mesopore volume during the annealing (stage III) facilitate reagent transport, when the second insertion of the modifier is realized (stage IV). This provides uniform filling of the fibrous matrix with SiC (Fig. 5.4). Regarding the sample after the stage IV, rather small porosity (0.25), which is caused mainly by meso- and macropores, has been found. Probably the residual porosity is affected by the pressure on the pore walls due to

Table 5.2 Categories of properties, which depends on porosity (adapted from [17])

Porosity dependence			Examples of properties
Category	Responsible phase	Character of dependence	
I		No dependence on porosity	Lattice parameter, unit cell volume, length of the elementary unit of the chain, thermal expansion, emissivity, melting, softening and boiling temperatures, ablation energy
II		Dependence only on the amount of porosity	Density, much dielectric constant data, heat capacity per unit volume
III		Dependence on both the amount and type of pores	
	Solid	Flux or stress dominant in the solid phase	Mechanical properties. Electrical and thermal conductivity under low and moderate temperatures
	Fluid in pores	All flux in the pore phase and filtration	Specific surface area and tortuosity, e.g. for catalysts
	Both solid and fluid	Flux in both pore and solid phases	Thermal conductivity, with larger and more opened pores at higher temperature

SiC deposition (swelling). This sample demonstrates higher bulk density ($1{,}680$ kg m^{-3}) in a comparison with unmodified fibers (650 kg m^{-3}). Silicon carbide, which possesses high density as well as significant thermal conductivity, keeps refractory function of the composite. Thus, the carbon plastic is related to starting materials, which allows one to obtain products with diametrically opposite properties: thermal insulator fibers (low density, inconsiderable thermal conductivity) or refractory composite (high density, significant thermal conductivity).

Thus, the MSCP allows one to identify the evolution of porous structure from the initial material to end-product, the functions of which can be essentially different. This possibility of the technique is caused by a wide diapason of pores, which can be recognized, and also by a nondestructive character of the measurements. Porous structure of thermal conducting materials determines their functional properties. This problem is considered further.

5.3 Functional Properties Affected by Porosity: Measurements of Thermal Conductivity

The porosity dependencies can be divided to three wide categories, which become more complex in a hierarchical manner (Table 5.2 [17]). There are properties, which demonstrate no dependence on porosity, at the lowest level. The characteristics, such as lattice parameter and unit cell volume, are affected only by the local atomic bonding. They determine macroscopic behavior (melting, softening, and boiling temperatures, thermal expansion, etc.).

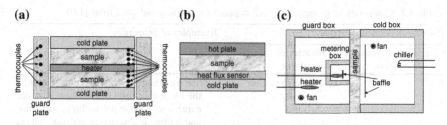

Fig. 5.5 Common steady-state techniques for measurements of thermal conductivity: **a** guarded hot-plate, **b** basic heat flow-meter, **c** guarded hot-box (adapted from [19])

The second category consists of the properties (density, heat capacity, and much dielectric constant data), which are dependant only on the pore volume per volume unit and not affected by the character of porosity. These characteristics follow a rule of relation for the volume fractions of pores and solid phase:

$$X = X_s(1 - \varepsilon) + X_p\varepsilon, \qquad (5.2)$$

where X is the property of porous material, X_s is the property of the nonporous solid and X_p is the property of the pore phase (it is often zero, but not always, $X_p \approx 1$ for the refractive index and dielectric constant).

The third, largest, most complex, and important category consists of the properties, which are dependant on the volume of pores per volume unit, their size, and shape. Three important subsets of this category are determined by a responsibility of solid phase as well as pore liquid and gas for a characteristic.

Since the main function of the insulating materials is to keep a constancy of temperature [18], further we will consider mainly their thermal conductivity. The order of its magnitude for the insulators is 10^{-2} W m^{-1} K^{-1}, this is much lower in a comparison with nonporous substances (10^{-1} W m^{-1} K^{-1} for polystyrene) and especially for metals (10^{1}–10^{2} W m^{-1} K^{-1}). The light concrete, which is characterized by the highest thermal conductivity, is often excluded from classifications of insulating materials [1, 2].

The devices for measurements are shown in Fig. 5.5 [19]. In owing to inconsiderable heat flow through the insulators, the determination of this characteristic is difficult and requires special accuracy as well as a large massive of experimental data. Thus, in spite of standardized techniques, new ways of measurements are still developed. The methods are based on measurements of a temperature gradient through the sample with known thickness, the heat flow from one side to the other is controlled. A one-dimensional flow approach is used, but also other geometrical arrangements are taken into consideration. The thermal conductivity (λ) is calculated using the Fourier equation [18]:

$$q = -\lambda A \operatorname{grad} T, \qquad (5.3)$$

where q is the steady-state flow (W m^{-2}), λ is the conductivity (W m^{-1} K^{-1}), A is the cross-section area of the sample (m^2), T is the temperature (K).

The guarded hot-plate technique involves uniformly wound heaters in a central metered section and in a thermally isolated guard area separated by a small coplanar gap. The temperature sensors are fitted tightly in all surfaces of the central and guard sections. Measured dc-power is applied to the hot plate, the temperatures of the cold plate and guard sections are adjusted and controlled. The zero temperature difference across the gap and the desired temperature difference across the sample are necessary. The device provides a very wide temperature interval of 173–473 K for measurements and allows one to analyze the sample with a conductivity of 7×10^{-3}–1 W m^{-1} K^{-1}.

Regarding the heat-flowmeter technique, a sample is placed between two plates. The heat flow through the sample is measured with sensors after stabilization of a temperature gradient. The measurement principle remains nearly the same, but the test section is surrounded by a guard heater, resulting in wider temperature diapason (from 93 to 1,273 K) and broader interval of thermal conductivity (1×10^{-4}–2 W m^{-1} K^{-1}).

The hot-box technique is normally used for investigation of the materials, the conductivity of which is 0.2–5 W m^{-1} K^{-1}. The overall thermal resistance, which includes air film resistances in the cold and warm sides as well as the sample resistance, can be obtained from these measurements. A large sample is placed between hot and cold chambers. The device provides the temperatures 253–313 K.

Dynamical [19] and pulsed [20] techniques as well as methods of hot wire [21] and hot ring [22] have been also developed for measurements of thermal conductivity of the insulating materials. The appropriated calculations have been also proposed [19–22].

5.4 Heat Transfer

Three heat transfer mechanisms, such as conductivity, convection, and radiation, are known [18, 23]. The first and the most effective conducting path is through the bulk of solid phase. Conductivity occurs, when the insulator particles are in a contact with each other and energy is transferred through these contacts. The second path is the transfer through the contacts (grain boundary or other interfacial phases). Sometimes the possible effects of the contacts is neglected, however it becomes important for composites. Another factors are convective and radiative transfer, which occurs exclusively and mostly via the pore phase. Convective heat transfer is a result of energy transfer by moving particles of gas or fluid. Radiation takes place due to the emission of electromagnetic waves, which transfer energy from a hot emitting object to a cold receiving surface. In opposite to conductivity cased by convection, this mode of heat transfer can be realized in vacuum. Some models, which are based only on empirical approaches, consider the thermal conductivity only as a function of porosity. In other words, the conductivity is plotted versus porosity or density, then the appropriated function is approximated. The functions obtained for matters with different pore shape are summarized in Table 5.3 [17].

Type of pores	Equation
Pores between spherical particles	$1 - 1.5\varepsilon$
Foams or pores between aligned cylindrical particles	$(1 - \varepsilon)^n$
Spherical pores	$(1 - \varepsilon)^{1.5}$
Cubic foam cells	$0.33(1 - \varepsilon^2)$
Various, including spherical pores	$(1 - \varepsilon)(1 + a\varepsilon)^{-1}$
Uniform cubic pores in simple cubic stacking	$(1 - \varepsilon^{0.66})(1 - \varepsilon^{0.66} + \varepsilon)^{-1}$

Table 5.3 Models for the porosity dependence of thermal conductivity of solids at ≈ 295 K (adapted from [17])

As seen, complication of pore geometry causes more complex type of the $\lambda - \varepsilon$ functions

Let us consider the models of heat transfer in more detail. Usually the first two paths of heat transfer are combined, thus, the total thermal conductivity (λ) can be divided to three terms [23]:

$$\lambda = \lambda_s + \lambda_g + \lambda_r \qquad (5.4)$$

where λ_s, λ_g, λ_r are the conductivity of gas, solid (including also contacts), and radiation, respectively. Since the largest of these factors is the solid conductivity, the insulation materials have to be highly porous with a small amount of solid phase. Regarding the materials with a small amount of solid, the importance of the radiation enhances (Fig. 5.6 [24]). This creates an optimal point from insulation perspective, for a certain material, where the sum of the contributions from radiation and solid conductivity is at a minimum. The gas conductivity can be considered as constant for conventional insulators. This gives a total thermal conductivity down to a minimum around 0.03 W m^{-1} K^{-1}, which can be compared to the air conductivity of 0.025 W m^{-1} K^{-1}.

There is also some difference in the solid conductivity for various materials. For instance, a number of models of heat transfer through the fibers has been developed. An exact theoretical formulation of solid conductivity for the fibrous geometry is a formidable task because of the tenuous nature of the myriad of different paths of varying fiber lengths and cross-sectional areas. A semi-empirical approach was used to model solid conductivity in unbounded fibers [25]:

$$\lambda_s = F_s \lambda_{s,0}(1 - \varepsilon)^b, \qquad (5.5)$$

which relates the solid thermal conductivity of fibrous insulator to the thermal conductivity of nonporous material ($\lambda_{s,0}$) and porosity. The F_s parameter characterizes the geometric effects of the fibrous matrix and accounts for the various fiber path lengths, arrangement of the fibers, and contacts between them. This parameter is assumed to be independent on temperature. The formula 5.5 looks similar to expressions 4.1, 4.2 (see Chap. 4), which describes the mass transport in pores (electric conductivity and diffusion). The deceleration factor of $(1 - \varepsilon)^b$ depends

Fig. 5.6 Heat transfer
mechanisms in conventional
fibers and foams (adapted
from [24])

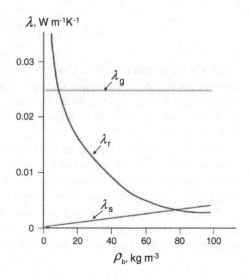

on tortuous path through pores. The F_s and b values were obtained from the thermal conductivity measurements in a vacuum at cryogenic temperatures (test condition with negligible gas conductivity and reduced radiation).

Regarding the polymer foams, the conductivity of solid, which includes the contacts and pore phase, is expressed as [26]:

$$\lambda_s = 0.33\,(1 - \varepsilon)\,(2 - F_s)\,\lambda_{s,0} \tag{5.6}$$

A very strong relation has been obtained for globular nanoporous materials like fumed silica [27]. It was suggested, that a characteristic size of globules (few nanometers) is very close to the phonons free path. The equation is as follows:

$$\lambda_s = \frac{\rho\,v_s}{\rho_b\,v_b}\lambda_{s,0}. \tag{5.7}$$

Here, v_s and v_b are the phonon velocities (m s^{-1}) in the monolithic and porous silica, they are assimilated to sound velocity in these materials.

The λ_b value is approximated as polynomial function of a temperature [23]. For instance, the following empirical equation has been proposed for silica [28]:

$$\lambda_{s,0} = -4.22 \times 10^{-6}T^2 + 4.36 \times 10^{-3}T + 0.44 \tag{5.8}$$

To minimize the solid conductivity, a proper solid material should be chosen. As shown in Fig. 5.6, the solid conductivity becomes negligible with a decrease of bulk density. However, the heat transfer due to radiation enhances. This constituent of the total heat flow is due to electromagnetic radiation, which is emitted by all surfaces. The net radiation is the difference between the radiation from the warm surface and the radiation from the cold surface. The rate of heat transfer by radiation is dependent on the temperature of a surface [23]:

$$\lambda_r = \frac{16n^2kT^3}{3\rho_b\varepsilon_{\text{ext}}} \tag{5.9}$$

where n is the refraction index, T is the temperature (K), k is the Stefan–Boltzmann constant (W m^{-2}K^{-4}) and ε_{ext} is the extinction coefficient (m^{-1}). In quite opened polyurethane foams, the contribution of radiation transfer to the total conductivity has been estimated as 5–40 % with a reducing of the foam density from 100 to 10 kg m^{-3} [29].

With an increasing temperature, the heat transfer by radiation increases rapidly. This increase can be counteracted by adding of an opacifier, for example, TiO$_2$, which scatters the radiation, or carbon soot, which absorbs the radiation [30]. Magnetit as well as SiC have been also proposed.

The conductivity of solid and gas was found to be proportional to the bulk density [31]. Based on this, the relation between different constituents of the opacified material has been developed [32]:

$$\lambda = \lambda'_s \left(\frac{\rho_b}{\rho'_b}\right)^{1.5} + \lambda'_g \left(\frac{\rho_b}{\rho'_b}\right)^{-0.6} + \lambda'_r \left(\frac{\rho_b}{\rho'_b}\right)^{-1} \left(\frac{T}{T'}\right)^3 . \tag{5.10}$$

where the " ' " index is related to the parameters of a nonopacified material. If the density as well as separated contributions of radiation, solid, and gas are known, it is possible to calculate the conductivity of any new material of the same type.

Convective effects are affected by the pore size and nature of fluid, in general, no convection occurs in pores, which size is less than 1–4 mm. Convection in porous materials can be divided to two types: the transfer inside the pores and through the material on a macroscale. For closed pore systems, there is no macroscale convection. At the same time, the convection in pores can be neglected, if their size is small. This is due to low temperature differences on the pore walls. Thus, the heat transfer caused by convection is often neglected for materials with a closed pores. In the case of materials with opened pores, the macroscale convection has to be taken into consideration.

The macroscale convection is either caused by natural or forced convection. For natural convection, the air movement is due to density differences as a consequence of temperature differences, while forced convection is a result of pressure difference induced by wind or fan.

The dimensionless Rayleigh number (Ra) characterizes natural convection in pores [33]:

$$Ra = \frac{g\beta\Delta T\delta^3}{a\nu} \tag{5.11}$$

where g is the acceleration due to gravity (m s^{-2}), β is the gas thermal expansion coefficient (T^{-1}), ΔT is the temperature difference between both sides of a pore (K), α is the gas thermal diffusivity (m^2 s^{-1}), ν is the kinematic viscosity (m^2 s^{-1}), δ is the characteristic size of pore (m), which can be interpreted as the distance

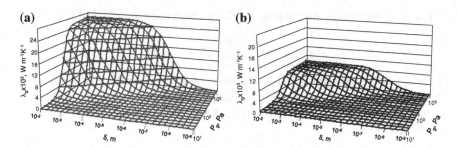

Fig. 5.7 Thermal conductivity of air (**a**) and krypton (**b**) as functions of the materials characteristic pore size and the gaseous pressure at a temperature of 300 K (adapted from [4]). The values have been retrieved from Eq. (5.12) to represent the Knudsen effect

between two parallel walls. The Rayleigh number is used to determine whether heat transfer through a fluid is mostly due to conductivity or natural convection. The computation of Rayleigh number for a 100 nm pore and temperature gradient of 1,000 K m^{-1} (the temperature difference between two pore walls is $\approx 10^{-4}$ K), gives Ra $\approx$ 10–17. No natural convection occurs for Ra < 2,000.

In order to reach lower λ_g value, the gas could be exchanged to a gas with lower conductivity (Fig. 5.7 [4]). Argon, the conductivity of which is lower in a comparison with nitrogen (0.016 and 0.024 W m^{-1} K^{-1} at 273 K, respectively) may be used for manufacture of foam polymers [34]. Inert gas can be trapped in closed pores or occupy a space between panels in the insulating structure.

Another way, which allows one to limit heat transfer through gaseous phase, is vacuum conditions [3, 4, 24]. The gas conductivity in porous media at lower pressure is determined by the number of gas molecules as well as by the number of obstacles for the gas, when it passes from the hot side to the cold one [33]. While reducing the gas pressure, the conductivity of the nonconvective gas remains almost unaffected until the free path of the gas molecules reaches values in the same order of size as the largest pores.

When the pore diameter is less, than the free length of path, the gas molecules will only collide with the pore surfaces without transferring energy by this elastic impact. Thus, the gas conductivity term has to be excluded from Eq. (5.4). However, to reduce the gas conductivity in conventional insulation materials as mineral wool, the pressure has to be reduced drastically to the range of 0.1 m bar or below and the thermal conductivity will rapidly increase with increasing pressure. Therefore, nanostructured core materials in a combination with the pressure reduction are more favorable. In such material, a fine vacuum is already adequate to reduce the gas thermal conductivity. Such a vacuum has a pressure around 10^2 Pa, a particle frequency of 10^{10} m^{-3} and a free length of path is 10^{-4} m. Thus, the thermal conductivity of the material is affected by the reduced gas pressure up to one-tenth of an atmosphere.

The gas conductivity in a porous media can be written as follows [33]:

$$\lambda_g = \frac{\lambda_{g,0}}{1 + 2\kappa Kn},$$

(5.12)

where $\lambda_{g,0}$ is the conductivity of the gas moving freely, κ is the constant for the effectiveness of energy transfer between gas molecules and pore walls, this value is commonly between 1.5 $\lambda_{g,0}$ and 2 $\lambda_{g,0}$, Kn is the Knudsen number, which is determined as:

$$Kn = \frac{l}{\delta},$$

(5.13)

where l (m) is the length of free path. The gas conductivity is strongly dependent on the ratio between the pore size and the free path of gas molecules inside pores. The free path is the average distance of travel of a molecule before colliding with another molecule. The l value can be calculated as:

$$l = \frac{k_B T}{\sqrt{2} s P_g}.$$

(5.14)

Here k_B is the Boltzmann constant (1.38×10^{-23} J K^{-1}), P_g is the pressure (Pa), s is the molecule cross-section area (m^2). For nitrogen and oxygen, which are the main components of air, the molecular cross-sectional area is around 0.4 nm^2. The mean free path becomes ≈ 70 nm at 293 K and 10^5 Pa. If the characteristic pore size is 100 nm, the Knudsen number is close to 1 [4].

The β constant depends on the gas, solid and temperature. Because of the high porosity of insulation materials, the contribution of the gas conductivity will play an important role under atmospheric pressure. However, the free air conductivity $\lambda_{g,0}$ in Eq. (5.12) will be strongly reduced due to the Knudsen effect in the case of nanoporous materials.

If we rewrite Eq. (5.12) for vacuum, a formula which accentuates the three main parameters for gas heat conductivity in porous medium is as follows [4]:

$$\lambda_{evac} = \frac{\lambda_{g,0}}{1 + C(T/\delta P_g)}.$$

(5.15)

Here λ_{evac} is the gas conductivity under vacuum, $C = \frac{2\beta K_B}{(\sqrt{2}s)}$.

Thus, there are several approaches to reduce the gas contribution to the thermal conductivity, The first one is to employ highly porous materials. Another way is to use low conductivity gases by means of their encapsulation into the materials or insulating structures.

As mentioned above, a modern trend among the structures is vacuum insulating panels [3, 4, 24]. If micro- or nanostructured materials with small pore size (for instance, fumed silica) are used as cores of these panels, only weak vacuum is required to reach a low thermal conductivity. To reduce the gas conductivity in

Fig. 5.8 Typical
compression stress–strain
curve of a low-density
elastoplastic foam

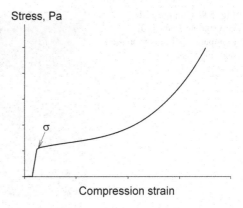

insulation materials with large pore sizes, the pressure has to be very low. It is difficult to maintain low pressure by envelopes made of organic materials. Therefore, a nanostructured core material in a combination with fine vacuum is preferred. A pore size of 10 nm or less would be ideal, in this case the gaseous conductivity reduces to zero even under atmospheric conditions. Polymer, metal, or metalized multilayer films are used as an envelope to provide vacuum conditions.

5.5 Compressive Strength

As follows from the previous sections of this chapter, the most of thermal insulating materials possess very high porosity, this is necessary to avoid high contribution of solid phase into the total conductivity. However, considerable porosity causes deterioration of their mechanical characteristics: compressive, tensile, and rupture strength, Young modulus, etc. In owing to this, it is important to focus special attention on mechanical properties in order to reach the consensus between thermal conductivity and mechanical strength. The optimal combination of the functional properties is necessary, since the constructions involving thermal insulators are always stressed due to squeezing, shock, tensile, and so on. In order to achieve the consensus, the effect of porosity on mechanical strength of the materials has to be considered.

One of the most frequently used parameter, which reflects mechanical durability, is the compressive strength (σ_b, here the "b" subscript refers to bulk of the material). The method of quantitative estimation of this characteristic is a determination of the maximum pressure, which corresponds to brittle failure of the sample (cracking) or dramatic change of a strain diagram, if no destruction of the sample occurs [35] (Fig. 5.8). The loading has to be perpendicularly to the sample surface. The compressive strength is not determined in the case of fibrous materials, since this parameter cannot be measured without their preliminary compaction.

Fig. 5.9 Compressed
polystyrene foam [10]
(compare with uncompressed
one, Fig. 5.3)

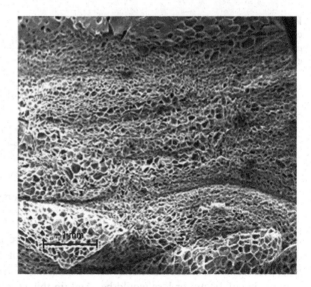

The structures, which involve the fibrous materials, must contain special devices (backup, unloading), which prevent compaction of the fibers.

The compressive strength is determined by means of a standardized method using special device. The σ_b value is calculated using the formula:

$$\sigma_b = \frac{F_{\text{des}}}{A},\tag{5.16}$$

where F_{des} is the failure load (N).

Foam polymers demonstrate pore destruction, since the strands lose steadiness, when the failure stress is reached. The rigid light polymer foams, which contain large pores, demonstrate the failure of the most impaired strands approximately at the height of one pore. The fine-pored materials show the strand collapse at an altitude of several pores (Fig. 5.9). Three fields are observed on the stress-strain diagram for elastic foams (see Fig. 5.8). The rapid ascent of the initial field, which corresponds to a stress growth, is related to the strand buckling before the loss of steadiness. The plateau reflects a bend of the strands after loss of steadiness. The third field is related to further increase of stress and caused by compaction until intercontacts of the intersecting strands.

Figure 5.10 shows a dependence of compressive strength on porosity both for polymer and inorganic foams [11, 35]. This parameter reduces nonlinearly with increase of porosity. Some models have been derived for the porosity dependence of strength based on mechanics analysis [17]. Pore stress concentration is assumed to increase the stress of the defects, which are located near the pore. As a result, the fissure appears in this field. The empirical approaches, which are mainly used for the thermal insulating materials, are commonly based on log–log or semilog plots giving straight lines.

Fig. 5.10 Compressive strength versus porosity of polystyrene foam (*1*), foam concrete (*2*). The *plots* are given based on analysis of [11] (*1*) and [35] (*2*). The density of polymer foam [11] were recalculated to porosity taking into consideration bulk density of polystyrene ($\approx 1,000$ kg m^{-3})

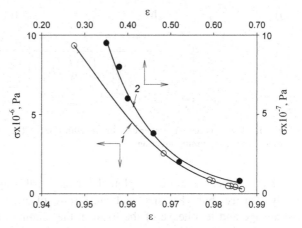

Regarding polymer foams, the equations for the relative compressive strength ($\frac{\sigma_b}{\sigma}$, where σ is the strength for the material with zero porosity) is [36]:

$$\frac{\sigma_b}{\sigma} = \tilde{N}\left(\frac{\rho}{\rho_b}\right)^n, \tag{5.17}$$

where C and n are the constants. The $\frac{\sigma_b}{\sigma}$ function is often plotted versus $\frac{\rho}{2r}$. As seen from Table 5.2, mechanical properties are strongly dependent not only on total porosity, but also on size and shape of pores. Thus, anisotropy of pores is often taken into consideration in a number of models.

In accordance to anisotropy, the foam structure can be subdivided to three types. The first type corresponds to foams with approximately spherical pores (Fig. 5.11). The pore shape of foams of the second type is ellipse-like, the ellipses are parallel to the foaming direction. The foams of the third type contain pores, which are oriented perpendicularly to this direction.

By means of image analysis, two parameters of pore distribution can be obtained. The first parameter is the average pore shape F in the plane of observation u (XY, YZ, XZ) [11]:

$$F_u = \frac{r_{u,\max}}{r_{u,\min}}, \tag{5.18}$$

where $r_{u,\max}$ and $r_{u,\min}$ correspond to the average maximum and minimum pore radius. The second parameter is the average cell orientation (θ), defined as the angle between the axis of maximum cell diameter and the testing direction in the plane of observation. These parameters are converted into anisotropy factor (ω):

$$\omega = F\theta \tag{5.19}$$

Here the F and θ parameters are the pore shape and pore orientation respectively attributed to one plane.

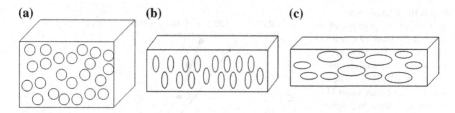

Fig. 5.11 Foam pores: **a**—spherical, **b, c**—ellipse-like. The pores are oriented parallel or perpendicular to foaming direction

The $\frac{\sigma_b}{\sigma}$ functions are often plotted versus $\frac{\rho}{2r\omega}$, this approximation involves not only density, but also size and anisotropy of pores. Anisotropy causes less shrinkage and less heave of the foam in the foaming direction than those in the perpendicular direction. Decrease of pressure drop inside and outside closed cells, spherical pores, increase of porosity as well as amount of opened pores provide mechanical stability of the foams.

Regarding inorganic foams like light concrete, the compressive strength is often given as a function of porosity, i.e., $(1 - \varepsilon)^n$ [35]. Exponential decrease, logarithmic as well as inverse proportional functions have been also proposed.

Thus, increase of porosity of foams leads to deterioration of their mechanical strength, however, the heat conductivity decreases. Depending on the application field, the insulator materials have to possess optimal combination of thermal and mechanical characteristics. This complex of properties is reached by means of formation of pretermined porous structure, which is viewed further.

5.6 Controlled Pore Formation in Thermal Insulating Materials

In order to impart necessary functional properties to the thermal insulators, special approaches are used. For production of polymer foams, such as polyethylene, polystyrene, polyurethane (Fig. 5.12), following methods are used: (i) gas formation directly in the polymer; (ii) foamer insertion to the polymer suspension; (iii) combination of these techniques [11, 29, 37–39]. The substances, which can be well distributed in the polymer, are used for the gas formation method. The requirements for the blowing agents involve no interaction with the polymer, gradual gas formation, when the temperature is close to that of the polymer softening. The heat generated during chemical reaction has not to cause the polymer destruction.

In accordance to physical state, the blowing agents are subdivided to solid, liquid, and gaseous. Some of them generate gas due to irreversible (diazoaminobenzene $C_6H_5-N = N-NH-C_6H_5$, benzenesulfonyl hydrazide $C_6H_5-SO_2-NH-NH_2$) and reversible (NH_4HCO_3, $(NH_4)_2CO_3$) thermal decomposition. The mixes of metal

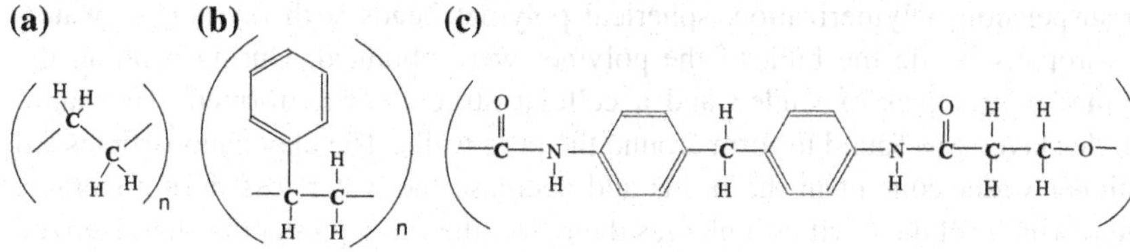

(a) **(b)** **(c)**

Fig. 5.12 Polyethylene (**a**), polystyrene (**b**), typical polymer related to polyurethanes (**c**)

powders (Zn, Al, Mg) with organic acids, for instance, oleic acid, are applied. In this case, the gas evolution occurs due to chemical interaction. Such adsorbents, as activated carbon and clays, silica, zeolites, which liberate gases adsorbed preliminary (thermal desorption), are also related to blowing agents. The substances, which become gaseous under heating (isopentane, freons), as well as gases (inert gases, air, N_2, CO_2) are also applied.

Usage of the blowing agents allows us to reach the highest porosity of the polymers (up to 85–95 %). Such foamers as soaps, proteins, pectins, saponins are often employed to produce polymer foams. Water also can be used as a blowing agent (water expandable polymers) [40]. Both the blowing agents and foamers allow one to obtain the polymer foams of honeycomb structure (see Fig. 5.2).

Strict requirements dealt to type of pores: they must be either opened (polyurethane for vacuum insulating panels) or closed (polymers containing inert gas). In owing to this, special attention is focused on the way of pore opening during foam formation [29]. This process is accompanied by heat evolution. Since the temperature of the foam bulk becomes higher, the foam top lose heat to ambient air. As a result of the heat loss, the temperature at the top surface of the foam is lower in a comparison with that of the bottom and center. The temperature profile yields slower reaction kinetics and retards phase separation. When the opening of the internal cells occurs, the top surface of the foam is still liquid. The expansion process is going on, if the internal pressure is enough to tear the liquid surface layer (skin blow-off time). The pores are opened continuously between the internal cell opening time and the skin blow-off time. Thus, the pore opening occurs during the foam expansion.

A number of ways for transformation of closed pores to opened ones have been developed also for the prepared foams. The destruction of pore walls can be caused by hydrolysis, oxidation, low or high pressure, thermal, mechanical, or sonical treatment. Blast or multiple compression are often used. As a result, the reticular foams are formed, their polymer phase is centered on edges (strands).

Last years the polymer nanocomposite foams focus the attention, since the addition of nanosized functional particles has been proved to be a viable strategy to reduce thermal conductivity of the materials [37].

For example, the nanocomposite based on water expandable polystyrene containing clay particles has been obtained [40]. The natural clay can be uniformly dispersed in water due to its hydrophilicity. It can be further inserted into the styrene monomer by the formation of water-in-oil inverse emulsion.

Via suspension polymerization, spherical polymer beads with myriads of water-clay droplets inside the bulk of the polymer were obtained. During heating, the polymer matrix was expanded and a cellular structure was formed. The nano-particles layer was found to form around the pore walls. The clay nanoparticles led to higher water content in the beads and decrease the water loss during storage. Water, which fill the cavities, enlarges their size, the foam product is characterized by ultra-low density (30 kg cm^{-3}) and low thermal conductivity.

The effects of incorporation of clay nanoparticles modified with organic sub-stances [37], vapor grown carbon nanofibers [41], natural fibers, and oil-based polyol [42] into polyurethane foams were studied. No cell opening and appearance of structural defects due to nanoparticles were found, moreover, the additions make pore structure more uniform. The nanoparticles decrease thermal conduc-tivity of the products and enhance their fire safety [41]. The amount of filler, which is enough for improvement of functional properties, is 0.5–5 mass %. Probably the improvement is caused by stretching of the pore walls under the influence of the filler. The dilatation leads to an increase of pore size and porosity, this causes reduction of bulk density and thermal conductivity.

In opposite to polymer materials, which demonstrate no chemical transforma-tion during the foaming process, the manufacture of foam concrete involves a change of chemical composition during the conversion of cement → concrete, the foaming occurs simultaneously. Cement is formed, for instance, under heating of the mixture consisting of $Ca(OH)_2$ and clay up to 1,723 K [43]. The mixture is melted partially, as a result, the clinker granules are formed. In order to obtain cement, some amount of gypsum is added to clinker, the mixture is milled finely. Gypsum controls the time of cement setting, it can be partially replaced by other forms of calcium sulfate. The composition of typical clinker, which contains usually several phases, is 67 % CaO, 22 % SiO_2, 5 % Al_2O_3, 3 % Fe_2O_3 and 3 % other components. During the reaction with water, dehydrated clinker minerals are transformed into hydrosilicates, hydroaluminates, and hydroferrites of calcium:

$$3CaO \cdot SiO_2 + H_2O \rightarrow 3Ca_2SiO_4 \cdot H_2O + Ca(OH)_2 \qquad (5.20)$$

$$Ca_2SiO_4 + H_2O \rightarrow Ca_2SiO_4 \cdot H_2O \qquad (5.21)$$

$$3CaO \cdot Al_2O_3 + 6H_2O \rightarrow 3CaO \cdot Al_2O_3 \cdot 6H_2O \qquad (5.22)$$

$Ca(OH)_2$ is also formed and gradually transformed to $CaCO_3$, this change affected CO_2 of air. Calcium hydroaluminates with gypsum form binary basic sulfates, for instance $Ca_6Al_2(OH)_{12}(SO_4)_3 \cdot 26H_2O$ and $Ca_4Al_2(OH)_{12}SO_4 \cdot 6H_2O$. As a result, very strength concrete consisting of calcium aluminate, silicates, and carbonates is derived from cement.

In order to obtain foam concrete, the air-entraining method is often used. The technique involves aluminum powder, hydrogen peroxide/bleaching powder, or calcium carbide, which are mixed into lime or cement mortar [7]. The gas (hydrogen, oxygen, or acetylene) escapes, leaving a porous structure. Among these

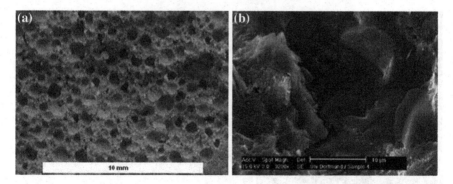

Fig. 5.13 Optical (**a**) [45] and SEM (**b**) [46] images of cross-section of the sample of gas concrete

blowing agents, aluminum powder is usually applied. The efficiency of aluminum powder process is affected by fineness, purity, and alkalinity of cement. The method allows one to obtain light material, which is called "air concrete" or "gas concrete." However, such terms as "foam concrete", "cellular concrete" can be used regarding to materials obtained by this manner.

The foaming method is reported as the most economical and controllable pore-forming process, since no chemical reactions between the components of the foaming agents occur. The pore formation is achieved through mechanical means either by preformed foaming (foaming agent mixed with a part of mixing water) or mix foaming (foaming agent mixed with the mortar). A number of foaming agents can be applied, such as detergents, resin soap, glue resins, saponin, hydrolyzed proteins such as keratin, etc.

Production of foam concrete by combining foaming and air-entraining methods was developed using aluminum powder and glue resin. Incorporation of poly-urethane foam into cement has been also proposed [44].

In aerated concrete, the method of pore formation (gas release or foaming) affect the structure and functional properties. The structure of this material includes solid microporous matrix and macropores [7]. The macropores are formed due to the expansion of the mass caused by aeration, the micropores appear in the walls between the macropores (Fig. 5.13 [45, 46]).

The macropores in the foam concrete are related to the porosity after hardening. High porosity caused by macropores means smaller density and better thermal insulation; however, the strength declines significantly [35]. Therefore, the unique characteristics of foam concrete arise from the reduction in strength as the price. The method, which is widely used and allows one to improve the mechanical properties of the cement-based materials, is adding of polymer fibers into ordinary cement-containing mixture [47].

Special requirements are for materials for vacuum insulating panels, which must have only opened pores to give a possibility to evacuate any gas from them. The products like "nanoporous silica" are the promising materials for this purpose.

Fig. 5.14 SEM image of agglomerate of fumed silica supported by fibers (**a**) [24], TEM image of aggregates of nanoparticles (**b**)

This term involves silica aerogel or xerogel as well as precipitated, pyrogenic (called also "fumed silica") and volatilized silica. The heat conductivity of fumed silica is close to 0.01 W m^{-1} K^{-1} up to 100 Pa and ≈ 0.02 W m^{-1} K^{-1} under ambient pressure under dry conditions [24]. No considerable increase of thermal conductivity occurs at relative humidity less than 50 %. Alternately, the heat conductivity of compact silica is 0.8–1.34 W m^{-1} K^{-1} in dry media.

These unique properties of fumed silica are mainly due to its porosity (0.85–0.9) and pore size (2–50 nm). Fumed silica is less sensitive to increasing of pressure after preliminary evacuation in a comparison with other silica materials. Obtaining of fumed silica involves interaction of tetrachlorosilane with hydrogen and oxygen:

$$SiCl_4 + 2H_2 + O_2 \rightarrow SiO_2 + 4HCl \qquad (5.23)$$

This reaction leads to the formation of silica particles with a mean diameter ≈ 20–30 nm, as seen in TEM image obtained by the authors (Fig. 5.14). These nanoparticles form grain chains (aggregates) with a typical length of few hundreds nanometers. Their agglomeration causes particles of micron size.

5.7 Conclusions

Low heat conductivity is the main functional property of thermal insulating material. This feature is provided mainly due to decrease of the contribution of a solid phase into the total conductivity by means of formation of significant porosity of the materials. For instance, in the case of polymer foams, the ratio of volumes of voids and solid can reach 0.98. High porosity is formed during foaming, which involves no change in chemical composition of the polymer. Regarding inorganic materials, such as light concrete, the foaming is realized simultaneously with cement → concrete transformation, which is provided by chemical reactions between the cement constituents and water. Significant porosity can be formed directly during synthesis of such inorganic material as fumed silica.

The order of magnitude of heat conductivity of highly porous insulators is 10^{-2} W m^{-1} K^{-1}. In the case of foam concrete, for which the thermal insulation is rather additional function, this value is higher approximately in 10 times. It is necessary to emphasize, that increase of porosity deteriorates mechanical durability of the insulators. Thus, the main task of the material engineering is to obtain the products with optimal combination of mechanical strength and thermal conductivity.

The modern approach to development of a new generation of insulating materials is to decrease a contribution of gaseous phase inside pores into the total thermal conductivity. This is achieved by encapsulation of inert gases into solid. Nanostructured materials, for instance, fumed silica, can be used as a core for vacuum insulating panels. Opacified fumed silica shows also a decrease of radiation contribution to the heat conductivity. Probably the development of nanotechnologies opens wide perspectives for obtaining of new thermal insulators as well as improvement of known ones.

References

1. Agham RD (2012) Int J Eng Innov Technol 2(6):97
2. Papadopoulos AM (2005) Energy Build 37(1):77
3. Schwab H, Heinemann U, Wachtel J, Ebert HP, Fricke J (2005) J Therm Envelope Build Sci 28(4):327
4. Baetens R, Jelle BP, Thue JV, Tenpierik MJ, Grynning S, Uvslokk S, Gustavsen A (2010) Energy Build 42(2):147
5. AbuBakr Bahaj S, James PAB, Jentsch MF (2008) Energy Build 40(5):720
6. Bobrov YuL, Ovcharenko EG, Shoikhet BM, Petukhova EYu (2003) Thermal isolating materials and constructions. UNFRA-M, Moscow (in Russian)
7. Narayanan N, Ramamurthy K (2000) Cem Concr Compos 22(5):321
8. Awang H, Mydin MAO, Roslan AF (2012) Adv Appl Sci Res 3(5):3326
9. Dubinin MM, Plavnik GM (1968) Carbon 6(2):183
10. Vaitkus S, Laukaitis A, Gnipas I, Kersulis V, Vejelis S (2006) Mater Sci (Medziagotura) 12(4):323
11. Gendron R (ed) (2005) Thermoplastic foam processing. Principles and development. CRC Press, Boca Raton
12. Talbor H, Lee T, Jeulin D, Hanton D, Hobbs LW (2000) J Microsc 200(3):251
13. Saadatfar M, Arns CH, Knackstedt MA, Senden T (2005) Colloids Surf A 263(1–3):284
14. Bakhtiyari S, Allahverdi A, Rais-Ghasemi M (2011) Asian J Civ Eng (Build Hous) 12(3):353
15. Yartsev DV, Lakhin AV, Vol'fkovich YuM, Manukhin AV, Bogachev EA, Timofeev AN, Sosenkin VE, Nikol'skaya NF (2010) Russ J Non-Ferrous Met 51(4):364
16. Bagotzky VS, Volfkovich YuM, Kanevsky LS, Skundin AM, Broussely M,Chenebault P, Caillaud T (1995) Power sources 15. In: Attewel A, Keily T (eds) Crowborough: international power sources symposium committee p 359
17. Rice RW (1998) Porosity of ceramics. Marcel Dekker, New York

18. Adkins CJ (1987) An introduction to thermal physics. Cambridge University Press, Cambridge
19. Bulent Y, Paki T (2007) Energy Build 39(9):1027
20. Jannot Y, Degiovanni A, Payet G (2009) Int J Heat Mass Transf 52(3–4):1105
21. Saito Y, Kanematsu K, Matsui T (2009) Mater Trans 50(11):2623
22. Coquard R, Baillis D, Quenard D (2008) Intern. J Therm Sci 47(3):324
23. Jeans J (1987) An introduction to the kinetic theory of gases. Cambridge University Press, Cambridge
24. Bouquerel M, Duforestel T, Baillis D, Rusaouen G (2012) Energy Build 54:320
25. Streed ER, Cunningtont GR, Zierman CA (1966) Thermophysics and temperature control of spacecraft and entry vehicles. Academic Press, New York, p 735
26. Placido E, Arduinischuster M, Kuhn J (2005) J Infrared Phys Technol 46(3):219
27. Fricke J, Hümmer E, Morper H-J, Scheuerpflug P (1989) J Phys Colloques 50(C4):87
28. Kamiuto K (1990) Int J Solar Energy 9(1):23
29. Lee ST, Ramesh NS (eds) (2004) Polymeric foams. Mechanics and materials. CRC Press, Boca Raton
30. Caps R, Fricke H (2000) Int J Thermophys 21(2):445
31. Fricke J, Lu X, Wang P, Buttner D, Heinemann U (1992) Int J Heat Mass Trans 35(9):2305
32. Hümmer E, Rettelbach T, Lu X, Fricke J (1993) Thermochim Acta 218:269
33. Lienhard JH IV, Lienhard JH V (2012) A heat transfer textbook. Phlogiston Press, Cambridge
34. Groover MP (2002) Fundamentals of modern manufacturing. Wiley, USA
35. Kearsley EP, Wainwrigh PJ (2002) Cem Concr Res 32(2):233
36. Gibson LJ, Ashby MF (1997) Cellular solids: structure and properties. Cambridge University Press, Cambridge
37. Mittal V (ed) (2014) Polymer nanocomposite foams. CRC Press, Boka Raton
38. Eaves D (ed) (2004) Handbook of polymer foams. Rapra Technology, Shawbury
39. Klempner D, Sendijarevic V (2004) Polimeric foams and foam technology. Carl Hanser, Munich
40. Shen J, Cao X, Lee LJ (2006) Polymer 47(18):6303
41. Harikrishnan G, Singh SN, Kiesel E, Macosko CW (2010) Polymer 51(15):3349
42. Kuranska M, Prociak A (2012) Compos Sci Technol 72(2):299
43. Hewlett P (ed) (2004) Lee's chemistry of cement and concrete. Elsevier, Oxford
44. Mounanga P, Gbongbon W, Poullain P, Turcry P (2008) Cem Concr Compos 30(9):806
45. Struharova A, Rousekova I (2007) Slovak J Civ Eng 15(2):35
46. Just A, Middendorf B (2009) Mater Charact 60(7):741
47. Jones MR, McCarthy A (2005) Mag Concr Res 57(1):21

Chapter 6
Characteristics and Structure of Powdered Medical Substances Used in the Pharmaceutical Industry

Mikhail L'vovich Ezerskiy and Vladimir Sergeevich Bagotsky

Abstract Medical substances (MS) used in the pharmaceutical technology are microheterogeneous solid-phase powder materials that underwent dispersion procedures in the gas phase. Many properties of these powders, particularly their stability and ability for subsequent processing into medications in forms acceptable to the consumer such as tablets, capsules, or suspensioms depend not only on their chemical composition, but also on their structure. MS have a high specific surface which results in increased inter-particle interactions and therefore to a great extent influences their chemical and technological properties. In order to optimize their further processing it is important to evaluate their structural properties by well-defined standard measuring procedures. In the pharmaceutical industry, besides MS different secondary, auxiliary substances are used which per se have no medical influence but help to produce the final medication. From the point of view of evaluating their structural properties these auxiliary substances are equivalent to the MS.

6.1 Structural Parameters of Medical and Auxiliary Substances and Methods for Their Measurements

MS are characterized by their physical–chemical parameters such as the size and shape of individual particles and the size and the properties of their surface. The sum total of these primary particle parameters determine the properties of the powder as a whole, such as dispersity degree, specific surface area, wetting by different liquids, adsorption of humidity, etc. From the technological point of view the main properties are those connected with their rheological behavior: flowability (free-running property) and with their packing properties.

Y. M. Volfkovich et al., *Structural Properties of Porous Materials and Powders Used in Different Fields of Science and Technology*, Engineering Materials and Processes, DOI: 10.1007/978-1-4471-6377-0_6, © Springer-Verlag London 2014

6.1.1 Particle Size (Degree of Dispersion)

The most important structural parameter of MS is their degree of dispersion. In general, MS-particles have a broad range of sizes, beginning from 2–5 μm up to 150–200 μm. The dispersion degree is a statistical parameter which can be un-ambiguously defined only for pore-less particles of regular shape. It is characteristic for MS that this parameter depends on the measurement method. Different methods are used in the pharmaceutical industry. Microscopic and electron microscopic methods are mainly used at the development stage of the MS. Mesh-screen analysis with sieves of different mesh size is widely used in technological procedures. The smallest opening in such sieves is about 50 μm, therefore this method is mainly used for bigger sized particles. In the pharmacopeias of different countries the dispersion degree of MS is classified according to the sieve's mesh size.

For measuring the dispersion degree sedimentation methods are also used. Computerized sedimentographs automatically fix the results as particle size dis-tribution curves. The lowest particle size detectable by this method is 2–5 μm. For smaller particles this method can be used in combination with a centrifugal machine.

The dispersion degree can also be measured by a conductometric method, which is based on a liquid's conductivity change when solid particles enter the gap between two electrodes [1]. This method is sometimes used in the pharmaceutical industry for dispersity degree control. In the development stage photometric methods with different laser light source are sometimes used in biological sus-pensions with particle sizes in the μm range (e.g., for amorphous and crystalline insulin). For all these methods a correct choice of the suspension medium and surface active substances is essential.

The dispersion degree of MS is characterized not only by particle size, but also by the particle shape, which influences the subsequent processing of the material and also properties of the final product, such as stability of tablets.

Up till now no method for quantitative determination of the particle shape has been described. For comparative assessment of the shape the sedimentation rate in comparison with that of regular spherical particles can be used. In the pharma-ceutical industry particles of regular shape are called *isometric*. The isometric coefficient is defined as the ratio of the minimal particle dimension to its maximal dimension. As a rule a lowering of this coefficient leads to increased technological difficulties.

A convenient averaged parameter of MS particle's dispersion degree is their specific surface S_M referred to as mass unit. The S_M value can be found by measuring the air permeability of an MS layer in a viscous flow mode. A special automated device PSH [2] based on this method allows rapid and reproducible determination of the S_M value, which is widely used for this purpose. The value of S_M correlates very well with the main technological properties of MS. In particular it correlates with one of the important biopharmaceutic properties of sulfamid medications—the rate of in vitro dissolution. In 1997 the parameter S_M was included into the European pharmacopeia as a parameter characterizing the dispersion degree of MS. The overall specific surface S_T (including the internal surface of open pores) can be measured by the BET method of low temperature nitrogen adsorption or by the more rapid gas chromatographic method. In 1995 this parameter was introduced into the USA pharmacopeia.

6.1.2 Adsorption and Wetting Properties

Another important characteristic parameter of MS is their ability to absorb humidity from the surrounding medium. This parameter determines their stability and their quality. Humidity adsorption isotherms can be measured by gas-chromatic methods. The wetting properties of powdered MS can be characterized by their wetting angles with liquids. A method for measuring the wetting angles for antibiotics was described in [3]. Some insoluble or poorly soluble substances have water wetting angles θ from 73 to 90°. Some substances are hydrophobic with $\theta > 90°$. The wetting angle characterizes the surface purity degree of MS (absence of adsorbed impurities). The value of this parameter influences the stability of MS suspensions. It was shown that suspensions of Ampicillin hydrate with wetting angles $\theta > 85°$ are not stable and prone to aggregation of the solid phase.

6.1.3 Technological (Rheological and Packing) Properties

Technological properties of MS are secondary properties which characterize the behavior of a large number of individual particles. They influence the subsequent processing of MS, in particular their consumer-friendly wrapping into tablets or capsules. The main rheological parameter is the MS powder's flowability which is measured by the powder's flow rate through a standard-size funnel. Commonly, glass funnels with a cone angle of 60° and an outflux opening of 10 mm are used. The funnels are placed on a vibrating device with a vibration frequency of 50 s^{-1} and a vertical amplitude of 0.08 mm. The reproducibility of such measurements is very high.

Another parameter, which characterizes the powder's flowability, is the slope angle after free strewing. In order to increase the accuracy of such measurements, strewing is performed with the help of the above-mentioned vibrating device. Slope angle measurements correlate fairly well with measurements of the flow rate through a funnel.

Important and frequently used parameters for MS powders are their packing properties, in particular, their mass after being strewn into a cylindric receptacle of a predetermined volume (either freely with the help of the mentioned vibrating device, or after application of a predetermined pressure).

In the pharmaceutical industry for filling of gelatine capsules and of glass phials with MS fast-acting automated machines are used. The accuracy of the filled amount must be very high. As different batches of MS-powders often have different rheological and packing properties it is necessary to make for each batch a control filling and to adjust the automated machine correspondingly.

Difficulties in accurate filling arise for powders containing particles of irregular form (needles, flat particles, etc.) and also containing large particles. The normal particle size for MS powders is 2–5 μm, but sometimes particles of size up to 200 μm can be found. Therefore, for smooth operation of the filling devices it is necessary to have the possibility of a disintegration of large particles. For this purpose mostly jet mills are used. These mills do not contaminate the powders and yield powders with particle size 2–4 μm. Independently of their previous form the powders are practically isometric.

6.2 Biopharmaceutical Properties of Some Sulfanilamide Powders, Suspensions, and Tablets

The therapeutic effect of medications in vivo can be controlled by processes in vitro. This is important for peroral use medications. Processes of medication dissolution in vitro and their resorption in vivo are called *biopharmaceutical*. An important property of peroral medications is their absorption rate by blood. This process can be (in simplified form) described as proceeding in two stages: (1) dissolution of the medication in the stomach-bowel tract (SBT) and (2) penetration through cell membranes into blood. For control of these processes in vitro it is necessary to determine the limiting stage in vivo. The lower the rate of the MS dissolution in water or aqueous solutions imitating the gastric juice, the higher the possibility that the dissolution stage is limiting the medication absorption. As the dissolution rate depends on the experimental conditions [4], for this measurement a standard device "rotating basket" is often used, the parameters of which are indicated in the pharmacopeias of many countries. After a predetermined time interval the medication must completely dissolve and the active MS must freely enter into the solution.

While developing new peroral medications (suspensions, tablets, or capsules) the influence of parameters of the initial MS on the medication's dissolution rate must be measured. The main parameter which determines this rate is the dispersion degree characterized by the above-mentioned parameter S_M.

In [5] the biopharmaceutical properties of poorly soluble sulfanilamid-type peroral medications such as sulfadimethoxin, sulfadimizine a.o. were investigated in detail. A part of these medications was disintegrated in jet mills. The disintegration led to a 7-fold increase in the S_M value from 0.24 to 1.67 m^2/g (corresponding to a decrease of the averaged particle diameter from 20 to 2.4 μm). The in vivo adsorption of the initial and disintegrated medications was studied on rabbits.

Fig. 6.1 Dependence of the sulfodimizine concentration in rabbits blood on time: (*1*) Initial medication; (*2*) medication after disintegration

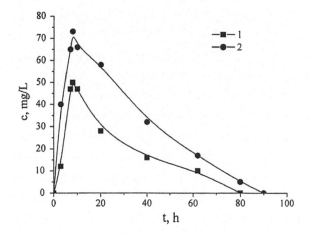

The absorption curves (dependence of medication concentration in blood vs. time) are shown in Fig. 6.1 for the initial medication (curve 1) and for the disintegrated one (curve 2).

These curves have a typical shape for biopharmaceutical curves—at the beginning the medication's concentration in the blood rises, then reaches a maximal value, then gradually decreases, and finally declines to a zero value. It can be seen from the figure that the absorption rate (the slope of the rising part of the curves) for the disintegrated medication is higher than that for the initial one. But the value of the maximal concentration (which mainly influences the therapeutical effect) remains almost the same. This is an indication that the disintegration process does not lead to mechano-chemical changes (e.g., of the medication's crystalline structure).

A thorough control and optimization of structural and wetting properties of the initial MS powders and of all auxilliary components is a prerequisite for the development and production of high-quality medications.

References

1. Orr NA (1977) Analyst 102:329
2. Khodakov CH (1968) Main methods for analysis the dispersion degree of powders. Publisher of building literature, Moscow (in Russian)
3. Ezerskiy ML, Pis'mennaya GM (1972) J Chem Pharm 6:50 (in Russian)
4. Marques M, Brown W (eds) (2010) The dissolution technology: questions and answers. Dissolution Technologies Publisher, Hockosin
5. Tentsova LI, Per'kova NN, Ezerskiy ML (1984) Pharmatsiya 2:13 (in Russian)

Chapter 7
Free Volume and Microporosity in Polymeric Gas Separation Membrane Materials and Sorbents

Vladimir Vasilievich Volkov and Yury Pavlovich Yampolskii

Abstract Free volume is an important property of polymers. When the size and connectivity of "microcavities" in polymers increase they often indicate the microporosity of polymers. In this review, an attempt is made to consider in a systematic manner the nature of free volume in polymers and different methods of their estimation. Several original approaches for evaluation of microporosity of polymeric gas separation membrane materials and sorbents are also discussed. It is important that the first industrially produced gas separation membrane was based on poly(vinyltrimethyl silane), the first glassy polymer with relatively high free volume. At present, the attention of researchers is attracted to several classes of membrane materials distinguished by high free volume. In turn, hypercrosslinked polystyrene sorbents are manufactured by several companies now, and they are widely used in chemical, food, and water treatment industries. In conclusion, the authors tried to emphasize common features and differences between free volume and microporosity in polymers.

7.1 Introduction: Concepts of Free Volume and Porosity in Polymers

Condensed state of matter that comprises amorphous polymers and also liquids, inorganic glasses, etc., includes voids of static or dynamic nature. In the case of polymers these voids form free volume. Often, instead of the notion of free volume, porosity (or inner porosity) in polymers are formulated, and the boundary between these two notions is somewhat uncertain.

The concept of free volume is of paramount importance for physical chemistry of polymers. The free volume in polymers determines their viscosity, relaxation and mechanical properties (deformation), thermal expansion, plasticization, as well as the rates of interdiffusion of polymers and diffusion of low-molecular-mass species (gases and vapors) in polymers.

Y. M. Volfkovich et al., *Structural Properties of Porous Materials and Powders Used in Different Fields of Science and Technology*, Engineering Materials and Processes, DOI: 10.1007/978-1-4471-6377-0_7,
© Springer-Verlag London 2014

When considering the free volume in polymers, one should keep in mind that this is not only an abstract physical parameter (although it can, of course, be described quantitatively), but also a real physical object characterized by an average size and shape of the "holes" (a free-volume element, FVE), FVE size distribution, topology, and connectivity (closed or opened internal porosity). The FVE in glassy polymers are treated as "frozen" in the polymer matrix. For rubbers, where the free volume has a fluctuation nature like in liquids, both the size of randomly formed FVE and the rate of their diffusion in the polymer matrix are essential.

According to Askadskii [1] there are three definitions for free volume.

1. Free volume is the difference between the real mole volume of a polymer body (specific volume) and the van der Waals mole volume of the chemical groups that form this polymer.

   Such value sometimes is called the empty volume.

2. Free volume is the difference between the volume of the polymer body at absolute zero and at certain temperature; that is, such definition of free volume characterizes the excess volume due to thermal expansion of the polymer body.
3. Free volume is the difference between the volume of the polymer body at certain temperature and the volume of the ideal crystal formed from the polymer of the same chemical structure.

Sometimes, the contributions within free volume that correspond to the main chains and small-scale movements of side groups are distinguished and discussed. There are other definitions of free volume that will be considered later in this chapter. Uses of different definitions are applicable regarding various problems of polymer science.

When speaking on porosity in polymers in most cases one means relatively large voids which: (1) have sizes larger than molecular sizes; (2) form infinite clusters in polymer bodies, i.e. pierce through the whole polymer plate, film, membrane, etc. However, there are "boundary" examples, when it is possible to speak on both free volume and porosity. They will be considered in this chapter.

It is traditional to evaluate free volume by the density of packing of chains, and sufficiently developed methods are available based on either experimental data (PVT curves) or using different computational models. On the other hand, during the last decades, researchers use more and more actively experimental methods for estimation of free volume on microscopic (atomistic) level, as well as detailed experimental study of porosity in polymers.

7.2 Historical Background

The concept of free volume was, apparently, first advanced by Frenkel [2] for interpretation of viscosity and diffusion in liquids. He generalized numerous observations made earlier on various liquids and was able to formulate different properties of "holes" in liquids: typical residence times, energy of formation (overcoming intermolecular forces), typical size, etc.

The works on free volume and porosity were being developed in independent directions since the 1960s. One of these directions implied studies of the group contributions into density via which it was possible to find empty volume and free volume as a function of chemical structure of polymers and temperature dependence of the specific volume above and below the glass transition temperature (Bondi et al. [3–6]). Interesting series of works was performed by Tager (see e.g. [7]), who considered contributions of van der Waals volume, empty volume, and maximum accessible for solute molecules pore volume into the total volume (or specific volume) of polymers. Interesting achievements of these works was demonstration that large "frozen" porosity can be characteristic for glassy polymers.

Another direction of research was focused on polymeric sorbents, such as macroporous copolymers on the basis of cross-linked polystyrenes, macroporous poly(divinylbenzenes), and hypercrosslinked copolymers. These materials are similar in chemical composition and properties but differ in the structure of the polymer matrix and pore structure. Hypercrosslinked polystyrene (HCL-PS) networks and sorbents, for the first time described in the scientific literature in the early 1970s [8], eventually came into routine practice by the end of the1990s. At present, hypercrosslinked sorbents are manufactured by several companies. They are widely used in chemical and food industries. All hypercrosslinked sorbents are materials with very fine pores at nanoscale level. From the polymeric viewpoint, they belong to the group of high free volume glassy polymers, which are presented in the following sections. Some of these polymers additionally display larger transport channels that have been introduced to enhance mass transfer. The latter heterogeneous products are often referred to as bi-porous.

An entirely different and rapidly growing family of nanoporous polymers is represented by membrane materials based on high free volume glassy polymers. Poly(vinyltrimethylsilane) (PVTMS), which was synthesized in 1962 [9], was the first representative of this group of glassy polymers. The production of the asymmetric gas separation PVTMS membrane was realized in the USSR in the mid-1970s [10]. The next breakthrough in the field of high free volume glassy polymers was related to the synthesis of poly(1-trimethylsilyl-1-propyne) (PTMSP) in 1983 [11]. Because of long term and numerous efforts by synthetic chemists directed at the design of glassy polymers with high free volumes, the scope of these materials has been substantially widened. Among them are poly(4-methyl-2-pentyne) (PMP) [12–14], poly[(1-trimethylgermyl)-1-propyne] (PTMGP) [15, 16], the additive poly(5-trimethylsilylnorborn-2-ene) [17, 18], random copolymers of 2,2-bis(trifluoromethyl-4,5-difluoro-1,3-dioxalane) and tetrafluoroethylene (e.g., AF-2400 [19, 20]), and polybenzodioxane PIM-1 [21, 22]. All high free volume glassy polymers comprise intrinsic microporosity, which is created in the material during film preparation from polymer solutions in a good solvent.

An important point is that although it is generally agreed that both HCL-PSs and extra high free volume glassy polymers are porous materials having high BET surface area (more than 1,000 m^2/g), however, their porosity has little in common with the porosity of conventional adsorbents such as alumina, silica, and carbon molecular sieves. There is no real interface in the polymeric materials considered; "pores" have no "walls" in common sense and they simply represent voids between loosely packed polymer chains. The high BET surface area is a result of high sorption capacity of the high free volume polymers towards nitrogen, rather than their surface area as such.

7.3 Free Volume Above and Below T_g

7.3.1 Simha-Boyer Concept

A traditional and, maybe, the most accepted model of free volume in a wide range of temperature was proposed by Simha and Boyer [5]. These authors paid attention to the fact that temperature dependence of the specific volume of polymers undergoes a break at T_g, and the slopes α_g (below T_g) is much less steep than α_l (above T_g). It was assumed that α_g is the same as the corresponding slopes for completely crystalline polymer and that at absolute zero the specific volume of the equilibrium liquid (glass) and of the crystalline state are equal. A graphical representation of this model is shown in Fig. 7.1.

Numerous data for different polymers both above and below T_g form an experimental basis for such dependencies [23]. Several empirical and approximate relations are the consequences of Simha-Boyer model, and they are often used for various estimates. Among them,

$$\alpha_l \cdot T_g = \text{const} \tag{7.1}$$

and

$$(\alpha_l - \alpha_g) \cdot T_g = \text{const.} \tag{7.2}$$

However, this model gives satisfactory results only for qualitative interpretation. One of the most important and strictly criticized [24] points of the Simha-Boyer model is the concept of so-called iso-free volume. It was assumed that the fractional free volume FFV $= V_f/V_{sp}$ where V_f is free volume and V_{sp} is the specific volume ($V_{sp} = 1/\rho$, ρ is the polymer density) at T_g has a universal value equal to 0.025. First, it was found that in many polymers the value FFV at T_g varies in a wider range 0.02–0.10. If the concept of iso-free volume is valid, then the diffusion and permeability coefficients at T_g should have constant values. An analysis of the available data for D and P for polymers with widely varying T_g indicated that $D(T_g)$ were not constant but increased with growing glass transition temperatures [25]. The same trend was noted for the size of free volume element at T_g according to the positron annihilation lifetime spectroscopy (PALS) [25, 26].

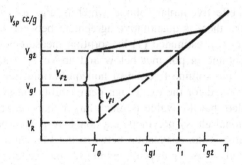

Fig. 7.1 Temperature dependence of specific volume in polymers above and below glass transition temperatures: T_o temperature of measurement, T_{g1} and T_{g2} glass transition temperatures of the polymers 1 and 2, V_R specific volume at T_o of a polymer in hypothetical equilibrium state, V_{g1} and V_{g2} specific volumes of two polymers 1 and 2 that are in nonequilibrium state, V_{f1} and V_{f2} excess free volume of polymers 1 and 2 at T_o

7.3.2 Dual Mode Sorption: Evaluation of Nonequilibrium Free Volume

Unexpected support to the concept of Simha-Boyer was provided by the results of numerous studies of gas sorption in glassy polymers. Sorption isotherms in such systems have a shape of curves concave to the pressure axis. The common approach for interpretation of such isotherms is an application of so-called dual mode sorption model [27]. It is assumed that in the population of solute molecules in a glassy polymer two fractions can be partitioned: the one sorbed in more densely packed part of polymer matrix described by Henry's law ($C_D = k_D p$) and another sorbed in excess nonequilibrium free volume (or in more loosely packed part) described by the Langmuir isotherm ($C_H = C_H' bp/(1 + bp)$) The superposition of these two populations results in the observed sorption isotherms:

$$C = k_D \cdot p + C_H' \frac{bp}{1 + bp} \tag{7.3}$$

Among the three parameters of this model, k_D and b are common equilibrium constants of sorption (cm^3(STP)/cm^3(polymer) atm), while Langmuir capacity parameter C_H'(cm^3(STP)/cm^3(polymer)) characterizes the quantity of gas sorbed in so-called "unrelaxed" ("frozen") free volume elements (FVE) or "microvoids." Observations on sorption isotherms (usually at the standard temperature $T_o = 35$ °C) made for polymers with widely varying T_g showed that the larger the difference $T_g - T_o$, the greater the C_H' values. Langmuir capacity parameters C_H' measured at 35 °C linearly increase with T_g of different polymers and can be extrapolated to zero at $T_g = 35$ °C [28]. Even quantitative agreement between these values can be reached if one takes into account the equivalent density of solute molecules in the Langmuir population ρ^* [29, 30]. The latter value can be estimated using an effective molar volume, which is, e.g., for CO_2 49 cm^3/mol. It was shown [30] that there is quantitative agreement between Langmuir sorption capacity parameters C_H' estimated (1) by sorption measurement and (2) via the plot for specific volume in polymer below and above T_g. Using the procedure introduced in [29], the volume fraction of nonequilibrium free volume, equal to 0.24, was first estimated for the extra high permeability glassy polymer PTMSP [31]. It was revealed that this value is much larger than similarly obtained estimates for conventional glassy polymers, e.g., 0.04 for polycarbonate [29].

Of special interest is the correlation between the Langmuir sorption capacity parameter C_H' of a glassy polymer proposed in [32] and the temperature gap $(T_c - T)$ between the measurement temperature T and the critical temperature T_c of a sorbate (or a penetrant in the case of permeability), which is shown in Fig. 7.2. As follows from this figure even though the content of nonequilibrium free volume of a glassy polymer is virtually constant at a given temperature [36] according to the Langmuir mechanism, this volume accommodates more molecules of such sorbate which has the minimum temperature gap between the critical temperature and the experimental temperature.

For substances with $T_c > T$ (the vapor region), sorption in a glassy polymer can be accompanied by condensation and formation of a dense sorbate layer at the Langmuir sites. In this case, dimension of the adsorbed layer is similar to those of the sorbate in liquid state at temperatures below the critical temperature. The bigger the dimensions of the penetrant, the lower the number of molecules accommodated in a given nonequilibrium free volume of a glassy polymer under its complete occupation. Therefore, in this interval, C_H' should correlate with the sorbate dimensions. Monotonic decrease in C_H' with increasing $(T_c - T)$ takes place as the sorbate dimensions increase in the same order. A similar trend not with $T_c - T$ but with the critical volume V_{cr} of solutes was noted also for C_H' values in PTMSP [37].

For penetrants with $T_c < T$ (the gas region), sorption in the nonequilibrium free volume of a glassy polymer does not seemingly lead to the formation of the dense layers. In any case, according to [38] for highly supercritical gases, the saturation density within the Langmuir sites is low. In other words, packing density of the penetrant sorbed in the Langmuir microvoids is lower compared with its packing density at temperatures below T_c. As the experimental temperature increases to T_c, packing density of the sorbed gas also increases; when the temperature approaches T_c, this parameter becomes equal to the liquid phase density at T_c or, in other words, some condensation of the penetrant upon its sorption takes place and this process is accompanied by the formation of dense pseudo-liquid layers at the Langmuir sites. Therefore, upon sorption of gases in a glassy polymer at $T_c < T$, as T_c (or ε/k) increases, the C_H' values increase due to the increase in the packing density of the gas sorbed under saturation conditions according to the Langmuir mechanism.

Thus, the parameter C_H' can serve as a measure of nonequilibrium free volume in glassy polymers. This statement is confirmed by the results of a review [39] where it was shown that permeability coefficients depend on $1/C_H'$ in the same way as they depend on 1/FFV.

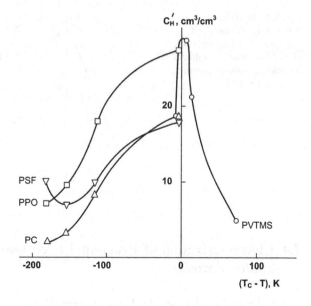

Fig. 7.2 Langmuir capacity sorption parameter versus temperature gap $(T_{cr} - T)$ between the ambient temperature and the critical temperature of a sorbate (adapted from [32]); the data for PC, PPO, and PSF were taken from [33–35], respectively

7.3.3 Interrelation Between Free Volume, T_g and Selective Permeability of Polymers

Glassy polymers, in contrast to rubbery ones, for a long time were regarded as low permeable and highly selective materials. For the time being, both the least permeable and the most permeable materials are glassy polymers, and the family of high permeability glassy polymers has been growing in recent years [40]. As shown in Sect. 7.3.2, this is because some glassy polymers have a significant contribution of nonequilibrium free volume into the total FFV.

The interrelation between three important for membrane gas separation parameters, namely permselectivity α_p, V_f and T_g is illustrated in Fig. 7.3 [41].

It had been shown earlier that the extreme character of the dependences on T_g for the diffusion coefficients (at room temperature) of light gases (curves with a minimum at T_g) [42] and for the activation energies of diffusion (curves with a maximum) [43] led to the assumption that the dependence of FFV on T_g had also to be described by a curve with a minimum in the vicinity of transition from rubbery to glassy state of a polymer. It follows from Fig. 7.3 that the same free volume fraction can be characteristic for two polymers (e.g., 1 and 2), each of which is in different physical (rubbery and glassy) state, and is therefore distinguished by different kinetic rigidity of its polymer chains. This results in a greater value of permselectivity for a glassy polymer having the same level of gas permeability as compared to a rubbery polymer. The conclusion was made that glassy polymers with high values of T_g were promising materials for gas separation membranes [41].

Fig. 7.3 Interrelation
between α_p, V_f and T_g of
polymers: R—Rubbery state,
G—Glassy state, 1—rubbery
polymer, 2—glassy polymer
(adapted from [41])

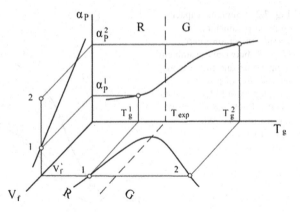

7.4 Characterization of Fractional Free Volume and/or Porosity

7.4.1 Fractional Free Volume Estimations

Let us discuss in more detail the notion of fractional free volume (FFV). It is an empirical parameter that is frequently used for characterization of integral free volume in polymers. It is based on a conception that free volume V_f (cm^3/g) is the difference between total (specific) and occupied volume in polymers. FFV then is defined as the ratio V_f/V_{sp}, where V_{sp} (cm^3/g) is the specific volume of a polymer and is defined as reciprocal density ρ (g/cm^3). Different approaches to finding V_f and FFV depend on the way the occupied volume is estimated. The most popular way for finding FFV has been proposed by Bondi [44] who suggested to estimate the occupied volume as follows: $V_{oc} = 1.3\,V_w$, where V_w is van der Waals volume of the repeat unit of the polymer:

$$FFV = \left(V_{sp} - 1.3V\right)/V_{sp} \qquad (7.4)$$

The factor of 1.3 is estimated based on packing density of molecular crystals at 0 K, while V_w can be easily evaluated via the tabulated increments, e.g., in the Van Krevelen book [45]. Such an estimate gives the FFV values of most polymers in the range 10–25 %, however, there are examples when FFV is as large as 35 % (e.g., in the case of PTMSP). In spite of its wide usefulness such definition of FFV has many disadvantages and is being often criticized. The coefficient 1.3 in calculating V_{oc} is estimated from packing density of molecular crystals at 0 K and cannot adequately describe polymers at room temperature.

It can be noted [1] that the parameter 1.3 is $1/k_{cr}$, where $k_{cr} = 0.769$ is the universal coefficient of molecular packing of ideal molecular crystals. In fact, k_{cr} value should depend on the chemical structure of polymers and the type of elementary crystallographic cell. It is suggested [1] to use another value $k_{cr} = 0.74$ (i.e., the factor 1.35) for calculation of FFV.

In addition, calculation of V_w of a repeat unit as the sum of the increments of van der Waals volumes of atoms that form this repeat unit can lead to inaccuracies because it poorly takes into account the dead volume which depends on conformation of the chain and is different for different gases. More reliable results based on quantum chemical corrections were reported by Ronova et al. [46].

A similar approach was proposed by Park and Paul [47]. Since FFV is frequently used in correlations with the permeability coefficients, it was suggested to find the empirical corrections in FFV values that take into account the differences in the occupied volume that are "sensed" by penetrants of different molecular size. In other words, instead of using the universal FFV value, the characteristics of a polymer, it is proposed to use, e.g., FFV(He) and FFV(CH_4) as more suitable for correlations.

Another practical point is related to the circumstance that in the case of glassy polymers the density may depend on the prior history of the sample and, to some extent, on the method of determination, which is discussed in detail in Sect. 7.4.3. So FFV found as it is described above can be considered as a semi-quantitative parameter. Nevertheless, the Bondi's formula is so widely accepted that it forms a convenient basis for comparison of different polymers.

7.4.2 Probe Methods for Determination of Free Volume (Advantages and Limitations)

In the previous sections we discussed different approaches for estimation of free volume in polymers as an integrated property of a material. However, it is equally important to analyze free volume on a microscopic, atomistic scale. This is an achievement of the last decades that such information became available due to the appearance of so-called probe methods and extensive studies of polymers by mean of the Molecular Dynamics (MD) and Monte Carlo (MC) techniques. We first consider the outcomes of probe methods widely used now, whereas the results of the application of computer simulations are only briefly considered at the end of this section.

Common features of the probe methods for investigation of the free volume in polymers consist of introducing certain particles (probes) into the polymer and monitoring their behavior (Table 7.1). The results of observations allow one to make conclusions about the size of free volume elements. In these methods probes of different nature and size are used, and they provide more or less detailed information. For instance, in ^{129}Xe NMR spectroscopy the only probe (^{129}Xe atom) "explores" the free volume in different polymers. In studies of polymers by PALS, an electron-positron pair (Positronium atom) serves as a probe. In inverse gas chromatography (IGC) and some other probe methods, the role of probes is

Table 7.1 Probe methods for investigation of free volume in polymers

Method	Probe
Positron annihilation lifetime spectroscopy (PALS)	o-Ps (e + e$^-$)
^{129}Xe-NMR	^{129}Xe
Inverse gas chromatography (IGC)	n-alkanes C_3–C_{16}
Spin probes method	Stable nitroxile radicals
Photochromic probes method	Azobenzenes, substituted stilbenes
Conformation probes method	Polar organic compounds
Electrochromic probes method	Aromatic dyes
Sorption probes (Dual mode sorption model)	Gases and vapors

played by series of structurally similar compounds having different molecular sizes.

It is important to be sure that different probe methods give reliable and consistent information on the size of FVE in polymers. Table 7.2 shows that there is reasonable agreement between the results of different and independent probe methods regarding the size of FVE in various polymers.

It is seen that in spite of some discrepancy between specific values of the radii, three methods give the results on the size of FVE which are in agreement. It has been also shown that the size of FVE in various polymers correlates with the permeability coefficients: according to the PALS method the largest radius of FVE (6.8 Å) is observed in PTMSP that has the permeability coefficients $P(O_2) = 8,000$–$12,000$ Barrer, while the smallest radius has been reported for copolyester Vectra (2.1 Å) that is a barrier material ($P(O_2) = 0.0005$ Barrer) [48]. It is important also to mention that application of several probe methods (photochromic probes, Inverse gas chromatography, ^{129}Xe-NMR, PALS) and use of computer modeling gave consistent values of FVE in a perfluorinated polymer [49].

The most detailed information on free volume in polymers, especially on its size distribution and concentration of FVE in polymers was provided by the PALS method. So we shall focus mainly on this method. The lifetime spectra of positrons before their annihilation in polymers consist of three components. Two components with shorter lifetimes τ_1 and τ_2 (0.2 ± 0.1 ns and 0.4 ± 0.1 ns) are caused by annihilation of p-Positronium atom and free positrons, respectively. Positronium atom is a hydrogen-like particle formed by a positron and an electron. In p-Positronium their spins are anti-parallel, and this results in faster annihilation. In o-Positronium total spin is one, so its lifetime in vacuum is very long—140 ns. However, when it gets into a condensed phase its lifetime τ_3 is shortened, though it remains much longer than τ_1 and τ_2. The PALS method is based on a paradigm that o-Positronium atoms get in a polymer into FVEs. The larger the size of FVE the longer the τ_3 lifetime [50]. In conventional polymers τ_3 lifetimes are in the range 1.5–3.0 ns, and this corresponds to the radii of FVE of 2–3.5 Å. In such polymers, the size distribution of FVE can be represented by a single Gauss curve [48, 50].

Table 7.2 Radii R, $\overset{\circ}{\text{A}}$ of FVE in some polymers according to the PALS, ^{129}Xe NMR and IGC methods

Polymer[a]	^{129}Xe-NMR		PALS		IGC
	R_{sp}	R_c	R_{sp}	R_c	
AF2400	8.04	5.12	5.95	6.33	6.4
AF1600	6.66	4.43	4.89	5.43	5.8
AD 80X	6.12	4.16	4.6-5.2[b]	–	
AD 60X	6.00	4.10	4.5-4.8[b]	–	
PTFE	5.69	3.94	4.2	4.9	
PPO	2.92	2.56	3.4	4.2	3.4
	2.84	2.52			
	2.80	2.50			
PC	2.48	2.34	2.9	3.8	
	2.50	2.35			
PS	2.52	2.36	2.88	3.76	
PEMA	2.60	2.40	3.0	3.9	

[a] *AF* amorphous Teflons, *AD* Hyflons AD80 and AD60, *PTFE* polytetrafluoroethyelene, *PPO* polyphenylene oxide, *PC* bis-phenol A polycarbonate, *PS* polystyrene, *PEMA* poly(ethylene metacrylate)
[b] Different values correspond to solution cast or melt pressed samples
R_{sp} and R_c correspond to spherical or cylindrical geometry of FVE, respectively [48, 49]

However, it has been shown [51] that polymers with larger free volume and higher gas permeability are characterized by bimodal size distribution of free volume and strongly increased lifetimes. This is characteristic of PTMSP [51] and other polyacetylenes, amorphous Teflons AF; recently it was shown also for additive Si-substituted norbornene polymers [52] and polymer of intrinsic microporosity PIM-1 [53]. So in highly permeable polymers *o*-Positronium lifetime is split into the two components τ_3 and τ_4. In some strongly porous materials (sorbents), e.g., in HCL-PS it is possible to discern even larger lifetimes (20–40 ns) and, respectively, additional components of size distribution of free volume of size FVE in the range 9–12 $\overset{\circ}{\text{A}}$ [54].

If the radius R of a spherical FVE is known and, hence, it is possible to find the volume of an average single "hole" as $4/3\ (\pi R^3)$ one can find FFV if the hole number density N_h is known:

$$FFV = \frac{4}{3}\pi R^3 N_h \qquad (7.5)$$

Several approaches are available to find N_h values [48, 50, 55]. It is interesting that different methods gave N values in quite different polymers within a relatively narrow range $(2–10)\ 10^{20}\ cm^{-3}$. FVE in polymers are located very close to each other, so the average distance between neighbor holes is approximately equal to the length of diffusing jump of gas molecules in polymers [56]. Approximately constant values of N_h explain a fulfillment of correlations of the type shown in Fig. 7.4.

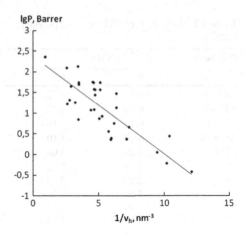

Fig. 7.4 Correlation of the permeability coefficient $P(O_2)$ with the unit volume v_f (nm^{-3}) of FVE in glassy polymers [57]

It should be noted that the FFV values found in this way are smaller than FFV according to Bondi. The reason for this is as follows: in the Bondi method the whole size distribution of FVE is taken into account, while the estimation of FFV via only its part is accounted in the PALS method (only those FVE with size larger than 1.06 Å, the size of o-Positronium atom). Computer simulations made by Hofmann et al. [58] showed that extrapolation to zero size of "virtual" probe particle results in extensive increase in the found FFV.

A disadvantage of the probe methods is their inability to provide information on so-called connectivity of free volume. Meanwhile, many important properties of polymers, and first and foremost their permeability, depend on close or open character of free volume in polymers. The MD method showed that polymers having large and similar free volume (as can be judged by their FFV) can be different in this respect. Thus, it was shown that PTMSP [58] and the polymer PIM-1 [59] are characterized by open porosity, while amorphous Teflons AF have close porosity [58] in spite of similar FFV and the sizes of FVE in these polymers. This is reflected in their selectivity of gas separation.

It can be added that the size distribution of free volume obtained using MD calculations for various polymers are in reasonable agreement with the results of the PALS method.

7.4.3 Estimation of Polymer Porosity and Free Volume Fraction by Density Measurements

The total pore volume (W_0) (porosity) of porous polymer materials may be determined by several methods. However, it is believed that more precise values result from measuring true, ρ_{tr}, and apparent, ρ_{app}, densities of a material. Based on these measurements, the porosity can be calculated by means of the following relationship:

$$W = \frac{\rho_{tr} - \rho_{app}}{\rho_{tr}} \qquad (7.6)$$

To our best knowledge, for polymer materials this approach was first applied for Amberlyst 15 (a sulfonic acid-type resin based on a styrene-divinylbenzene copolymer) [60]. The apparent density was measured by mercury displacement (mercury densitometry), and the true density was measured by helium displacement (helium densitometry). The porosity was defined as follows: (pore volume)/ (pore volume + polymer volume).

The same approach was also applied for characterization of porosity in high free volume glassy polymers [41]. Among them were PVTMS and PTMSP; the pore ("microvoids") volume fraction was found to be equal to 0.20 and 0.07 for PTMSP and PVTMS, respectively. As mercury does not wet the polymer, the mercury densitometry provides a way of estimating the total volume of the sample including pore volume. The same results can be obtained by measurements of geometric dimension of the sample.

Later, an extended method of hydrostatic density measurement was developed as an efficient approach for the evaluation of microporosity and nonequilibrium free volume of membrane materials based on highly permeable glassy polymers [61]. The principle of this method is based on comparing weights of a sample in air, non-wetting liquid, and wetting liquid. Using non-wetting liquids, the apparent or geometric density ρ_g, which is the density of the membrane sample as a whole, is measured. With wetting liquids, a pycnometric density ρ_p is measured. This density is related to the densest regions of a high free volume glassy polymer ("true" or "skeletal" density). The total free volume fraction (microporosity) W of the dense films can be calculated similar to Eq. 7.6, by the following equation [61]:

$$W = \frac{\rho_p - \rho_g}{\rho_p} \qquad (7.7)$$

For hydrophobic polymers such as PTMSP, PTMGP, and PMP water and ethanol was used as non-wetting and wetting liquids, respectively [61, 62]. For example, the dense films of PTMSP, PTMGP, and PMP were characterized by the hydrostatic density measurement method before and after 30 h of heat treatment at 100 °C [62]. As seen from Table 7.3 this method confirms that membrane densification takes place after annealing. The geometric density ρ_g was higher for treated membranes, while pycnometric density ρ_p was almost constant. In other words, the heat treatment results in partial relaxation of excess free volume. For example, the microporosity W of PMP drops from 0.19 to 0.16 after exposure at 100 °C for 30 h; meanwhile, the highest densification effect was observed for PTMSP (prepared in the presence of the TaCl5/TIBA catalyst)—from 0.27 to 0.20 (Table 7.3).

Table 7.3 Hydrostatic density measurements for PTMSP of different prehistory: geometric density ρ_g, pycnometric density ρ_p and microporosity W (adapted from [62])

Membrane	Initial sample			After 30 h at 100 °C		
	ρ_g, g/cm^3 (H$_2$O)	ρ_p, g/cm^3 (EtOH)	W	ρ_g, g/cm^3 (H$_2$O)	ρ_p, g/cm^3 (EtOH)	W
PTMSP TaCl$_5$/TIBA	0.75	1.02	0.27	0.83	1.03	0.20
PTMGP	1.05	1.33	0.21	1.09	1.36	0.20
PMP	0.77	0.94	0.19	0.79	0.94	0.16

Polymers exposed to high pressure undergo compression, densification, or even collapsing of free volume, as PALS studies showed [63, 64].

It should be pointed out that the estimation of porosity of polymer sorbents by using individual wetting and non-wetting liquids is a rather convenient approach; however, the drawback of this simple method is related to the fact that the polymer usually swells in wetting liquid environment. It is practically impossible to find out an individual wetting liquid in which the polymer exhibits zero swelling. To overcome this problem, a binary solution of wetting and non-wetting liquid (e.g., ethanol aqueous solutions for hydrophobic PTMSP), instead of pure wetting liquid, was first proposed as a wetting liquid medium in the extended method of hydrostatic density measurement [61]. Gradual increase in the content of wetting component in the binary solution makes it possible to change continuously the wetting ability of liquid medium, as a result of which the desired combination of "wetting liquid" and "no polymer swelling" can be easily attained. For instance, wetting of PTMSP was observed at the ethanol content in water of 5 % wt. and more while there was no polymer swelling up to 15 % wt. of ethanol aqueous solution [61].

7.4.4 Dynamic Desorption Porometry

One of the most important characteristics of porous materials (and polymers in general) is concerned with their pore size distribution (PSD). At atomistic level, the PALS technique [55], a single among the probe methods, provides size distribution of FVE, while the same aim in a more general way can be achieved by MD simulation [58]. For porous objects (sorbents), methods for the estimation of PSDs are traditionally based on sorption studies, namely on the measurements of adsorption isotherms [65]. Usually, the above-mentioned traditional techniques are known to be time-consuming.

Some time ago, a novel dynamic desorption technique for obtaining vapor isotherms was developed [66]. This so-called dynamic desorption porometry (DDP) method was illustrated in particular by desorption of benzene vapors from mesoporous silica MCM-41 [67]. To calculate the pore size distribution, the

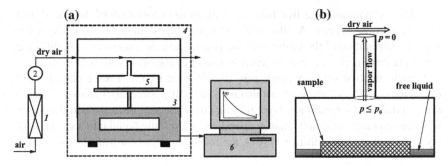

Fig. 7.5 a Schematic representation of a dynamic desorption technique: (*1*) air drier; (*2*) flow meter; (*3*) balance; (*4*) thermostat; (*5*) measuring cell; (*6*) control system (PC) and **b** Schematic of measuring cell with a sample. Vapor pressure of an adsorbate $p = 0$ at the cell outlet. The liquid evaporation rate varies proportionally with the vapor pressure p over the sample [67]

Derjaguin–Broekhoff–de Boer theory in its combination with the Wheeler model for capillary condensation was used. The mean mesopore sizes estimated by DDP were shown to be in fair agreement with the calculations through the geometrical method based on the X-ray diffraction data. Since this method has been employed only for several objects, we shall describe it here in some detail.

The key feature and advantage of DDP is that no vapor pressure measurements are necessary. To reach this goal, a special experimental setup was designed (Fig. 7.5a). Dry gas (air) is passed over the top end of the cell to ensure the complete removal of adsorbate vapors from the cell outlet (Fig. 7.5b). Changes in the weight loss of the sample with time were recorded.

Thus, the DDP method is based on the analysis of the kinetics of liquid evaporation in the quasi-steady state regime. This regime is provided by optimization of the rate of evaporation from the cell containing the sample. With decreasing the measured rate of vapor removal from the cell, the vapor pressure over the sample tends to reach its equilibrium value (Fig. 7.5b). Therefore, the rate of liquid evaporation is proportional to the vapor pressure p over the sample. Using geometrical parameters of the off-take tube (internal diameter and lengh) of the measuring cell, one can estimate the pressure p from the evaporation rate. To avoid any estimation of equilibrium vapor pressure and to reduce possible errors from uncontrolled changes in the experimental conditions, the principle of self-calibration is used. To this end, excessive amounts of the adsorbate (wetting liquid) are supplied into the cell containing the sample saturated with the liquid (Fig. 7.5b).

Then, it is evident that until all free liquid fully evaporates from the cell, vapor pressure in the cell is equal to the equilibrium pressure p_0 over its surface. This pressure is associated with the maximum rate of evaporation W_0 from the cell. In this case, the time dependence of evaporation rate is characterized by the plateau region. This region is the necessary validation criterion for all experiments, and the W_0 parameter is the reference value for the calculation of the relative pressure in each experimental run. The end of the initial plateau region corresponds to complete evaporation of a free liquid as well as to evaporation of the liquid from the interparticle spaces. As the content of the liquid in the sample decreases further, evaporation of the probe from the porous particles takes place; in this case, vapor pressure is $p < p_0$ and the corresponding evaporation rate is $W < W_0$.

Current relative vapor pressure p/p_0 is controlled by the relative rate of liquid evaporation W/W_0. One can expect that $p/p_0 \approx W/W_0$. More strict analysis requires the account for a stagnant gas layer (dry air) over the sample in the cell. Using the well-known model of steady-state evaporation of a volatile liquid into a stagnant gas layer (so-called Stefan diffusion) [68], the following equation was obtained [66]:

$$p/p_0 = \left[1 - (1 - x_0)^{W/W_0}\right]/x_0, \qquad (7.8)$$

where $x_0 = p_0/p_{atm}$ is the mole fraction of saturated vapor in the measuring cell, p_{atm} is the overall (atmospheric) pressure of the gas phase. Equation (7.8) is valid when adsorbate vapor pressure outside the cell is equal to zero.

Continuous weighing of the sample allows reproducible measurements within a moderately short time period. The experiment provides nearly infinite number of experimental points together with a detectable random noise of the weighing readings. Therefore, the data should be carefully smoothed by averaging and cubic spline. The accuracy of the temperature control exerts a significant influence on the experimental errors; this value does not exceed 6 % [67] and 0.5 % [69] at the temperature control within ±0.1 and ±0.03 K, respectively.

For polymeric materials, the DDP was used to study the sample of HCL-PS CPS-100 swollen in different solvents [70], butanol-PTMSP system [71, 72] and the properties of water in polystyrene sulfo acid gels with various degrees of cross-linking [69]. For all systems studied, the desorption isotherms, obtained by the dynamic DDP method and conventional static methods, were in good agreement.

The desorption isotherms of hexane, benzene, methanol, and water for the sample of HCL-PS CPS-100 are presented in Fig. 7.6.

Regardless of the polarity of adsorbates (except for water), the starting state of the polymer in the swollen state includes an approximately equal (abnormally large) total volume of adsorbates. Studies by DDP demonstrated that this volume is ~ 2.8 cm^3 g^{-1} for CPS-100, which is substantially larger than the volume of pores in the dry polymer (~ 0.43 cm^3 g^{-1}) and is indicative of a large swelling in the polymer samples (spherical granules of sizes from ~ 0.2 to 2 mm).

Cumulative curves of volume distributions for adsorbates with respect to the potentials of adsorption RT ln(p$_0$/p) (liquid–solid bond energies according to Rehbinder) calculated from the desorption isotherms are shown in Fig. 7.7.

The series of the curves in Fig. 7.7 from left to right correlates with the series of dielectric permeabilities (i.e., the polarities) of the adsorbates: water (81), methanol (33.7), benzene (2.23), and hexane (1.89). As can be seen from Fig. 7.7, methanol desorption occurs at substantially lower potentials of adsorption (~ 3 times lower) compared to hexane desorption. This observation agrees with the

Fig. 7.6 Isotherms of desorption from the polymer CPS-100: hexane (*1*), benzene (*2*), methanol (*3*), and water (*4*) [70]

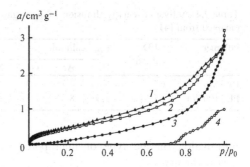

Fig. 7.7 Volume distributions for adsorbates with respect to the potentials of adsorption for the polymer CPS-100: hexane (*1*), benzene (*2*), methanol (*3*), and water (*4*) [69]

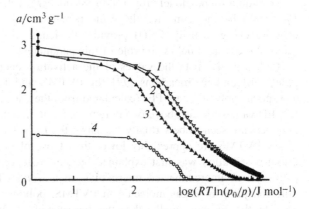

difference between the integral heats of wetting and swelling of HCL-PS by methanol and hexane [73].

Water desorption differs radically from desorption of other adsorbates. The initial volume of "adsorbed" water in CPS-100 is 0.98 cm^3 g^{-1}, which is three times smaller than the retained volumes of organic solvents but is twice as large as the volume of pores of the dry sample. Approximately the same volume (0.9 cm^3 g^{-1}) of nitrogen is condensed in this polymer at liquid nitrogen temperature. Evidently, these volumes correspond to equilibrium swelling of the material in water and in liquid nitrogen. The nonmonotonic character of water desorption from HCL-PS observed by DDP may be indicative of a bimodal distribution of micropores in the samples under study. This observation is in agreement with the results of PALS measurements [74].

7.4.5 Nonequilibrium FVE or Micropore Connectivity

For selective transport of gases and vapors in highly permeable glassy polymers, key parameters include not only the nonequilibrium free volume content or microporosity but also micropore (nonequilibrium FVE or microvoids) connectivity.

Table 7.4 Activation energy diffusion of light gases and hydrocarbons for PVTMS and PE (adapted from [41])

Polymer	CED cal/cm^3	E_D, kcal/mol							
		He	Ar	O_2	N_2	C_2H_6	C_3H_4	C_3H_6	C_3H_8
PVTMS	67	2.8	4.2	4.3	4.4	6.0	6.2	9.0	10.2
PE (100 %)	65	5.4	8.8	8.3	8.4	12.3	10.9	12.5	13.8

Information on the type of micropores can be extracted from the diffusion measurements.

According to the model proposed by Meares [75] activation energy of diffusion E_D should be the same for the same penetrant in the polymers with similar cohesion energy density (CED), provided the length of the diffusion jump λ is the same. We consider below the role of this parameter.

Table 7.4 [41] lists the values of the activation energy of diffusion for light gases [76] and hydrocarbons [77] for PVTMS and for a hypothetical 100 % amorphous polyethylene (PE) (calculated from the data collected by [78]) as well as CED values for the above two polymers. Even though the CED values for both polymers are seen to be virtually the same, the E_D values of gases and vapors for glassy PVTMS are appreciably lower than those of elastomeric PE. This interesting observation was first explained by the existence of pre-existing diffusion pathways, as interconnected microvoids, in glassy PVTMS [77] Then, for any diffusion jump of a gas molecule in PVTMS, polymer chains should be drawn apart by the distance smaller than the diameter of a diffusing molecule.

This analysis is based on an assumption of existence of pre-existing diffusion pathways between FVEs. This is apparently true for some high free volume polymers as will be seen from the subsequent discussion. A more general analysis of Meares equation [79] for polymers with large and small free volume indicated that average distance between the neighbor FVE is identical or very close to the length of the diffusion jump λ, in Meares equation. Recently, using the experimental values of the hole number density of FVE N_h [55, 80] it was shown [57] that the following equations hold for the λ parameter:

$$\lambda = N_h^{-1/3} \cdot \left[1 - (6FFV/\pi)^{1/3}\right] = N_h^{-1/3} - 2R \qquad (7.9)$$

where R is the radius of FVE and for the diffusion coefficients

$$\ln D = \ln D_0 - \frac{\pi}{4} N_A d^2 \left(N_h^{-1/3} - 2R\right) \cdot CED \qquad (7.10)$$

as seen in Fig. 7.8.

Note that the N_h values used in these equations decrease when the radius of FVE increases. An alternative assumption that N_h is constant and independent of the R values often considered in previous works (see Ref. [47]) leads to

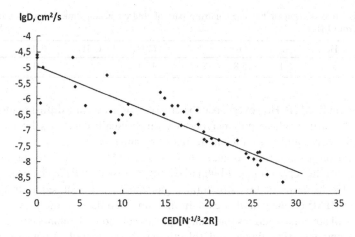

Fig. 7.8 Dependence of $D(CO_2)$ $(cm^2 s^{-1})$ on the parameter $CED[N_h^{-1/3} - 2R]$ [56]

inconsistent conclusions. The fractional free volume FFV can be regarded as $FFV = N_h v_h$, where v_h is the volume $(Å^3)$ of average FVE. If one assumes that N_h is independent of the v_f value and the same as for some low free volume polymers (e.g., polyimides or polycarbonate) then for PTMSP, the polymer with the greatest measured R values and v_f of about 1,200 $Å^3$, the resulting FFV value would be about 100 %, which is absurd. Decreasing values of N_h predict rather close location of neighbor FVE but not their complete overlapping.

Let us return now to the analysis of the data presented in Table 7.4. In terms of the Meares model it is possible to estimate for some polymers the diameter of the diffusion pathway bottleneck (the narrowest cross section of interconnected micropores) which imposes limitations on the diffusion rates. In general, this relationship can be presented as [41, 77]:

$$E_D(GP)/E_D(RP) = (d^2 - a^2)/d^2 \qquad (7.11)$$

where $E_D(GP)$ and $E_D(RP)$ are the gas diffusion activation energies, respectively, for a glassy polymer and for a rubbery polymer with the same CED values; d^2 is the cross section area of a gas molecule, and a^2 is the cross section area of the bottleneck part of the micropore. The calculated a^2 values are summarized in Table 7.5. As follows from Table 7.5, the averaged a value is equal to 2.6 Å for PVTMS. The similar approach, based on the Meares model, allowed Srinivasan et al. [81] to conclude that, in PTMSP, pore-like continuous diffusion pathways exist where main chains are loosely packed and spaced by a chain-to-chain distance of, at least, 3 Å.

Therefore, in highly permeable glassy polymers, molecular transport of light gases and vapors does not require significant separation of polymer chains because of micropores (or nonequilibrium FVEs) connectivity. This is the main reason behind the extra high permeability and diffusion coefficients and low ideal selectivities for

Table 7.5 Cross section of the most narrow part of interconnecting micropores for PVTMS (adapted from [41])

Gas	He	Ar	O_2	N_2	C_2H_6	C_3H_4	C_3H_6	C_3H_8
a^2, $Å^2$	3.2	6.1	5.8	6.5	10.1	7.8	7.0	6.9

light gases in PTMSP. However, once the cross-section area of a diffusing molecule exceeds a^2 of a given glassy polymer, permeability critically decreases. For example, for PTMSP, permeability coefficient decreases by an order of magnitude from 2.3 to 0.21 Barrer on passing from *n*-butane to *iso*-butane [41, 82]. As a result, an ideal selectivity for the separation of *n*-butane/*iso*-butane vapors in PTMSP appears to be equal to 14. In terms of this structural organization of the nonequilibrium free volume in PTMSP, one can also explain the published data on the pervaporation separation of aqueous solutions for binary aqueous solutions of butanol isomers [83]. Reported pervaporation fluxes and selectivities of *n*-butanol, 2-butanol, and *i*-butanol were 6.4–8.4 mol/m^2 h and 61–63, respectively. At the same time, substantially lower partial flux (0.729 mol/m^2 h) and selectivity (9.4) of *t*-butanol is the evidence that gas separation and organophilic pervaporation behavior of PTMSP is similar [84]. This behavior is consistent with the view that PTMSP acts as molecular sieves rather than conventional glassy polymers [41, 84–86].

It is known that a crosssection of a molecule controls penetration of the molecule into zeolite pores [87]. The crosssection of *n*-butane and isobutane is equal to 4.3 and 5.0 Å [87], respectively. Thus, to the conclusion that PTMSP voids may be linked through chain-to-chain gaps at least 3 Å wide [81] could be added so that the widths of these chain-to-chain gaps do not presumably exceed 5 Å [84]. Since the crosssection of *n*-butanol, 2-butanol, and *i*-butanol are smaller than the narrowest sections of interconnected micropores in PTMSP, the permeability coefficients of these compounds are high and do not depend strongly on the structure of alcohol. Crosssection of *t*-butanol, presumably, exceeds the narrowest section in pore network of PTMSP causing a tenfold decrease in permeability. Taking into account that PTMSP is an amorphous glassy polymer, one should expect less pronounced sieving effect due to larger PSD and possible polymer swelling in gas separation and pervaporation effect observed for zeolites.

7.5 Concluding Remarks

In this chapter we tried to consider common features of and differences between free volume and porosity in polymers. Both notions characterize "empty volume" in polymer matrices. The difference between them refers to two circumstances.

1. The first implies that FVE or elementary free volumes are relatively small, comparable with the sizes of individual low molecular mass molecules (intrinsic size of the order 1 nm). In contrast to liquid and rubbers, in glassy

polymers FVE are frozen in solid matrix and their residence time is relatively long in comparison with the intrinsic times of measurement of the experimental parameters dependent of free volume. However, even in the glassy state some movements of FVE are possible, and it is manifested in aging phenomena [88] or densification of some larger free volume polymers [89] within intrinsic times of months or sometimes even weeks. Porosity in polymers is rather an unusual phenomenon and can be induced by certain processing such as sol–gel technology, treatment by super-critical fluids, or formation of radiation-induced defects like in track etched membranes [90]. Typically, the size of pores in polymers is 1–3 orders larger. In some cases inner porosity is formed due to rigid structure of polymeric chains.

2. Another, and maybe more important difference is related to connectivity. A traditional (and apparently correct) view is that free volume in polymers can be regarded as a sort of "closed" porosity. In conventional glassy polymers individual FVEs are separated by virtually "non-porous," non-permeable walls. On the contrary, "free volume" in porous polymers has high degree of connectivity, i.e., forms macroscopic clusters and in limiting cases pierces the whole size of the sample, like it takes place in sorbents or catalysts. Visualization of such type of porosity in PTMSP was demonstrated by Hofmann et al. [58]. However, the openness of pore structure depends on a size of probe penetrant molecules. As has been discussed in Sect. 7.4.5 even PTMSP does not reveal "open" porosity for larger penetrant molecules like isobutene and, presumably, larger gaseous penetrants like neopentane or SF_6. On the other hand, Greenfield and Theodorou showed [91] that even densely packed polymer matrix like atactic polypropylene reveals opened porosity for a hypothetical (nonexisting) penetrant of radius ca. 0.9 Å (that is smaller than the smallest gas molecule He with the radius of 1.28 Å). It means that connectivity of free volume in polymers depends on a specific type of transport or sorption problem.

Thus, both concepts, free volume and porosity, can be helpful in understanding and interpreting mass transfer problems in highly permeable glassy polymers and polymeric sorbents.

Acknowledgments The support of the proposal of Russian Science Foundation No 14-29-00272 is greatly acknowledged.

References

1. Askadskii AA (2012) Vysokomol Soed A 54:11
2. Frenkel YI (1946) Kinetic theory of liquids. Clarendon Press, Oxford
3. Bondi A (1968) Physical properties of molecular crystals, liquids and glasses. Wiley, New York
4. Van Krevelen DW (1972) Properties of polymers. Correlations with chemical structure. Elsevier, New York

5. Simha R, Boyer RF (1962) J Chem Phys 37:1008
6. Boyer RF (1963) Rubber Chem Technol 36:1303
7. Tager AA, Askadakii AA, Tsilipotkina MV (1975) Vysokomol Soed A 17:1346
8. Rogozhin SV, Davankov VA, Tsyurupa MP (1969) Patent USSR 299165
9. Nametkin NS, Topchiev AV, Durgaryan SG (1964) J Polym Sci Part C 4:1053
10. Yampolskii YP, Volkov VV (1992) J Membr Sci 64:191
11. Masuda T, Isobe E, Higashimura T, Takada K (1983) J Am Chem Soc 105:7473
12. Masuda T, Kawasaki M, Okano Y, Higashimura T (1982) Polymer J 14:371
13. Morisato A, Pinnau I (1996) J Membr Sci 121:243
14. Khotimsky VS, Matson SM, Litvinova EG, Bondarenko GN, Rebrov AI (2003) Polym Sci A 45:740
15. Litvinova EG, Khotimskiy VS (1994) Proceedings of 2-nd international symposium progress in membrane science and technology. June 27–July 1. Enschede p 57
16. Khotimsky VS, Tchirkova MV, Litvinova EG, Antipov EM, Rebrov AI (2001) Polym Sci A 43:577
17. Finkelshtein ES, Makovetsky KL, Gringolts ML, Rogan YV, Golenko TG, Yampolskii YP, Starannikova LE, Platé NA (2007) Patent RF 2296773
18. Finkelshtein ES, Makovetsky KL, Gringolts ML, Rogan YV, Golenko TG, Starannikova LE, Yampolskii YP, Shantarovich VP, Suzuki T (2006) Macromolecules 39:7022
19. Nemser SM, Roman IC (1991) US pat. 5,051,114
20. Alentiev AYu, Yampolskii YP, Shantarovich VP, Nemser SP, Platé NA (1997) J Membr Sci 126:123
21. McKeown NB, Budd PM, Msayib KJ, Ghanem BS (2004) PCT/GB2004/003166
22. McKeown NB, Budd PM (2006) Chem Soc Rev 35:675
23. Zoller P, Walsh DJ (1995) Standard pressure-volume-temperature data for polymers. Technomic Publishers, Basel
24. Lipatov YuS (1978) Usp Khim 47:332
25. Yampolskii YP, Kamiya Y, Alentiev AYu (2000) J Appl Polym Sci 76:1691
26. Srithawatpong R, Peng ZL, Olson BG, Jamieson AM, Simha R, McGervey JD, Maier TR, Halasa AF, Ishida H (1999) J Polym Sci Part B Polym Phys 37:2754
27. Paul DR (1979) Ber Bunsenges Phys Chem 83:294
28. Toi K, Morel G, Paul DR (1982) J Appl Polym Sci 27:2997
29. Koros WJ, Chan AH, Paul DR (1977) J Membr Sci 2:165
30. Koros WJ, Chern RT (1987) Separation of gaseous mixtures using polymer membranes. In: Rousseau RW (ed) Handbook of separation technology. Wiley, New York, pp 862–953
31. Srinivasan R, Auvil SR, Burban PM (1994) J Membr Sci 86:67
32. Volkov VV, Bokarev AK, Durgaryan SG, Nametkin NS (1985) Dokl AN SSSR 282:641
33. Koros WJ, Chan AH, Paul DR (1977) J Membr Sci 2:165
34. Toi K, Morel G, Paul DR (1982) J Appl Polym Sci 27:2997
35. Erb AJ, Paul DR (1981) J Membr Sci 8:11
36. Barrer RM, Barrie JA, Slater J (1958) J Polym Sci 27:177
37. Merkel TC, Bondar V, Nagai K, Freeman BD (1999) J Polym Sci Part B Polym Phys Ed 37:273
38. Chern RT, Koros WJ, Sanders EH, Hopfenberg HB, Chen SH (1983) Industrial gas separation. In: Whyte TE, Yon CM, Wagener EH (eds) ACS Symposium Series No 233, Washington DC
39. Kanehashi S, Nagai K (2005) J Membr Sci 253:117
40. Yampolskii Yu (2012) Macromolecules 45:3298
41. Volkov VV (1991) Polymer J 23:457
42. Yampolskii YP, Durgaryan SG, Nametkin NS (1982) Vysokomol Soedin Ser A 24:536
43. Volkov VV, Durgaryan SG (1983) Vysokomol Soedin Ser A 25:30
44. Bondi A (1968) Physical properties of molecular crystals, liquids, and gases. Wiley, New York
45. Van Krevelen DV (1990) Properties of polymers, 3rd edn. Elsevier, New York

46. Ronova I, Rozhkov E, Alentiev A, Yampolskii Yu (2003) Macromol Theory Simul 12:425
47. Park JY, Paul DR (1997) J Membr Sci 125:23
48. Yampolskii YP (2007) Russ Chem Rev 76:59
49. Jansen JC, Macchione M, Tocci E, De Lorenzo L, Yampolskii Yu, Sanfirova O, Shantarovich V, Heuchel M, Hofmann D, Drioli E (2009) Macromolecules 42:7589
50. Bartos J (2000) Positron annihilation spectroscopy of polymers and rubbers. In: Meyers RA (ed) Encyclopedia of analytical chemistry. Wiley, Chicheste, p 7968
51. Shantarovich V, Kevdina I, Yampolskii Yu, Alentiev A (2000) Macromolecules 33:7453
52. Gringolts M, Bermeshev M, Yampolskii Yu, Starannikova L, Shantarovich V, Finkelshtein E (2010) Macromolecules 43:7165
53. Budd PM, McKeown NB, Ghanem BS, Msayib KJ, Fritsch D, Starannikova L, Belov N, Sanfirova O, Yampolskii Yu, Shantarovich V (2008) J Membr Sci 325:851
54. Shantarovich VP, Suzuki T, He C, Davankov VA, Pastukhov AV, Tsyurupa MP, Kondo K, Ito Y (2002) Macromolecules 35:9723
55. Dlubek G (2008) Positron annihilation spectroscopy. In: Seidel A (ed) Encyclopedia of polymer science and technology. Wiley, Hoboken, pp 1–23
56. Alentiev AYu, Yampolskii YP (2002) J Membr Sci 206:291
57. Ryzhikh V, Alentiev A, Yampolskii Y (2013) Vysokomol Soed A 55:4 (in press)
58. Hofmann D, Entrialgo-Castano M, Lerbret A, Heuchel M, Yampolskii Yu (2003) Macromolecules 36:8528
59. Heuchel M, Fritsch D, Budd PM, McKeown NB, Hofmann D (2008) J Membr Sci 318:84
60. Kunin R, Meitzner EF, Oline JA, Fisher SA, Frisch N (1962) I & EC Prod Res Dev 2:140
61. Volkov VV, Khotimsky VS, Gokzhaev MB, Litvinova EG, Fadeev AG, Kelley SS (1997) Rus J Phys Chem 71:1396
62. Trusov A, Legkov S, van den Broeke LJP, Goetheer E, Khotimsky V, Volkov A (2011) J Membr Sci 383:241
63. Dong Q, Jean YC (1993) Macromolecules 26:30
64. Hong X, Jean YC, Yang H, Jordan SS, Koros WJ (1996) Macromolecules 29:7859
65. Gregg SJ, Sing KSW (1982) Adsorption, surface area and porosity, 2nd edn. Academic Press, New York
66. Shkolnikov EI, Volkov VV (2001) Dokl Phys Chem 378:152
67. Shkolnikov EI, Sidorova EV, Malakhov AO, Volkov VV, Julbe A, Ayral A (2011) Adsorption 17:911
68. Bird RB, Stewart WE, Lightfoot EN (1960) Transport phenomena. Wiley, New York
69. Gagarin AN, Tokmachev MG, Kovaleva SS, Ferapontov NB (2008) Russian. J Phys Chem 82(11):1863
70. Pastukhov AV, Davankov VA, Sidorova EV, Shkolnikov EI, Volkov VV (2007) Russ Chem Bull Intern Ed 56:484
71. Malakhov AO, Volkov VV (2000) Polym Sci Ser. A 42:1120
72. Malakhov AO, Volkov VV (2008) Polym Sci Ser. A 82:1863
73. Tsyurupa MP, Volynskaya AI, Belchich LA, Davankov VA (1983) J Appl Polym Sci 28:685
74. Shantarovich VP, Suzuki T, He C, Davankov VA, Pastukhov AV, Tsyurupa MP, Kondo K, Ito Y (2002) Macromolecules 35:9723
75. Meares P (1954) J Amer Chem Soc 76:3415
76. Teplyakov VV, Durgaryan SG (1984) Vysokomol Soed A 26:2159
77. Volkov VV, Nametkin NS, Novitskii EG, Durgaryan SG (1979) Vysokomol Soed A 21:927
78. Michaels AS, Bixler HB (1961) J Polymer Sci 50:413
79. Alentiev AYu, Yampolskii YP (2002) J Membr Sci 206:291
80. Yampolskii Yu (2010) Macromolecules 43:10185
81. Srinivasan R, Auvil SR, Burban PM (1994) J Membr Sci 86:67
82. Platé NA, Bokarev AK, Kaliuzhnyi NE, Litvinova EG, Khotimskii VS, Volkov VV, Yampolskii YP (1991) J Membr Sci 60:13
83. Fujii Y, Fusaoka Y, Aoyama M, Imazu E, Iwatani H (1989) Proceedings of 6th International Symposium. Tubingen, p 71

84. Fadeev AG, Selinskaya YA, Kelley SS, Meagher MM, Litvinova EG, Khotimsky VS, Volkov VV (2001) J Membr Sci 186:205
85. Pinnau I, Toy LG (1996) J Membr Sci 116:199
86. Pinnau I, Casillas CG, Morisato A, Freeman BD (1997) J Polym Sci Polym Phys Ed 35:1438
87. Breck DW (1974) Zeolite molecular sieves. Wiley/Interscience, New York
88. McCaig MS, Paul DR, Barlow JW (2000) Polymer 41:639
89. Nagai K, Masuda T, Nakagawa T, Freeman BD, Pinnau I (2001) Prog Polym Sci 26:721
90. Mulder M (1996) Basic principles of membrane technology, 2nd edn. Kluwer, Dordrecht
91. Greenfield ML, Theodorou DN (1993) Macromolecules 26:5461

Part II
Natural Materials

Chapter 8
Oil and Gas Bearing Rock

Anatoly Nikolaevich Filippov and Yury Mironovich Volfkovich

Abstract This section is devoted to characterization of a porous structure of hydrocarbon-bearing fields using methods of standard contact porosimetry and atomic force microscopy. Multidimensioned peculiarities of the fields and of the oil production process are also discussed. The model for calculation of concentration profiles of the suspended particles along the depth of the filtration bed in running time is developed. The model simulates the oil reservoir with retarded proppant particles. A good agreement between experimental and theoretical values of numerical concentration of the proppant particles as functions of coordinate and time is established. The model can be applied for evaluation of the dynamic porosity of rocks as well as for changing their specific permeability. The experimental data describing change in saturation of displacing fluid in the pore space of a rectangular porous block filled with more viscous fluid are obtained. The mathematical model of instability of a two-phase flow in the form of propagation of "fingers" of displacing fluid is investigated. Modeling results are compared with the experimental data. It is shown that the influence of interfacial forces must be taken into account for adequate description of two-phase filtration.

8.1 Structure of Oil and Gas Deposits

Oil and natural gas are most often localized near the lift and fold layers of clastic and carbonate sedimentary rocks (sandstone, limestone, siltstone, shale), which are accumulations of mineral grains connected with cementing material and transformed by geological processes [1].

The pore space of clastic rocks is a complex irregular system of communicating and sometimes isolated inter-granular voids with a typical pore size of several to tens of microns (Fig. 8.1).

In carbonate rocks (limestone and dolomite), the pore system is more inhomogeneous, in addition, they have a more fully developed system of secondary voids occurring after the formation of the stock itself. This includes cracks caused

Y. M. Volfkovich et al., *Structural Properties of Porous Materials and Powders Used in Different Fields of Science and Technology*, Engineering Materials and Processes, DOI: 10.1007/978-1-4471-6377-0_8, © Springer-Verlag London 2014

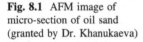 **Fig. 8.1** AFM image of micro-section of oil sand (granted by Dr. Khanukaeva)

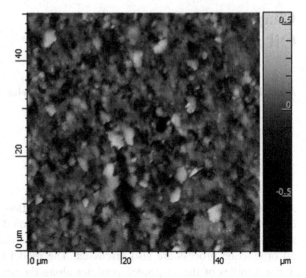

by tectonic stresses, as well as channels and cavities caused by the dissolution of rock skeleton with water, which may be accompanied by a chemical reaction. In this case, the length of cracks and size of the cavity can be much greater than the initial size of the pores. Liquid or gaseous hydrocarbons whose density is less than the density of water are collected in the uplift of rocks (so-called traps), superseding water which was earlier there. In order to preserve oil or gas, reservoirs must be isolated from permeable higher and lower layers with top and bottom: the layers of impermeable rocks, mostly of clays or salts (Fig. 8.2). The structure of the oil and gas deposits is complicated by considerable heterogeneity and, above all, the multi-layering of the constituent species. Oil and gas bearing formations often cross major tectonic faults—breaks in the continuity of the rocks.

8.2 Multidimensional Peculiarities of Hydrocarbon-Bearing Fields

Oil and gas production, exploration and research of layers of hydrocarbon deposits are conducted through individual wells with diameter of 10–20 cm, separated by hundreds of meters. These facts imply the features of filtration of oil and natural gas through reservoirs. One of them is the need to consider processes in different areas, the characteristic dimensions which differ by orders of magnitude: the pore size (a few tens of micrometers) the hole diameter (tens of centimeters), the thickness of layers (a few tens of meters), the distance between wells (hundreds of meters), the length of fields (up to tens or even hundreds of kilometers). In addition, the heterogeneity (thickness and area) has virtually characteristic dimensions of any size. Information about the reservoirs in all their diversity is

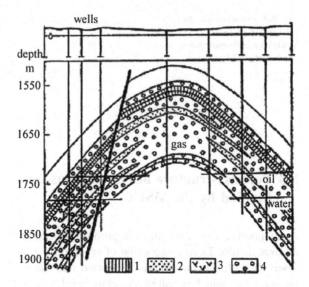

Fig. 8.2 Scheme of oil and gas deposits: *1* clays, *2* clay sandstones, *3* dense layers, *4* sands, sandstones (adapted from [1])

always limited. It consists of geological and geophysical data: surveying data of rock samples and well testing, the analysis of samples taken from the wells of oil, gas, and produced water, and, finally, from the history of the field development, i.e., the aggregate data on the dynamics in pressure, selection, or injection of oil and water for individual wells and in the whole object. Even if you have all the listed amount of information that is not always possible, it is not enough for the unique construction of the reservoir model. Therefore, the main objective of the study is to establish high-quality patterns, consistent trends, and quantitative relationships that are resistant to change in the source data. The most important characteristic of oil or gas reservoir is its hydrodynamic permeability. Recent technological developments in oil and gas production, designed to increase the percentage of recoverable hydrocarbons, are connected with the possibility to adjust the permeability by injection fluid (water), enriched by surfactants, solid particles of micron and submicron range or gas, into the borehole. This raises the need for modeling the properties of these dynamic porous media with variable permeability. In the literature, such media are called complex porous media.

In the most general case the cores, drilled in various directions and then tested in vitro, are used to determine the hydrodynamic permeability of oil and gas bearing rock. As gas and oil formations due to extended stratification are almost always transversely isotropic porous media, the results of Chap. 13 can be used to calculate the specific permeability tensor (13.10):

$$\tilde{k}_{ij} = \begin{pmatrix} \tilde{\mu}^0 \tilde{L}_{11}^{\perp} & 0 & 0 \\ 0 & \tilde{\mu}^0 \tilde{L}_{11}^{\perp} & 0 \\ 0 & 0 & \tilde{\mu}^0 \tilde{L}_{11}^{\parallel} \end{pmatrix}. \tag{8.1}$$

Fig. 8.3 Porosimetric curves
for various rocks of
Vuktylsky carbonate gas
condensate field

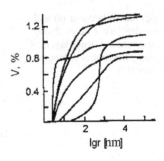

8.3 Porous Structure of Oil and Gas Bearing Rocks Studied by the MSCP

As mentioned above, oil stores in porous oil-bearing strata of considerable length
at great depths. During extraction of crude oil the latter is displaced from the oil
reservoir with pressurized water injected into the wells. The minimum capillary
pressure Δp^c, which is required to extrude oil with water can be obtained from the
modified Laplace equation:

$$\Delta p^c(v) = \left| p_w^c(v) - p_o^c(v) \right| = \left| 2(\sigma_w \cos \theta_w - \sigma_o \cos \theta_o)/r \right| \qquad (8.2)$$

where θ—the contact angle of wetting, v—the degree of flooding of the oil reservoir, and the subscripts «w» and «o» refer, respectively, to water and oil.
Usually in physics of oil reservoir, octane simulates crude oil [2]. Therefore, the
dependence of $\Delta p^c(v)$, which is important for petroleum production, can be
obtained by measuring porosimetric curves for cores from the oil reservoir using
MSCP with water and octane as measuring liquids.

Equation (8.2) shows that the value of Δp^c is inversely proportional to the pore
radius. Therefore, in practice, oil is extracted only from sufficiently wide pores
with radii of $r > r_{cr}$, because oil is held by large capillary pressures (capillary
forces) inside very narrow pores. According to [2], the critical pore radius r_{cr} for
common methods of oil production is approximately 1 micron. Hence the volume
fraction of pores with $r > r_{cr}$ can be estimated from porosimetric curves built for
cores. Further, estimating the volume of the oil reservoir through wells located on
its perimeter, one can evaluate the total oil reserves on the basis of the total
porosity and then calculate the proportion of recoverable oil basing on the value of
the volume of pores with radii $r > r_{cr}$. Based on these data we can better target
investment in the oil industry.

Using MSCP we investigated the porous structure of a number of fields: Yareg
oil and Vuktylsky gas condensate (Timan-Pechora region, Russia) (Fig. 8.3) [3, 4],
Salem (West Siberia, Russia), and Uzen (Mangyshlak Peninsula, Kazakhstan) oil
fields (Fig. 8.4) [5], as well as deposits in Texas in the United States (Fig. 8.5).

Figure 8.4 shows the integral porograms (obtained by measurement with
octane) for different depths of Uzen field.

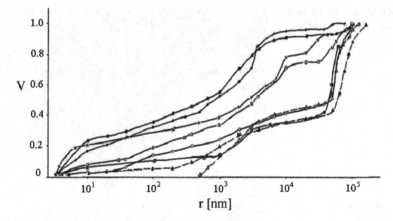

Fig. 8.4 Porograms for different wells of Uzen field

Fig. 8.5 Comparison of differential porosimetric curves for oil field in Texas, measured by MSCP (2) and by mercury intrusion (1)

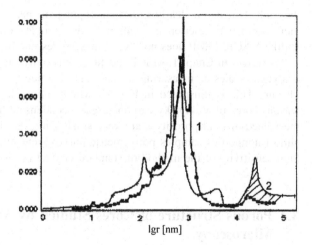

As one can see, the variety of porous structures in this field is very wide and therefore it is necessary to average porograms. We also see that the proportion of the volume of recoverable oil varies considerably at $r > r_{cr}$ for different locations in the field. It should also be borne in mind that the value of r_{cr} depends essentially on the specific process and parameters of oil extraction. For example, value of r_{cr} increases by heating the water injected into the well and applying surfactants.

For comparison, Fig. 8.5 shows the differential porosimetric curves for oil-bearing rock in the state of Texas, measured by MSCP and mercury intrusion (method of mercury porosimetry–MMP).

It is seen that a quite satisfactory agreement of both curves occurs within a wide range of r values from 1 to $\sim 10^4$ nm. However, in the region of tens of microns MSCP shows and MMP does not show the presence of macropores. The image obtained by scanning electron microscopy (SEM) was taken (Fig. 8.6) to find out

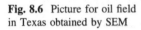

Fig. 8.6 Picture for oil field
in Texas obtained by SEM

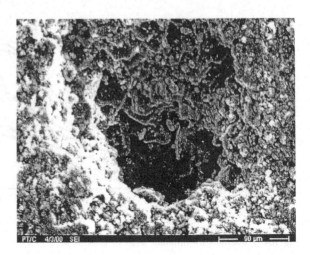

the truth. This photo clearly shows macropores of specified dimensions. However, their volumetric fraction in relation to the sample volume is only about 1 %. Unlike MSCP, MMP does not "see" these pores due to insufficient sensitivity.

As shown in Chap. 1, MSCP has high sensitivity due to the use of electronic analytical scales as measuring device, and also due to control of capillary equilibrium. This is illustrated in Fig. 8.3 which shows the porosimetric curves for various cores of Vuktylsky carbonate gas—condensate field [3]. As one can see, the values of total porosity ε are very small: only 0.8–1 %. And even for those almost nonporous samples porosimetric curves were measured that indicates the high sensitivity of the method of standard contact porosimetry.

8.4 Porous Structure of Cores Studied by Atomic Force Microscopy

Core samples, provided by the Department of Lithology at Gubkin Russian State University of Oil and Gas, were studied at the Department of Mathematics by Associate Professor Dr. Daria Khanukaeva using a scanning probe microscope SmartSPM® 1000 manufactured by AIST-NT (Russia). Scanning was performed in a tapping mode, and silicon cantilevers with the probe curvature radius of order of 10–30 nm were used. Low resolution optical microscope was used to select the scan area so that a region of 1 × 1 cm of the sample gets into the field of vision. In that region the investigator can select different areas to study.

(i) *medium-gravelly sandstone*

The sandstone sample was presented as a micro-layer supported on glass. Resolution of the supplementary optical microscope allows to find the existence of

Fig. 8.7 AFM-image of the dark area of the medium-gravelly sandstone (granted by Dr. Khanukaeva)

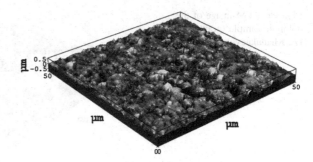

Fig. 8.8 AFM-image of the bright area of the medium-gravelly sandstone (granted by Dr. Khanukaeva)

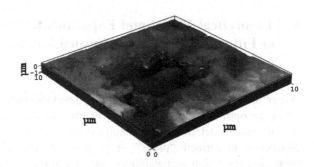

the grain structure, as well as to distinguish areas respond differently to external lighting. The sample regions which intensively reflect falling light were allocated, and thus they were visible as bright formations and, conversely, the darker areas. Figure 8.7 shows the image of a dark region with the size of 50×50 µm. Its surface is fairly evenly covered with grains and pores; size, shape, and distribution of which can be subjected to statistical processing and mathematical description by making use of the method developed in [6].

Similar formations of irregular shape, having inner relief, are also clearly visible in the image of the light area of size 10×10 µm (Fig. 8.8), indicating the submicron pore size.

(ii) *dolomite*

The sample of carbonate rock was presented in the micro-section form, the grain structure of which was visible in the optical microscope as a fairly homogeneous one without abrupt changes in brightness and color. Sharply protruding heterogeneities of a nanometer range, which may be individual crystals of dolomite, are clearly seen in the image of size of 3×3 µm (Fig. 8.9). It should be noted that the surface of the dolomite samples studied is not porous to the same scale as the surface of the sandstone shown in Fig. 8.7.

Fig. 8.9 AFM-image of the
dolomite sample (granted by
Dr. Khanukaeva)

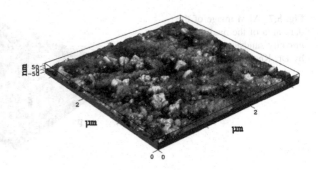

8.5 Theoretical Model and Experimental Study of Filtration of Aqueous Suspension Through a Porous Medium

One of the main problems in the rational development of hydrocarbon fields is to select the optimal mode of intrusion of the particles (proppant) in the hydraulic fracture and to determine the conditions of their uniform filling, as well as the destruction of formed "plugs" of proppant. Cracks filled with large particles (proppant) have high permeability compared to the surrounding rocks, which can significantly increase the recovery of oil from their surfaces.

In filtration experiments (Fig. 8.10) [7] we used aqueous suspension containing solid particles of silicon carbide (moissanite) of 10 micron average diameter and concentration of 84 g/l that corresponds approximately to 0.5×10^6 particles per cm^3. Sheets of 0.9 mm $\times$ 24 cm $\times$ 15 cm foam rubber, compressed between plates of polished glass mirrors, which is a braiding of polyurethane filaments with thickness of 100 μm with an average pore size of 600 μm (Fig. 8.11), were used as porous medium (initial thickness of the sheet was 5 mm and initial porosity was 0.98). The porosity of the compressed foam rubber decreased to 0.93, and its permeability to 150 Darcy. Before each experiment, the porous layer was filled with water. Dissemination of the suspension into the reservoir model occurs under differential pressure created at the ends of the reservoir model using a vacuum pump. Monitoring the area of penetration of the suspension into the porous medium was conducted using a camera that records the light coming through different parts of the medium with different concentrations of filtered particles. The density distribution of the particles deposited in the pores was correlated with the distribution of brightness on the obtained photographs. The experimental dependences of the volumetric filtration rate $Q(t)$ which is monotonically decreasing with time t and the concentration profiles $n(x, t)$ of the particles deposited in the foam rubber at different time were obtained.

The probabilistic-sieve model of microfiltration [8, 9] is applied to describe the pore blocking process, which allowed to determine physicochemical parameters of nonstationary filtration process of aqueous suspension (see also Chap. 13). According to the model the linear filtration rate $u(t)$ is of the following form:

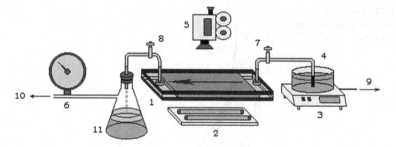

Fig. 8.10 The experimental setup. *1* Hele-Shaw cell with a sheet of foam rubber, *2* backlight, *3* weighing-machine, *4* container with aqueous suspension, *5* camera, *6* gauge, *7, 8* tubes with taps, *9* to the computer interface, *10* to the vacuum pump, *11* buffer reservoir

Fig. 8.11 Foam rubber cells with retained particles: polyurethane fibers diameter –100 μm, average pore size –0.6 mm, average particle size of silicon carbide –10 μm

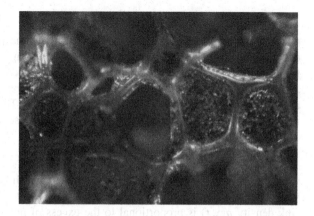

$$u(t) \equiv \frac{Q(t)}{S} = \frac{Q_\infty/S}{1 - (1 - Q_\infty/Q_0)e^{-\omega t}}, \qquad (8.3)$$

where $Q_\infty = \lim_{t\to\infty} Q(t)$, $Q_0 = Q(0)$, S is the cross-sectional area of the medium and ω–parameter which depends on the particle concentration and Q_∞. In our experiments we had $Q_0 = 12.96\,\text{cm}^3/s$ and $S = 2.16\,\text{cm}^2$. The values of $Q_\infty = 3.67 \times 10^{-3}\,\text{cm}^3/s$ and $\omega = 2.7 \times 10^{-6}s^{-1}$ were fitted using (8.3) and the least square method. Figure 8.12 presents experimental (symbols) and theoretical (curve) dependencies of $Q(t)$.

In order to describe the behavior of numerical particle concentration $n(x,t)$ we need to employ the law of mass balance in the form:

$$\frac{\partial n}{\partial t} = -\frac{\partial J}{\partial x} + q(x,t), \qquad (8.4)$$

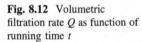

Fig. 8.12 Volumetric filtration rate Q as function of running time t

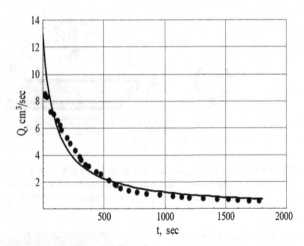

where x is the coordinate which is directed along the pressure gradient (in our experiments the pressure difference $\Delta p = 0.6$ bar was applied at 15 cm depth of the filtration bed), $q(x, t)$ is the sink density of the particles due to blockage of pores and J is the flux rate of the particles. We evaluated the particle diffusion coefficient as $\sim 10^{-10}$ cm^2/s, so characteristic Peclet number $\sim 10^8 \gg 1$. It means that we may neglect the diffusive part of the flux and can consequently write

$$J = \alpha u(t) n(x, t), \tag{8.5}$$

where α– correcting factor which takes into account the difference between mean velocity of the particles in the pores and filtration rate. We assume also that the sink density $q(x, t)$ is proportional to the excess of number of particles (at given point and on given time) over some critical value n_* ($H(x)$—the Heaviside function, $\beta = const$—coefficient of proportionality):

$$q(x, t) = -\beta(n(x, t) - n_*)H(n - n_*). \tag{8.6}$$

On substitution of (8.4) and (8.5) into (8.6) we get the transport equation and its solution takes the form:

$$n(x, t) = n_*\left(1 + e^{-\beta t} \cdot f\left(x - \alpha\frac{Q_\infty}{S\omega}\ln\left(e^{\omega t} - \left(1 - \frac{Q_\infty}{Q_0}\right)\right)\right)\right), \tag{8.7}$$

where $f(x)$ is an auxiliary function which should be determined from initial and/or boundary conditions. In our case, experimental values of numerical concentration of the particles can be approximated by the following simple formula:

$$n(x, t) = n_0(t) + n_1(t) \cdot e^{-x/x_0(t)}, \tag{8.8}$$

Table 8.1 Time-dependent coefficients in formula (8.8)

t, s	$n_0(t) \times 10^{-6}$, 1/cm^3	$n_1(t) \times 10^{-6}$, 1/cm^3	$x_0(t)$, cm
100	0.34	3.08	2.50
200	0.87	6.58	1.55
300	1.18	9.18	1.40
400	1.41	11.01	1.43
500	1.55	12.26	1.67
1000	1.92	15.10	2.44

Fig. 8.13 Profiles of numerical concentration at different moments of time

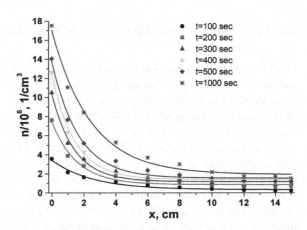

So, in order to match (8.7) and (8.8), we should choose for $f(x)$ an appropriate decay law:

$$f(x) = A + Be^{-\gamma x} \ (A, B, \gamma = \text{const}). \tag{8.9}$$

Table 8.1 gives dependencies of coefficients in formula (8.8) as functions of running time, which were fitted by the least square method. Figure 8.13 shows experimental (symbols) and theoretical (curves) behaviors of these coefficients. It is seen that there is good agreement between them.

It is necessary to mention that from (8.4–8.6) we can extract the stationary distribution of the particles concentration $n_\infty(x)$ along the depth of the filtration layer. Assuming that $t \to \infty$, we arrive at the ordinary differential equation for $n_\infty(x)$:

$$\alpha u_\infty \frac{dn_\infty}{dx} = -\beta(n_\infty - n_*), \tag{8.10}$$

where $u_\infty = Q_\infty/S = 1.7 \cdot 10^{-3}$cm/s. General solution of Eq. (8.10) is of the form:

$$n_\infty(x) = n_* + C \cdot e^{-\frac{\beta}{\alpha u_\infty}x}, \tag{8.11}$$

where C is unknown constant. We can evaluate n_* and C, as well as ratio β/α comparing Eqs. (8.8) and (8.11) assuming that $t = 1000\,s$ is "almost infinite" characteristic time for the system under consideration. From Table 8.1 we conclude that $n_* \approx 1.92 \cdot 10^6\,\text{cm}^{-3}$, $\beta/\alpha \approx 7 \cdot 10^{-4} s^{-1}$ and $C = 15.1 \cdot 10^6\,\text{cm}^{-3}$.

Therefore, using the model proposed here we can predict the profiles of particles retarded by porous medium and evaluate changes in its specific permeability and local porosity with time.

8.6 Laboratory Modeling of Two-Phase Jet Flows

Problems of studying the stability of multiphase filtration flows are certainly relevant in the development of oil and gas fields [10]. Tasks associated with the need to consider the instability of the water–oil interface arise in the development of methods for stimulating oil production by flooding, as well as when investigating the movement of reservoir waters in the process of exploitation of fields. Similar problems are first considered by Muskat, Leverett, Rapoport, Leibenson, and then later in the works of Barenblatt, Nikolaevskij, Ryzhik, Zheltov, Basniev, Entov, Dmitriev, Lyubimov, and others [1, 11–17]. But issues of determining the large-scale structure of unstable multiphase flows, the influence of rheological properties of liquids and forces of capillary interaction on the character of development of the filtration instability still remain poorly understood. Experimental studies of Saffman and Taylor [18] and Saffman [19] and Chouke et al. [12] have shown that the development of perturbations of a plane front of displacement of a viscous fluid in a porous medium occurs in violation of stability in the form of indefinitely proliferating "fingers" or "tongues" of the displacing fluid. Mathematical model of propagation of "tongues" of the displacing fluid with exclusion of interfacial forces was proposed by Barenblatt et al. [1]. It was assumed that this process can be described by equations of the Buckley-Leverett model, with relative permeabilities linearly dependent on the respective saturations. In this paragraph, using laboratory simulation, the problem about the impact of capillary forces on the developing process of filtration instability in the form of jets of the displacing fluid is investigated.

To study the filtration instability of the two-phase flow, we used the same experimental setup with the cell of the Hele-Shaw type shown in Fig. 8.10. We performed a series of experiments to displace more viscous fluid, filling the reservoir model, using a less viscous liquid immiscible with the first. Water with viscosity of 1 cP, and mineral oil VM-1 with viscosity of 120 cP were used as immiscible liquids. A foam sheet of thickness of 5 mm was used as the filtering medium. The initial porosity of the foam sheet was 0.98. The sheet of foam rubber was compressed between the planes of polished rectangular glass with dimensions of 125×400 mm to reach the final thickness of 0.9 mm. The average pore size was reduced by compression 1.7 times, and porosity—to the value of 0.93. Specific permeability of the compressed foam was 150 Darcy.

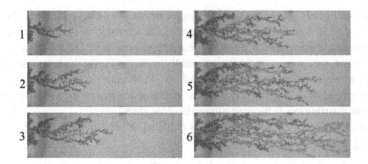

Fig. 8.14 Successive stages of displacement at a differential pressure of 0.385 bar. Time step between frames is 15 s, the flow is directed from left to right

Before the beginning of each experiment, the porous array was filled with oil. In this case the actual permeability of the array and the total pore volume were measured. Because of similar refractive index of oil and foam material (polyurethane) the model of porous medium is optically transparent. Displacement of oil with water occurs under the pressure drop caused at the ends of the reservoir model using a vacuum pump. Continuous in time measurements of total flow rate of the injected fluid in the reservoir model were carried out using precision electronic scales that measure the mass of the cell with the pumped liquid. Observation of the penetration area of displacement liquid into a porous medium was conducted using a camera. The displacing liquid was colored to increase the contrast. Water saturation distribution of the porous array along filtration flow was correlated with the distribution of the brightness of different portions of pore space in the photographs obtained.

To establish the relation between a particular brightness and the corresponding saturation, a series of calibration measurements was conducted. The pore space of the model reservoir was filled with a mixture of liquids (water and oil) with known volume ratio and was photographing after filling. During calibration, the pictures were taken at the same exposure settings and in the same conditions as for the conduct of basic experiments on fluid displacement. The brightness of the obtained image was averaged across the region occupied by a mixture of liquids. As a result of a series of measurements with different volume ratios of liquids, the dependence, which allows finding the water saturation of a certain area of the model collector taking brightness of this area on the photograph into account, was obtained. Experiments have shown that this dependence is approximately linear in the range of values of water saturations of interest.

During the study, we obtained experimental data on changes in the distribution of saturation of displacing liquid in the process of more viscous liquid displacement at fixed pressure drop. It was found that displacement front is unstable at all velocities of displacement used in the experiment, when the ratio of viscosities of the displacing and displaced fluid is 1:120. Development of perturbations of initially flat front of displacement occurs in the form of sprawling fingers or jets of displacing liquid (Fig. 8.14). Advancing rate of fingers is proportional to the total flow rate of liquids. In the process of displacement at constant pressure drop, the total flow rate of liquids increases, while the characteristic thickness of displacing jets decreases tending in the limit to the average pore size. Distribution of water saturation of the porous reservoir, formally averaged over its cross-section, is complex along the direction of filtration flow. In the later stages of displacement with the spread of displacing jets over the entire space reservoir, a profile of water saturation distribution is aligned along the flow direction.

In order to describe a process such as considered here, authors of [1] made the following assumptions:

(i)   the length of the fingers in the direction of flow is much greater than their width;
(ii) the flow on average is regarded as one-dimensional, so the filtration rate of each liquid, averaged over some representative cross-section, was directed along the axis of filtration; and
(iii) the saturation was assumed to be constant within each finger.

It was assumed that under such an approach, ordinary two-phase filtration equations are valid for averaged flow, but with relative permeabilities which are linearly dependent on the respective saturations.

Let us consider the process of displacement occurring in a horizontal array of unit thickness. The array is presented by homogeneous and isotropic porous medium with porosity ε and specific permeability k. The coordinate x is measured along the sample, direction of flow is horizontal (Fig. 8.10). The width ω of the array is considered sufficiently small compared with its length L, so that the pressure and saturation can be regarded constant over sections. Liquids are assumed as incompressible ones; the pressure P in the displaced and displacing phases are considered the same. Let at the initial time $t = 0$ the liquid with viscosity μ_1 be on the left up to the coordinate $x = 0$ of the porous medium and liquid with viscosity μ_2 be on the right of $x = 0$ up to the coordinate $x = L$. For definiteness, we assume that $\mu_0 = \mu_1/\mu_2 < 1$. Starting with the moment of time $t = 0$ liquid 1 at a flow rate $Q(t)$ is injected in the porous collector, which is initially occupied by liquid 2, thereby displacing it.

We assume also that the displacement occurs in the form of "fingers" pushing in the direction of flow. Displacing phase saturation within the finger is considered equal to 1, the saturation outside the finger is 0, and that is, we suppose that the piston displacement of fluids occurs within the pores on the passage of the liquid interface. This assumption allows us to think that liquid flow in each phase obeys the Darcy law:

$$u_1 = -\frac{k}{\mu_1}\frac{\partial P}{\partial x}, \; u_2 = -\frac{k}{\mu_2}\frac{\partial P}{\partial x} \tag{8.12}$$

Here $u_{1,2}$ are velocities of the respective phases of fluids. Let us introduce $\sigma(x)$ as the ratio of the total cross-sectional area $x = \text{const}$ of all "fingers" of the displacing liquid to the total cross-sectional area ω of the porous array of unit thickness. Formally, the relative area of filtration σ can be regarded as the saturation of the displacing phase, averaged over the cross-sectional area of the porous array.

In order to know how σ changes in the displacement process let us consider the balance of each phase as a homogeneous fluid, written for elementary micro-volume. Then we obtain the continuity equation for the displacing liquid:

$$\varepsilon \frac{\partial}{\partial t} \sigma(x,t) + \frac{\partial}{\partial x} [\sigma(x,t) u_1(x,t)] = 0. \tag{8.13}$$

A similar continuity equation can be written for the second phase:

$$-\varepsilon \frac{\partial}{\partial t} \sigma(x,t) + \frac{\partial}{\partial x} [(1 - \sigma(x,t)) u_2(x,t)] = 0.$$

Adding the continuity equations and integrating, we obtain:

$$\sigma u_1 + (1 - \sigma) u_2 = \frac{Q(t)}{\omega},$$

that is, the total volumetric flow rate of both phases, as expected, does not depend on the coordinate x.

The system of Eqs. (8.12–8.13) allows obtaining the following equation for σ by exclusion of other (dependent) variables [10]:

$$\varepsilon \frac{\partial \sigma}{\partial t} + \frac{Q(t)}{\omega} [g(\sigma) + \sigma g'(\sigma)] \frac{\partial \sigma}{\partial x} = 0, \tag{8.14}$$

where the notation is introduced

$$g(\sigma) = \frac{1}{\sigma(1 - \mu_0) + \mu_0}.$$

Function $g(\sigma)$ is similar to the Buckley-Leverett function used in the classical theory of two-phase flow [1, 11]. For each section of the porous array, it is equal to the ratio of the filtration rate of the displacing phase to the average speed of the entire flow. If we consider, as assumed above, the total relative area σ of filtration flow of the displacing phase as its saturation, averaged over the cross section of a porous array, then the resulting Eq. (8.14) coincides with the classical Buckley-Leverett equation for two-phase flow, in which the relative permeabilities of both phases are assumed to be equal to their corresponding saturations.

Let us consider such trajectories $x(t)$ on the phase plane (x, t), in which the total area of "fingers" of displacing fluid takes a given constant value $\sigma(x, t) = \text{const}$. The equality $d\sigma = 0$ is valid on these lines and the position x_σ of specified value σ as a function of time is

$$x_\sigma(t) = \frac{V(t)}{\omega \varepsilon} \frac{\mu_0}{[\sigma(1 - \mu_0) + \mu_0]^2} + x_\sigma(0). \tag{8.15}$$

Here $x_\sigma(0)$ is the initial position of the front with the corresponding value σ at $t = 0$; $V(t) = \int_0^t Q(\tau)d\tau$ is the total volume of liquids passed through the porous array since the beginning of displacement. Since the inlet to the array is fed only with phase 1, the total volume $V(t)$ is equal to the total volume of phase 1. Furthermore, the initial conditions give $x_\sigma(0) = 0$ for all values of σ.

With the law of motion of the displacement front for each fixed σ, we can get the dependence of the jet tip coordinate of the displacing fluid on time. To do this, tending σ to 0 in the expression (8.15), we obtain for the coordinate x_f of the leading edge of displacement

$$x_f(t) = \frac{1}{\mu_0} \frac{V(t)}{\omega\varepsilon}. \tag{8.16}$$

Similarly, taking σ equal to 1 in the expression (8.15), we obtain the time dependence of the coordinate of the trailing edge of a full displacement:

$$x_1(t) = \mu_0 \frac{V(t)}{\omega\varepsilon} = \mu_0^2 x_f.$$

For all values $x_1 < x < x_f$ due to monotonicity of the function x_σ with respect to σ we obtain the following relationship from Eq. (8.15):

$$\sigma(x,t) = \frac{\left(\frac{\mu_0}{x} \frac{V(t)}{\omega\varepsilon}\right)^{\frac{1}{2}} - \mu_0}{1 - \mu_0}.$$

Using (8.16), we arrive at the final solution for σ in the area $0 \le x \le L$ and $t \ge 0$:

$$\sigma(x,t) = \begin{cases} 1 & , \quad 0 \le x < \mu_0^2 x_f \\ \left(\left(\frac{x_f}{x}\right)^{\frac{1}{2}} - 1\right) \frac{\mu_0}{1 - \mu_0} & , \quad \mu_0^2 x_f \le x < x_f \\ 0 & , \quad x_f \le x \le L \end{cases} \tag{8.17}$$

The function $\sigma(x,t)$ can be considered as the volume fraction of a unit length of the array of the pore space occupied by the displacing fluid. It is easy to verify that the result of integration of the function $\sigma(x,t)$ in respect of x within the segment $0 \le x \le x_f$ will be equal to $V(t)/\omega\varepsilon$.

Our observations of the coordinate of the displacement front showed that the velocity of areas with low values of σ is significantly inferior to the rate predicted by the model. Measured travel times of the displacement front from the mouth of a flow to its drain are an order of magnitude longer than the times calculated from the abovementioned theoretical estimates. Apparently, this is due to the influence of surface tension forces arising at the interface. In formulating the mathematical model it was assumed that the relative sectional area σ of the jet tends to 0 upon approaching the front of displacement. Experiments have shown that the movement of the displacing liquid occurs in the form of individual jets, the width of which is smaller, the more their velocity. At the ends of jets of the displacing fluid a zone of rapid growth of their cross-sectional area is forming, where regions with different relative area σ are moving at the same velocity. The length of this zone decreases with increasing the jet velocity and has the size of the order of its characteristic width.

To describe such a zone in the classical model, the equation, known as the equation of Rapoport-Leas [20], is applied. It is a mass balance equation written for the phases, each of which moves under the influence of "own" pressure gradient [1, 20]. Pressures at interfaces differ by a capillary pressure jump $P_c = P_1 - P_2$, depending on local saturation [15]. In addition to the surface tension, preferential wetting of the porous solid skeleton by one of the phases also affects the magnitude of the pressure jump. It is believed that the motion of the fluids is going on in a quasi-equilibrium manner, i.e., under given saturation the liquids are distributed as in conditions of hydrostatic equilibrium. Thus, the curves giving the dependence of the capillary pressure on the saturation represent the integral characteristic of the pore structure.

A similar equation can be written for the case of a jet displacement. We assume that each mobile phase flows in its occupied space under the action of "own" pressure, i.e., as if it was limited only by rigid walls. Then the following equations are valid for one-dimensional two-phase flow:

$$u_i = -\frac{k}{\mu_i}\frac{\partial P_i}{\partial x}, \quad i = 1, 2 \quad \text{and} \quad P_1 - P_2 = P_c(x)$$

which implies that,

$$u_2 = \mu_0 u_1 + \frac{k}{\mu_2}\frac{\partial P_c}{\partial x}.$$

Using the condition of conservation of the total flux of liquids, we obtain for the velocity of the displacing phase:

$$u_1 = g(\sigma)\left[\frac{Q}{\omega} - (1 - \sigma)\frac{k}{\mu_2}\frac{\partial P_c}{\partial x}\right].$$

Substituting this expression in the mass balance equation for the first phase (8.13), we finally obtain partial differential equation for determination of σ:

$$\varepsilon\frac{\partial \sigma}{\partial t} + \frac{Q}{\omega}[g(\sigma) + \sigma g'(\sigma)]\frac{\partial \sigma}{\partial x} - \frac{\partial}{\partial x}\left[\sigma(1 - \sigma)g(\sigma)\frac{k}{\mu_2}\frac{\partial P_c}{\partial x}\right] = 0. \qquad (8.18)$$

In the above model we assumed that the saturation of the displacing phase inside the jet is constant and equal to the limiting value, that is, the magnitude of the capillary pressure jump must be constant. However, as seen from Eq. (8.18),

Fig. 8.15 Scheme for
derivation of the displacing
jet equation

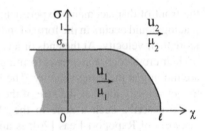

a constant pressure difference between phases results in the same mass balance
equation because the third term on the left side of (8.18) vanishes. For description
of the zone of constant velocity at the end of the jet of the displacing fluid within
the proposed new model it is reasonable to assume that the pressure difference in
phases depends on the local value of the relative cross-sectional area σ of the
filtration flow. This is possible if we take into account the finite time to establish
capillary equilibrium between phases. Furthermore, we assume that the pressure
drop at the interface is equal to 0 in equilibrium, that is, the pore surface of the
array has neutral wettability. Under these assumptions, the value of the capillary
pressure will be greater, the higher the velocity of propagation of the liquid
interface (front) relatively to the solid skeleton, tending in the limit to the value
determined by the surface tension:

$$P_c = P_1 - P_2 = \alpha(\frac{1}{R_1} + \frac{1}{R_2}),$$

where α—the coefficient of interfacial tension, and R_1 and R_2—the principal radii
of curvature of the interface at this point, which are close to the pore size.

Accounting for the strength of the surface interaction, which depends on the
propagation velocity of the fluid interface, gives the condition of the lowest possible "width" of the jet at the displacement front. Let us consider the motion of an
extended jet of the displacing fluid with constant relative cross-sectional area σ_0
(Fig. 8.15). We assume that the velocity of the displacing fluid in a jet is constant
and equals $u_1 = g(\sigma_0)Q/\omega$. The axis x is directed along the direction of motion.
The coordinate $x = 0$ as shown in the figure is compatible with the position of the
beginning of the transition zone at initial moment of time $t = 0$, so that $\sigma(0) = \sigma_0$.
Length of the transition zone is denoted by l, so that at the initial time we have
$\sigma(l) = 0$, which corresponds to the tip of the jet, as the function $\sigma(x)$ decreases
monotonically along the interval from 0 to l. We assume that when $x < 0$ the fluids
near the interface are in equilibrium, so that $P_c = P_1 - P_2 = 0$. The capillary
pressure has a maximum at the tip of the jet and is determined by the Laplace
formula: $P_f = 4\alpha/a$, where a is the average pore size.

Since the velocity of propagation of the displacing fluid is constant, we seek a
solution of Eq. (8.18) in the form of a traveling wave. To do this, let us make the
change of variables $\chi = x - Dt$, $D = g(\sigma_0)\frac{Q}{\omega\varepsilon}$, thereby going to the reference
frame which is relative to the position of the jet, where D is the jet velocity. After

integration and algebraic manipulation with (8.18) we obtain the ordinary differential equation for the capillary pressure:

$$\frac{dP_c}{d\chi} = \frac{\mu_2}{k} \varepsilon D(1 - \mu_0)\sigma_0 \left[\frac{1 - \frac{\sigma}{\sigma_0}}{1 - \sigma}\right]. \tag{8.19}$$

It is necessary to integrate the resulting expression (8.19) in order to determine the magnitude of the transition zone. For that we need to know the dependence of the capillary pressure on the value of σ, which is analogous to the capillary pressure curve for the unsteady distribution of saturation. However, to evaluate the magnitude of the transition zone we can use the fact that the dimensionless area of the filtration section on the integration interval ranges from $\sigma_0 < 1$ at $\chi = 0$ to 0 at $\chi = l$ and $\sigma(\chi)$ decreases monotonically. Therefore, in the integration of (8.19) the following estimation can be made:

$$\int_0^l \frac{1}{1 - \sigma}\left(1 - \frac{\sigma}{\sigma_0}\right) d\chi \simeq l.$$

Thus, taking into account that $P_c(0) = 0$, $P_c(l) = P_f$, we obtain the expression for the dimensionless width σ_0 of the jet:

$$\sigma_0 \simeq \frac{k}{\mu_2} \frac{P_f}{l} \frac{1}{\varepsilon D} \frac{1}{1 - \mu_0}. \tag{8.20}$$

In our experiments, the displacing phase propagated as "fingers" consisting of separate jets that moved at the same velocity. The value of the relative filtration area of the displacing phase was typically constant (in average) throughout the length of the finger, and was steady at a distance from its end which is comparable with the width of the finger. Formally, substituting the length of the transition zone, which is approximately equal to the width of a finger, that is $l \simeq \sigma_0 \omega$, in (8.20) and keeping in mind that $P_f = 4\alpha/a$ we obtain:

$$\sigma_0 = \left(\frac{1}{\omega} \frac{k}{\mu_2} \frac{4\alpha}{a\varepsilon D} \frac{1}{1 - \mu_0}\right)^{\frac{1}{2}}. \tag{8.21}$$

Figure 8.16 shows the curves of saturation of the pore space on the dimensionless distance x/x_f. Theoretical curve 1 is obtained from the condition of proportionality of the coordinate of the jet tip and the total volume of the displacing phase, measured in the experiment: $x_f = g(\sigma_0)V/\omega\varepsilon$. Curve 2 is calculated by the formula (8.21), using experimental data for the advancing velocity of the liquid interface, under an average pore size of $\alpha = 100$ μm and coefficient of surface tension $\alpha = 70 \times 10^{-3}$ (N/m). Both curves are calculated for the viscosity ratio $\mu_0 = 1/120$. Discrepancy between the curves 1 and 2 at low values of x/x_f is

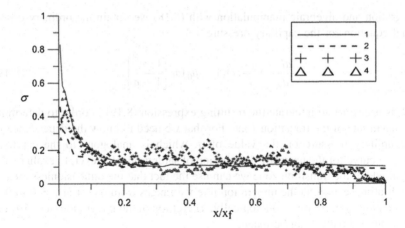

Fig. 8.16 Dependence of the saturation of the pore space on the dimensionless distance x/x_f at different stages of the displacement process: *curves 1* and *2* show simulation results, *curve 3* corresponds to 105 s from the beginning of displacement, *curve 4*, 65 s

due to the fact that the length of jets of the displacing liquid was much smaller than their typical "width" at the initial stage of displacement that contradicts the assumption concerning the length of the transition zone, which was used to obtain the formula (8.21). For other values of x/x_f the simulation results are in a good agreement with the experimental curves.

For more accurate description of the process of two-phase flow within the frame of the "jet" model with constant saturation of displacing fluid, it is required to establish the dependence of the capillary pressure on the local value of the velocity of propagation of a fluid interface.

Acknowledgment The support of the Russian Foundation for Basic Research (RFBR) is acknowledged (grant No 12-08-01091_a).

References

1. Barenblatt GI, Entov VM, Ryzhik VM (1990) Theory of Fluid Flows Through Natural Rocks. Kluwer Academic Publishers, Dordrecht
2. Gimatudinov ShK (1963) Fizika Neftyanogo Plasta (Physics of the Oil Reservoir). Gostoptekhizdat, Moscow (in Russian)
3. Matusevich VM, Popov VK, Rimskikh VF, Rodygin VR, Volfkovich YM, Shkolnikov EI, Sosenkin VE (1981) Izvestiya vuzov. Seriya: Neft i gas. In: Proceedings of the Universities. Series: Oil and Gas. vol 7. p 3 (in Russian)
4. Volfkovich Yu.M, Bagotzky VS, Sosenkin VE, Blinov IA (2001) Colloids Surf, A 187:349
5. Kovalev AG, Novokreshchenova AP, Chumakina MS, Volfkovich YM, Shkolnikov EI (1982) Sbornik nauchnykh trudov Vsesoyuznogo neftegazovogo nauchno-issledovatelskogo instituta. In: Proceedings of scientific papers of All-Union oil and gas research institute, vol 78. Moscow, p 122 (in Russian)

6. Khanukaeva DY, Filippov AN (2013) Membrany i Membrannye Technologii (Membranes and Membrane Technologies) 3(3):210 (in Russian)
7. Gabova A, Belyakov G, Baryshnikov N, Tairova A, Filippov A (2013) In: Proceedings of international conference on ion transport in organic and inorganic membranes, Kuban State University, Krasnodar-Tuapse, 2–7 June 2013, p 79
8. Filippov A, Starov V, Lloyd D, Chakravarti S, Glaser S (1994) J Membr Sci 89:199
9. Filippov AN, Iksanov RKh (2012) Pet Chem 52(7):520
10. Baryshnikov NA, Beljakov GV, Turuntaev SB, Filippov AN (2013) In: Proceedings of Gubkin Russian state university of oil and gas vol 3(272). p 15 (in Russian)
11. Basniev KS, Dmitriev NM, Kanevskaja RD, Maksimov VM (2005) Podzemnaja gidromekhanika (Underground Hydromechanics). Izhevsk (in Russian)
12. Chouke RL, van Meurs P, van der Poel C (1959) Pet Trans A.I.M.E. 216:188
13. Zheltov YP (1975) Mehanika neftegazonosnogo plasta (Mechanics of oil and gas reservoir). Moscow (in Russian)
14. Leibenson LS (1947) Dvizhenie Prirodnyh Zhidkostej I Gazov V Poristoj Srede (Motion of Natural Liquids and Gases in a Porous Medium). Gostehizdat, Moscow-Leningrad (in Russian)
15. Leverett MC (1939) Trans A.I.M.E. 132:381
16. Muskat M, Meres MW (1936) Physics 7:346
17. Nikolaevskij VN (1984) Mehanika Poristyh I Treshhinovatyh Sred (Mechanics of Porous and Fractured Media). Nedra, Moscow (in Russian)
18. Saffman PG, Taylor, Sir GI (1958) Proc R Soc Lond A245:321
19. Saffman PG (1986) J Fluid Mech 173:73
20. Rapoport L, Leas W (1953) Trans A.I.M.E. 198:139

Chapter 9
Determination of Active Porosity in the Field to Solve Problems of Protection of Groundwater Against Pollution

Alexander Vladilinovich Rastorguev

Abstract When justifying the remediation projects of aquifers from pollution, as a rule, field studies are in preference. In comparison with laboratory experiments, they cover a larger volume of rocks that contain groundwater, and therefore better reflect the average properties of the reservoir. The most important parameter determined in field trials is the porosity. It determines migration velocity of the contamination front and characterizes the amount of pollutants in the reservoir, etc. Depending on the objectives of the study, it can be determined under field conditions in different ways: injection of chemical or thermal indicators, or vice versa pumping or injection–pumping. There are examples of three studies presented below. In the first example, one well-experienced results of injection–pumping of clean water into the contaminated aquifer are presented. As a result, the active porosity of gravel and pebble reservoir with sand filling is equal to 0.15. The data obtained by the experience have been required to support the interception of contaminated flow. The second example shows the results of pumping heated water into the reservoir folded by fractured carbonate sediments (limestones and dolomites). The resulting value of the active porosity is 0.005. Field trial was needed to validate the use of groundwater for cooling of process equipment. The third example is related to the processing of man-made pumping oil. As a result, we obtained porosity of 0.39, residual water saturation of 0.05, and residual oil saturation of 0.25. The need to determine the parameters has been associated with the need to optimize the remediation works.

9.1 Determination of Active Porosity on the Basis of One Borehole Experience: The Injection–Pumping of Clean Water into the Polluted Horizon

To estimate migration parameters (active porosity is the main among them) one borehole experiments are often used [1]. In the course of the experience for one borehole, the injection starts, and then pumping of the indicator proceeds.

Y. M. Volfkovich et al., *Structural Properties of Porous Materials and Powders Used in Different Fields of Science and Technology,* Engineering Materials and Processes, DOI: 10.1007/978-1-4471-6377-0_9, © Springer-Verlag London 2014

In the professional literature such scheme is called "push-pull" experience. An example of conduction and interpretation of the one borehole expertise is given below to identify the active porosity of the alluvial deposits in eastern Kazakhstan. The need for such expertise was associated with forecasts of migration and interception of contaminated groundwater, forcing their way to the territory of the tailings storage in the valley of the Irtysh River. The cut of the well, in which the experiment was carried out, is shown in Fig. 9.1. Migration studies have focused on the determination of the parameters of pebble deposits drowned in a depth interval from 29.7 to 31.8 m.

Chloride ion was chosen as an indicator. In the alluvial horizon before the experiment, the chlorine ion concentration was high and equal to 235 mg/l. Therefore, it was decided to inject clean water with low content of chloride ion (5 mg/l). Experimental work is timing tracking changes in the concentration of chloride ions in the pilot hole. Injection flow rate was 0.19 l/s. A total of 200 l was injected, which is almost five times more than the volume of water contained in the borehole. In 62 min after injection pumping was launched with a flow rate of 0.09 l/s. The time span of pumping was 90 min. Thus, the observations of the concentration change were performed for 152 min after injection. During this time, 20 samples of water were selected. The concentration of chloride ions in the water samples was determined by spectrophotometry.

Interpretation of the results of the migration experience was carried out with the help of mathematical modeling of convective-dispersive transport using MT3DMS code [2]. The equation of convective-dispersive transport is as follows [3]:

$$\theta \frac{\partial C}{\partial t} = \frac{\partial}{\partial x_i}\left(\theta D_{ij}\frac{\partial C}{\partial x_j}\right) - \frac{\partial}{\partial x_i}(\theta v_i C) + q_S C_S \tag{9.1}$$

wherein θ—the active porosity; C—the content (concentration) of the component which is an indicator of pollutants in the groundwater, $[ML^{-3}]$; D_{ij}—hydrodynamic dispersion tensor, $[L^2 T^{-1}]$; v_i—the actual velocity of groundwater inside pores which is associated with the Darcy or averaged filtration velocity q_i as follows: $v_i = q_i/\theta$, $[LT^{-1}]$; t—time, $[T]$; x_i—Cartesian coordinates, $[L]$, $i = 1, 2, 3$; C_s—the content (concentration) of the indicator in injected-pumped water $[ML^{-3}]$; q_S—specific rate of the injection-pumping $[T^{-1}]$. The components of the hydrodynamic dispersion tensor are determined according to the formula [3]:

$$D_{ij} = \alpha_T |v|\delta_{ij} + (\alpha_L - \alpha_T)\frac{v_i v_j}{|v|},$$

$$|v| = \sqrt{v_x^2 + v_y^2 + v_z^2}, \tag{9.2}$$

$$\delta_{ij} = \begin{cases} 1 & i = j \\ 0 & i \neq j \end{cases}$$

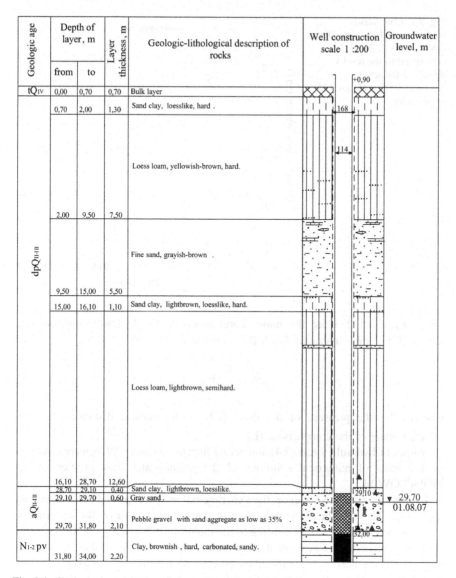

Geologic age	Depth of layer, m		Layer thickness, m	Geologic-lithological description of rocks	Well construction scale 1 :200	Groundwater level, m
	from	to				
tQ$_{IV}$	0,00	0,70	0,70	Bulk layer		
dpQ$_{II-III}$	0,70	2,00	1,30	Sand clay, loesslike, hard .	168	
	2,00	9,50	7,50	Loess loam, yellowish-brown, hard.	114	
	9,50	15,00	5,50	Fine sand, grayish-brown .		
	15,00	16,10	1,10	Sand clay, lightbrown, loesslike, hard.		
	16,10	28,70	12,60	Loess loam, lightbrown, semihard.		
aQ$_{II-III}$	28,70	29,10	0,40	Sand clay, lightbrown, loesslike.		
	29,10	29,70	0,60	Gray sand .	29,10	29,70 01.08.07
	29,70	31,80	2,10	Pebble gravel with sand aggregate as low as 35% .	32,00	
N$_{1-2}$ pv	31,80	34,00	2,20	Clay, brownish , hard, carbonated, sandy.		

+0,90

Fig. 9.1 Geological engineering section of experimental well

where α_L and α_T are the parameters of the longitudinal and transverse dispersivity, [L]. The components of the actual velocity vector needed to solve (1) are determined by Darcy's law:

$$v_i = \frac{q_i}{\theta} = -\frac{K_i}{\theta}\frac{\partial h}{\partial x_i} \qquad (9.3)$$

Fig. 9.2 Comparison between actual and simulated data. *Curves 1, 2 and 3* correspond to the results simulated for active porosities 0.05, 0.15 and 0.25, respectively

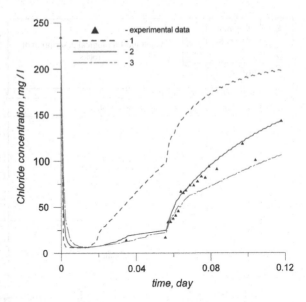

where K_i, $i = 1, 2, 3$ are the major components of the hydraulic conductivity tensor, $[LT^{-1}]$; h—hydraulic head, [L], is given as

$$h = \frac{P}{\rho g} + z \qquad (9.4)$$

wherein $\frac{P}{\rho g}$—the pressure of the fluid, [L]; z—the vertical distance from the selected plane of the comparison, [L].

Values of hydraulic head (9.4) and actual filtration velocity (9.3) were obtained on the basis of numerical solutions of the groundwater flow problem using MODFLOW [4].

In interpreting the experience, the experimental well was placed in the center of the model. The calculated discretization of the influence area on the experiment was assumed to be 10 cm. According to the monitoring, boundary conditions were set using the hydraulic head gradient which is equal to 0.04. Hydraulic conductivity of alluvial horizon represented by gravel with sand filling was assumed to be 30 m/day according to the previously conducted experiments.

The migration parameters were obtained by selection of their values in the process of mathematical modeling of a convection-dispersion flow. Model graphs of a time schedule tracking (Fig. 9.2) show a significant effect of active porosity on the level of change in concentration of chloride ions in the water pumped out of the pilot hole. The numerical solution of a series of migration problems gave the following values of alluvial aquifer parameters: active porosity—0.15; longitudinal dispersivity—3 cm, and transverse dispersivity—3 mm.

9.2 Determination of Active Porosity and Heat Transfer Parameters According to Injection of Heated Water into Fractured Rocks

Studies of heat transfer are now relevant to justify the use of low-grade heat with heat pumps. But the experiment described below was held in Moscow in the 1970s/1980s of the last century. At that time, the construction of large computing centers was carried out and underground water, as the primary option, was supposed to be used for cooling process equipment. It seemed comfortable because underground water in Moscow has a constant temperature of 10–11 °C throughout the year. The disadvantage of this approach was that, when cooling equipment, water is heated to 25 °C and the question that arises is what to do with hot water. The result was the adoption of a plan of experimental work, which included the drilling of experienced and observation wells and conducting experiments on the injection of hot water on the doublet scheme. Section showing the location of the experimental wells is shown in Fig. 9.3. From this figure, it follows that the doublet experiment was carried out in the upper 24 m pack of the Oka-Protvino aquifer. Borehole 2—evacuated well, 2ok—uploading well, 7—observation well. The main parameters that characterize the experimental conditions are given in Table 9.1.

The first 2 weeks of experiment were conducted in a relatively stable manner. Pumping and injection rate were constant, while the injection temperature was kept at 58 °C by a boiler. The results of the temperature change in the trunk of the observation wells are shown in Fig. 9.4, from which it follows that in a day after the start of the experiment there was an increase in temperature in the filter area of the observation well. After 13 days, a temperature of 25.4 °C was reached in the observation well. This rapid distribution of heat in the aquifer required an explanation. Interpretation of the experimental results was carried out on the basis of three models: a heterogeneous block model of concentrated capacity, a model based on the thermal conductivity of the formation and convection, and a model of unlimited capacity (Lauverier's scheme) [5]. The graphs (Fig. 9.5) give a comparison of the relative temperature $T^* = (T-T_0)/(T-T_c)$ obtained using all three models and the data in the observation well after 13 days from the start of the experiment. In all cases, the typical thermal parameters were used for calculations. The best agreement was obtained for the heterogeneous block model of concentrated capacity (see Eqs. 9.5–9.6), its parameters are given in Table 9.2:

$$c_w q_x \frac{\partial T}{\partial x} + c_w q_y \frac{\partial T}{\partial y} + c_w \theta \frac{\partial T}{\partial t} + c_b (1 - \theta) \frac{\partial T_b}{\partial t}$$
$$= \chi (T_b - T) \pm \sum_{j=1}^{n} c_w \frac{Q_j}{m} T_c \delta (x - x_j) \delta (y - y_j) \tag{9.5}$$

Fig. 9.3 Hydrogeological cross-section in the experimental area

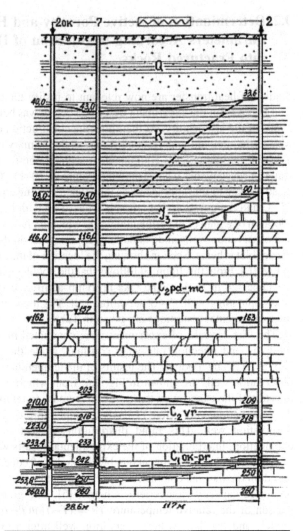

Table 9.1 The parameters of conducting the experiment

List of parameters	Value of the parameter
The distance between the injection well and pumping well, m	117
Distance from pumped well to the observation well, m	28
Injection-pumping rate, Q (m^3/h)	30
Thickness of the injection interval, m (m)	24
The temperature of formation water, T_0 (°C)	10.6
Temperature of the injected water, T_c (°C)	58
Hydraulic conductivity, k (m/day)	4–6
The gradient of the natural flow, i [-]	0.001

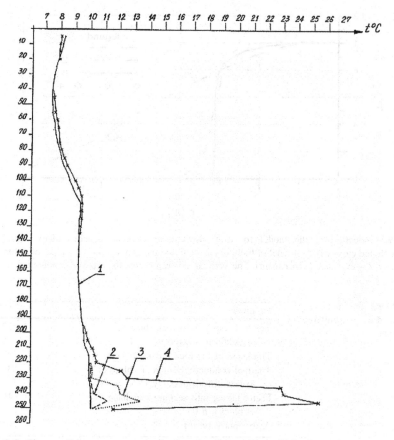

Fig. 9.4 Change of the temperature in depth at the observation well: *1* prior to the experiment, *2* 1 day after initiation of the experiment, *3–5* days after beginning of the experiment, *4* in 13 days after start of the experiment. The vertical axis represents the depth, the horizontal axis—the temperature

$$(1 - \theta)c_b \frac{\partial T_b}{\partial t} = \chi(T - T_b) \tag{9.6}$$

where x, y—spatial coordinates; t—time; T and T_b—the temperature of the water and rock blocks, respectively; c_w and c_b—heat capacity of water and blocks of rocks, respectively; χ—the coefficient of heat exchange between water and blocks, $(\chi = (\beta \lambda_b)/m_b^2)$; n—total number of wells; $\delta(x)$—Dirac delta function; q_x and q_y—average filtration velocity along the x and y axes, respectively. Definitions of other symbols are given in Tables 9.1 and 9.2.

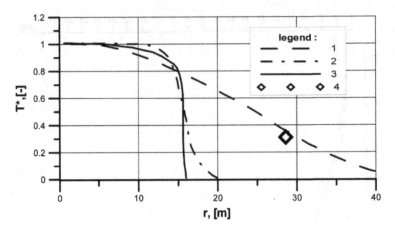

Fig. 9.5 Selection of the model for heat distribution: *1*—heterogeneous block model of concentrated capacity, *2*—model of thermal conductivity and convection, *3*—model of unlimited capacity, *4*—evidence (experiment). The vertical axis represents the relative temperature

Table 9.2 Parameters obtained by interpretation of the experiment

Parameter	Value
The heat capacity of rock blocks, c_b (kcal/(m^3 degree))	274
Thickness of the rock blocks, m_b (m)	28
Thermal conductivity of rock blocks, λ_b (kcal/(m degree day))	30
Factor taking into account the shape of blocks, β [-]	10
Active porosity, θ [-]	0.005

9.3 Determination of Porosity and Residual Oil Saturation When Pumping Technological Oil Occurs

Groundwater pollution with oil products is widespread throughout the world and in Russia as well. To indicate the liquid oil product polluters, LNAPL—light non-aqueous phase liquids (which can be interpreted as light water-insoluble liquid phase having a density less than the density of water) was accepted. Light hydrocarbons are sources of groundwater contamination at many sites around the world. These pollutants usually occur as a result of the strait of petroleum products, which are generally multicomponent organic mixture consisting of components of different water solubility. According to the EPA in January 1987 in the U.S., there were between 3 and 5 million underground storage tanks for liquid petroleum products and chemicals. Among them, according to the same agency 100–400,000 containers with inlet and outlet pipes allow leakage of their contents directly into the ground. An example of the extensive pollution of the urban area in

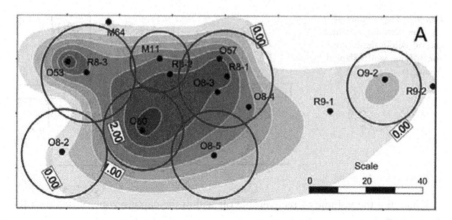

Fig. 9.6 The lens of oil pollution with the location of suction wells and gripping zones (adapted from [7])

the refinery can serve Novokuybyshevsk, Samara Region [6]. Here, even the reserves of man-made oil field of C1 category have been approved to the amount of 1,650,000 tons in more than 500 ha.

In the rehabilitation of the areas where there have been accidental contamination of soil and groundwater with petroleum products, it is necessary to predict the distribution of hydrocarbons to assess their inventories and, if possible, to extract these hydrocarbons. Justification of the extraction systems requires parameters that characterize porosity, residual oil saturation, and other factors determining the multiphase flow.

Below are the results of the processing of man-made kerosene which was pumped out of the lens in the Leningrad Region, Russia. Here for 4.5 months oil was pumped from man-made lens area of 1,500 m^2 using four wells. Oil pollution was formed due to leakage of kerosene. Pumping was carried out with the help of wells, the construction of which allow the pumping of kerosene and water separately. Water pumping is necessary to prevent flooding and create a capture area of the well. Average flow rate of water pumping from one well was 37.5 m^3/day. As a result of rehabilitation work, the volume of kerosene evacuated from four wells was 19.6 m^3. The following analysis of the referred experimental results of pumping in one of the wells to obtain the required parameters was performed.

The LNAPL Distribution and Recovery Model (LDRM) code [7], developed by the American Petroleum Institute (API), was used for processing the experience.

LDRM allows to calculate the efficiency of the extraction system consisting of the wells having capture radius R_c. Example of wells intercepting the lens of oil pollution is shown in Fig. 9.6.

Computing module LDRM allows to lead the calculation of the pumping of oil products and water separately with the flow rates Q_o and Q_w, respectively. The scheme of such oil pumping of a lens with capacity b_o and b_w of the aquifer is shown in Fig. 9.7.

Fig. 9.7 The scheme of
separate pumping of water
and oil within the capture
zone of one well (adapted
from [8])

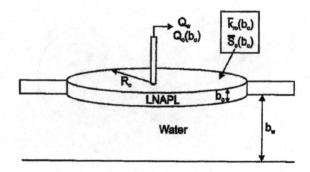

Changing the volume of recoverable oil within the capture zone of one well R_c
is given as

$$-\pi R_c^2 \frac{dV_o}{dt} = Q_o, \qquad (9.7)$$

where specific volume V_o of recovered oil is equal to

$$V_o = \theta \int_{z_{ow}}^{z_{ow}+D} (S_o(z) - S_{or})dz, \qquad (9.8)$$

where θ—porosity, S_o—oil saturation, S_{or}—residual oil saturation. So, for characterization of oil saturation in problems of migration of oil contamination, the
empirical dependence Van Genuchten-Parker [9, 10] is widespread as

$$S_o(z) = S_{wr} + (1 - S_{wr} - S_{or})(1 + (\beta_{ow}\alpha(h_{ow})^N)^{-M}, \qquad (9.9)$$

here, $h_{ow} = h_o - h_w = (1 - \rho_{ro})(z - z_{ow})$—capillary oil–water pressure in meters
of water column, [L]; ρ_{ro}—density of oil relative to water; z_{ao}—is the vertical
coordinate of the air/oil interface measured from the general plane of the comparison; z_{ow}—the vertical coordinate oil/water interface measured from the general
plane of the comparison; z—the vertical coordinate of the point from the general
plane of comparison,

$$\left\{ \begin{array}{l} h_w = \dfrac{P_w}{g\rho_w} = (1 - \rho_{ro})z_{ow} + \rho_{ro}z_{ao} - z \\[2mm] h_o = \dfrac{P_o}{g\rho_w} = \rho_{ro}(z_{ao} - z) \\[2mm] h_a = \dfrac{P_a}{g\rho_w} = 0 \end{array} \right\} \text{—hydraulic head of water, oil, air, [L];}$$

Here $\beta = \frac{\sigma_w}{\sigma_{ow}}$, σ_{ow}—surface tension at the interface between the hydrocarbon and water, σ_w—surface tension of pure water (72 dyne/cm), α and N—parameters of empirical relationship proposed by Van Genuchten [6] (the parameter α is directly proportional to the size of pores, and the parameter N is inversely proportional to the length of pore size distribution), $M = 1 - 1/N$, S_{wr}—residual water saturation.

Note that the zone of saturation with oil $S_o > 0$ in the soil is greater than the thickness of the petroleum lens observed in the wells $b_o = Z_{ao} - Z_{ow}$. The thickness of the area saturated by petroleum products can be found from the equation $\beta_{ao}\rho_{ro}(z|_{so=0} - z_{ao}) = \beta_{ow}(1 - \rho_{ro})(z|_{so=0} - z_{ow})$.

Denoting $z|_{so=0} - z_{ow} = D$—the thickness of the zone inside a layer which is rich in oil, $\beta_{ao} = \frac{\sigma_w}{\sigma_{ao}}$, σ_{ao}—surface tension at the air/hydrocarbon interface and, bearing in mind that $z|_{so=0} - z_{ao} = D - b_o$ we get:

$$D = \frac{b_o \beta_{ao} \rho_{ro}}{\beta_{ao}\rho_{ro} - \beta_{ow}(1 - \rho_{ro})} \qquad (9.10)$$

The expression (9.9) determines the true thickness of oil products in the reservoir and detects the upper limit of integration in the formula (9.7).

Oil production rate can be found on the basis of the proportionality of the mass flows of water and oil, as well as their water T_w and oil $T_o(b_o)$ transmissivity.

$$Q_o = \frac{Q_w T_o(b_o)}{\rho_{ro} T_w} \qquad (9.11)$$

Aqueous flow rate can be found on the basis of the Dupui formula:

$$Q_w = \frac{2\pi T_w S}{\ln(R_i/r_w)} \qquad (9.12)$$

where R_i—the radius of influence for pumping water, r_w—the radius of the filter, S—lowering the water level in the well during pumping.

Oil transmissivity T_o and water transmissivity T_w in (9.10) and (9.11) are evaluated according to the following formulas:

$$T_o(b_o) = \frac{\rho_{ro}}{\mu_{ro}} \int\limits_{z_{ow}}^{z_{ow}+D} K_{sw} K_{rn}(S_o) \, dz,$$

$$T_w(b_w) = \int\limits_{z_{aw}-b_w}^{z_{aw}} K_{sw}(z) \, dz.$$

Fig. 9.8 Graph comparing the calculated and actual volumes of oil extracted from the process well. *Curve 1* corresponds to the actual data, *curve 2*—the result of the calculations

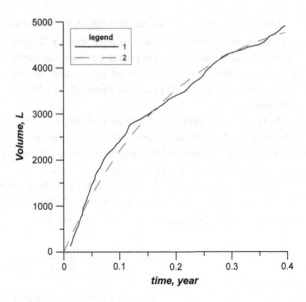

Table 9.3 The parameters obtained in the interpretation of the experimental pumping of man-made kerosene

Porosity	Van Genuchten's parameters		Residual saturation of water	Residual saturation of oil	Hydraulic conductivity (m/day)
	α (m^{-1})	N			
0.39	5	2.9	0.05	0.25	3

Here K_{sw}—the coefficient of water permeability at full saturation, μ_{ro}—the ratio of the viscosity of petroleum products to the viscosity of water. The relative permeability of the oil phase is estimated according to the relationship of van Genuchten-Mualem [9]:

$$k_{rn}(S_o) = S_o^{1/2}(1 - S_w^{1/M})^{2M}$$

Multiple approximate solution of (6) with the help of the LDRM program made it possible to determine the parameters that characterize the flow of kerosene to technological wells. An example of a comparison of actual and calculated data is shown in Fig. 9.8. Table 9.3 shows the values of the parameters for the variant where the best fit was obtained between the actual and estimated data.

References

1. Hall SH, Luttrell SP, Cronin WE (1991) Ground Water 29(2):171
2. Zheng C, Wang PP (1998) MT3DMS, a modular three-dimensional multispecies transport model for simulation of advection, dispersion and chemical reactions of contaminants in groundwater systems. Documentation and user's guide, Departments of Geology and Mathematics, University of Alabama, 1998
3. Zheng C, Bennett GB (2002) Applied contaminant transport modeling, 2nd edn. Wiley, New York
4. McDonald MC, Harbaugh AW (1998) MODFLOW, a modular three-dimensional finite difference ground-water flow model. U.S. Geological Survey, Open-file report 83-875
5. Lauverie HA (1955) Appl Sci Res Sect A 5(2–3):145
6. Kuranov PN, Rastorguev AV (2004) Solving groundwater problems. In: International conference "MODFLOW and more", Karlovy Vary, Czech Rep, p 207, Sept 2004
7. Charbeneau RJ (2003) Models for desing of free-product recovering systems for petroleum hydrocarbon liquids. Regulatory analysis and scientific affairs department, API publication number 4729, American Petroleum Institute, Washington DC, 2003
8. Charbeneau RJ, Johns RT, Lake LW, McAdams III MJ (2000) Ground Water Monit Rem 20(3):147
9. van Genuchten MTh (1980) Soil Sci Soc Am J 44:892
10. Parker JC, Lenhard RJ (1989) Transp Porous Med 5:187

References

Part III
Biological Materials

Chapter 10
Food Materials

Mikhail Yurievich Sidorenko

Abstract The current chapter is dedicated to the research on water state in food objects of various origins. It is shown that the bound water in such objects is not homogeneous. This phenomenon is of great importance for the food industry. Studies illustrating this problem in relation to several food products, such as various grains and its processing goods and sugar, including foodstuff based on those products, are given in this chapter. Modern analytical methods are used. Among them are NMR and DSC methods. Both methods have given a high correlation of results in the separating of bound water for "simply bound" and "strongly bound." Based on the data provided by the research of water state in various products of plant origin, moisture sorption isotherms are obtained and a mathematical model, describing the moisture state in such products, is developed.

10.1 Introduction

Commonly food products are of bioorganic nature. They can be classified into products of plant and animal origin.

Food products of plant origin mainly consist of natural biopolymers. The formation of tissues during vegetation involves an active participation of water, which simultaneously serves as a conveying medium and a part of formed tissues. The first type of water can conventionally be considered as free, since it serves mainly as a solvent and transports plant ingredients to the place of their synthesis or accumulation by means of capillary forces. The second type of water is bound water, which does not undergo phase transitions of the second order when changing the state of aggregation of free water. Bound water in plant tissues can be chemically, colloid-chemically, and physically bounded.

After the process of vegetation plant tissues remain porous. Thus, a bioorganic object in its entirety retains its ability to undergo chemical, biochemical, and microbiological processes with the participation of water as programmed by nature. The porous structure allows moisture to be moved sufficiently coordinated in a predetermined volume of plant tissue and initiates the various chemical reactions.

Y. M. Volfkovich et al., *Structural Properties of Porous Materials*
and Powders Used in Different Fields of Science and Technology,
Engineering Materials and Processes, DOI: 10.1007/978-1-4471-6377-0_10,
© Springer-Verlag London 2014

At rest, only a minimal amount of water remains in the living plant tissue which is sufficient for chemical reactions of life support in the *anabiosis* state. Examples of such objects, having bioorganic nature, are grain products (cereals, beans, buck-wheat, etc.), fresh vegetables, fruits, and spicy aromatic raw materials.

Products of industrial processing of vegetable raw materials tend to fully retain the physical properties of the source object to the full, but physicochemical and biochemical—only partially. According to the depth of processing, natural plant and animal raw materials can be divided into two groups:

- Raw materials subjected to processing without significant changes in the aggregate state (malt, getting flour and starch, etc.);
- Raw materials subjected to processing with a significant change in the state of aggregation (production of oil and butter, the preparation of the crystalline sugar, caramel syrup, glucose, the production of ethyl alcohol, and other fermentation products).

Foodstuffs of the first group, accordingly, retain their porous structure and physical properties, such as hygroscopicity, porosity, ability to oxidation, and adhesion characteristics. Therefore, sorption properties of the goods of the first type in the chain of transformations are comparable to each other (e.g., grain—flour—starch). Foodstuffs of the second type acquire physical properties which are different from those inherent in their native state. Thus, sorption properties of crystalline sugar, glucose, and sugar confectionery are different from those of sucrose contained in sugar beet or sugarcane.

In view of the above, the sorption characteristics of food products belonging to the second group require separate case studies and cannot be predicted a priori.

The sorption properties of bioorganic objects are mainly determined by their interaction with molecules of water. Under the partial pressure of water vapor of less than 1 the interaction is established by the chemical nature of the surface, and is described by moisture adsorption isotherm. The ability to bind vapor moisture is called "hygroscopicity". When the substance turns into a solution, it retains its identity in the interaction with the water molecules, but is characterized by another magnitude—"hydration". Hydration and water absorption (hydroscopicity) of the same substance are interrelated and allow qualitative assessment of each characteristic based on the value of the other.

10.2 Physicochemical Properties of Water

Structurally, the water molecule is a triangle whose vertices are two hydrogen atoms and one oxygen atom. In the steady state, the angle near the oxygen atom is $104°27'$, the distance between the oxygen and the hydrogen atoms is 0.9584 Å, and between the hydrogen atoms—1.5150 Å [1–3].

A water molecule in the aggregation state of ice is characterized by greater interatomic distances than that in the liquid state: for the O–H bond the distance is 0.99×10^{-10} m and for the H–H bond—1.62×10^{-10} m. The angle at the oxygen atom is equal to $109.5°$. Accordingly, the intermolecular distance of ice is 2.76×10^{-10} m, and the radius of the water molecule is 1.38×10^{-10} m. Water is characterized by a pronounced polarity: its molecules exhibit a significant dipole moment (1.86 D), due to which, as indicated by Rehbinder [4], the bond forms of water with various substances are very diverse.

Despite the considerable amount of research on the interaction of water with other substances, so far there is no consensus on the structural model of water and its mechanism of interaction.

10.3 Forms of Sorption Bonds in the Objects of Bioorganic Nature

Bioorganic objects are able to retain a significant amount of moisture. The moisture is held by the electrostatic forces of different nature and intensity. Moisture sorption of the first layer is the most tightly bound. The energy of the hydrogen bonds of such moisture can almost match the energy of the covalent bonds.

Conditionally, moisture bioorganic objects (MBOO) can be differentiated on the strength of its binding by bioorganic matrix. Moisture, whose interaction energy between molecules is higher than the energy of interaction with the matrix, can be attributed to the free water. This water is subjected to phase transitions of the second kind with cooling freezes, and when heated—turns to vapor at a temperature close to 100 °C.

The moisture having higher energy levels due to the matrix rather than between its molecules must be assigned to the bond (associated) moisture. In this, a part of this water turns into vapor at temperatures above 110 °C. This moisture can be called "weakly bound water" (WBW). The moisture that is not evaporated at a temperature exceeding 110 °C can be called "tightly bound water" (TBW). P. A. Rebinder proposed a classification of adsorbed moisture according to its forms of bond with the matrix and identified three groups of moisture held by the following types of bonds: (1) chemical bond, (2) physicochemical bond, and (3) physico-mechanical coupling [4].

10.4 Hygroscopicity of Grain, Flour, and Starch

10.4.1 General Patterns of Sorption

Cereal grains according to depth of processing are related to the first group and are characterized by the native porosity system. Pores of natural polymers have functional groups, which are primarily capable to form hydrogen bonding with water molecules. The degree of coupling of MBOO with the matrix depends on the partial pressure of water in the storage medium of grains. Moisture sorption capacity depends on the activity of water. The indicator A_w—"water activity" is dimensionless and represents the ratio of the water vapor pressure over a given material to the vapor pressure of pure water at the same temperature. Functional groups of different hydratable surfaces hold water molecules with different power, which affects their mobility and the probability of their detaching from the surfaces (desorption) at any fixed time. Therefore, the water activity allows evaluating the hygroscopic properties of the object. Water activity is numerically equal to the equilibrium relative humidity of the air, immediately adjacent the surface of the product, and divided by 100. In the literature, this value is often called the partial water pressure. The partial pressure of water is numerically equal to its activity under condition provided to establish equilibrium in the entire volume of air when the water content of boundary layer becomes equal to the moisture of air, which is infinitely remote from the interface.

When water activity is less than the partial pressure of water vapor in the ambient air then sorption occurs, i.e., grain will be wetted. Conversely, if the water activity at the surface of the grain is more than the partial vapor pressure of the ambient air, then moisture will be desorbed from the interface.

Graphic dependence of the equilibrium moisture content of grain w_p on the relative humidity φ of air, obtaining in the same temperature, called sorption or desorption isotherm. Sorption and desorption isotherms were not the same (except for the extreme points of $\varphi = 0$ and $\varphi = 1$), i.e., the phenomenon of sorption hysteresis takes place. At the same temperature, the desorption isotherm is higher than the sorption isotherm. The largest discrepancy between the two is in a plot with a relative humidity of 20–80 %. The difference between the equilibrium moisture according to sorption and desorption isotherms in this area reaches 1.2–4.0 % for crops [5].

Hygrothermal equilibrium is extremely slow and the biological system tends to a natural state indefinitely long, resulting in a balance which will not be true. Therefore, the value of w_p defined by sorption is less than the true value, and in the desorption—more than the true value.

Maximum equilibrium moisture content of grain, corresponding to the maximum relative humidity $\varphi = 100$ %, is called hygroscopic moisture and is denoted as w_g. Hygroscopic point on the grain sorption curve is the limit value for w_p under grain moisture absorption of water vapor from the environment. Further hydration of grain is only possible with direct contact of its surface with liquid. According to Egorov [6, 7] the maximum hygroscopic moisture (w_p) at $t = 25$ °C is for wheat—36.3–38.5 %, oats—36.5 %, rye—36.5 %, buckwheat—32.5 %, and rice—30.5 %.

For humidity of the storage environment $0.1 \leq \varphi \leq 0.75$ and the temperature 0 °C $\leq t \leq 60$ °C, the dependence of the equilibrium moisture content of grain w_p on temperature t and the relative air humidity φ can be described by the following empirical dependencies:

For wheat:

$$w_p^c = 4.0 - 0.035t + (19.7 - 0.075t)\lg\left(\frac{1}{1-\varphi}\right)^{0.5} \tag{10.1}$$

Table 10.1 Values of the scaling index of beer wort ingredients [8]

The object of bioorganic nature	Scaling index
Ethanol	0.735
Maltose syrup	0.609
Hop extract	0.515
Pale malt wort	0.493

for maize:

$$w_p^c = 1.8 - 0.1t + (22.5 - 0.03t)\lg\left(\frac{1}{1-\varphi}\right)^{0.5} \tag{10.2}$$

Caryopsis, from the chemical point of view, consists of a set of chemical compounds. Among the major ingredients of grain starch, gluten and cellulose should be mentioned. It is found that the cellular tissue, as compared with the other components of grain, has reduced hygroscopicity and hydrophilicity of starch at moderate values of relative humidity (up to $A_w = 0.75$) exceeds hygroscopicity of gluten mass. However, at higher relative humidities, sorption capacity of gluten mass increases at a faster rate relative to starch.

This character of the moisture content ratio of starch and gluten can be explained in terms of the scaling index. Scaling index is an exponent of the polymer concentration in the equation describing the dependence of the bound water content on the polymer concentration. The physical sense of the scaling parameter is to account for the availability of sorption (hydrated) surface from the total surface of the polymer. Obviously when swollen, scaling index of gluten mass grows more rapidly in relation to starch. Table 10.1 shows the values of the scaling exponent for a number of objects of bioorganic nature—fermented ingredients and hopped beer wort [8].

As seen in Table 10.1, even in the liquid phase hydration is available for only a specific portion of the surface molecules of studied ingredients. Taking into consideration the condition of the identity of the properties of water absorption and hydration, we can predict incomplete availability of biopolymers, which are in the solid state.

The scaling exponent of grain biopolymers depends on the degree of development of the capillary system: the growth of the total surface area increases the availability of surface for adsorbed moisture. Egorov [7] showed that the capillaries of wheat grain can be attributed to the mesopores, their diameter is 2.5×10^{-9} m (for corn—1.1×10^{-9} m), which is two orders of magnitude less than 1.0×10^{-7} m that conventionally defines the boundary between micro and macrocapillaries. The total amount of active surface of the wheat grain lies in the range of 200–250 m^2/g and exceeds the visible surface area of about 200,000 times. In this connection, in hygroscopic area a grain can absorb up to 37–38 % of water by weight of dry substances.

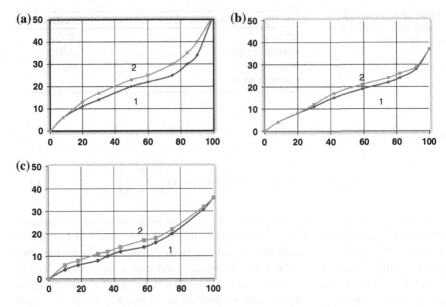

Fig. 10.1 Sorption (*1*) and desorption (*2*) isotherms of starches: **a** potato; **b** corn; **c** wheat. The horizontal axis—relative air humidity, (%). The vertical axis—the equilibrium moisture of starch, (%) [5]

10.4.2 Sorption and Desorption Isotherms of Starch of Different Origin

Hygroscopic properties of starch depend on the nature and chemical composition. Moisture sorption isotherms were determined by a strain gauge at 20 °C. Moisture was determined by drying at the temperature of 130 °C. Investigations were carried out for wheat, potato, and corn starches, both through their gradual hydration and by drying. Suitable adsorption–desorption isotherms of moisture for mentioned starches are shown in Fig. 10.1 [5].

Sorption and desorption isotherms of all studied starches have a sigma-like character with a strong sorption hysteresis. At the same relative humidity of air, the equilibrium moisture of the samples is different: it is higher to the isotherms obtained by desorption method in comparison with that obtained by successively increasing the concentration of absorbent. Hysteresis is associated with capillary-porous structure of starch. During storage, capillaries of starch absorb air, which being localized on the inner walls of the capillaries, provides some resistance to the penetration of water vapor for moistening (sorption) of the product, which lowers the equilibrium moisture content of the product.

During dehydration (desorption) of the product, moisture does not experience internal resistance, thus drying of the starch removes more moisture and, consequently, equilibrium moisture content of the product is higher.

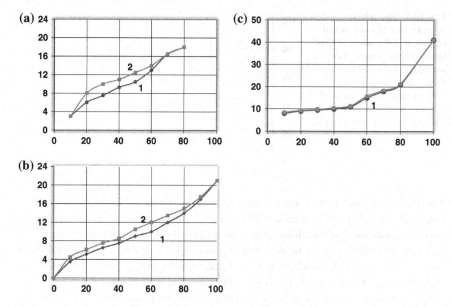

Fig. 10.2 Sorption (*1*) and desorption (*2*) isotherms at 20 °C: **a** buckwheat; **b** rice; **c** wheat flour. The horizontal axis—relative air humidity, (%). The vertical axis—the equilibrium moisture of starch, (%) [5]

Analyzing the sorption and desorption curves, storage conditions of the finished product and humidity to which the product must be produced can be defined. For example, if stored at room temperature of 20 °C at relative humidity of 75 %, the equilibrium moisture content of corn starch should be 16.7 % based on air-dry substance. When $\varphi = 70$ % starch humidity on air-dry basis is equal to 16.1 %. Similarly, for potato starch it is equal to 21.5 % based on air-dry substance during storage at room temperature of 20 °C and relative humidity of 75 %, and 19.6 %, based on air-dry substance at $\varphi = 70$ %.

10.4.3 Hygroscopic Properties of Products of Grain Processing

Figure 10.2 shows the sorption (1) and desorption (2) isotherms of premium wheat flour received by a strain gauge method [5].

One of the main factors affecting the storage of grain processing products (wheat and barley) is moisture with which the product is sent to storage. If the humidity of the product is below 14 %, perhaps its long-term storage is possible. Reducing the storage temperature to 10 °C increases the shelf life by 3–6 months (for rice cereals), which is associated with the value of the equilibrium moisture content of the product. The practical meaning of the adsorption hysteresis effect is

Fig. 10.3 The relation between free (*2*) and bound (*1*) water in the whole loaf of bread during storage. The horizontal axis—relative humidity of air, (%); the vertical axis—percentage of free and bound water, (%), in the total amount of water [5]

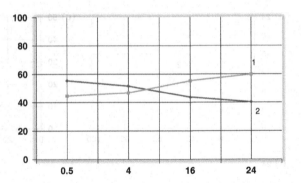

that objects of bioorganic nature which are brought during the drying process have values of humidity lower than equilibrium one; during subsequent storage under adverse conditions of humidity they can adsorb more moisture from the environment without the threat of destructive processes. Destructive processes should include all chemical, biochemical, and microbiological processes, which are available only when free water is present: chemical hydrolysis of proteins, carbohydrates, and fats; activation of enzymatic processes, the development of the microbiota, especially fungi.

In the production of bread, water condition plays a crucial role. The kinetics and extent of hydration of the polymer in flour in the process of dough preparation affects the rheological properties of semi-finished and finished products. [5] The kinetics of swelling of polymers is associated with the intensity of their homogenization in the preparation of dough that in turn affects the processing characteristics of the dough. The depth of hydration and the amount of water bounded by starch grains and proteins have a decisive impact on the bread during storage. Total moisture of wheat bread is reduced from 55 to 34 % during storage for 24 h, whereby the proportion of free water is reduced from 80 to 40 %, and a fraction of bound water decreases from 64 to only 60 % [5]. Figure 10.3 shows the dynamics of changes in the ratio of free and bound water in storage of wheat bread.

As can be seen from Fig. 10.3, the proportion of bound water in bread increases during storage, so the conditions for high-quality hydration of flour biopolymers in the preparation of dough is a prerequisite for the prevention of retrograded starch.

10.4.4 Assessment of the State of Water in the Bioorganic Objects by NMR

One of the promising methods to assess the state of water in the solid-phase bioorganic objects is the method of nuclear magnetic resonance (NMR). The most sensitive, due to the strength of signals induced and changes in magnitude of NMR—parameters depending on the environment of the water molecules in the sample, was hydrogen proton spectroscopy (^{1}H NMR).

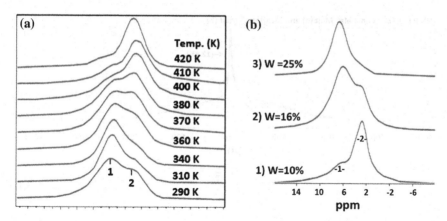

Fig. 10.4 ¹H NMR—spectra of narrow lines corresponding to the mobile protons: **a** buckwheat in dependence on the storage temperature; **b** gluten having moisture content, % wt 1–10, 2–16, 3–25. Numbers *1* and *2* denote the maxima corresponding to the loosely coupled and tightly bound water, respectively [5]

Almost all of the ¹H NMR—spectra of the samples of food products [9] have a spectrum consisting of broad (line width at half maximum of the peak is 20,000–60,000 Hz) and narrow (500–2,000 Hz) lines. The intensity of absorption lines in the ¹H NMR spectrum (peak area) is proportional to the number of the sample protons inducing such lines. Previously, a narrow spectral line corresponding to the mobile protons of hydrogen was allocated. The mobility of protons is caused by reorientation—translational motions of water molecules in the structure of biological molecules and proton exchange between H_2O and terminal OH—groups of bioorganic molecules. The integrated intensity of the narrow line is proportional to the moisture content of the sample. Figure 10.4a shows the ¹H NMR—spectra of a narrow line corresponding to the mobile protons of the buckwheat in dependence on the storage temperature. As can be seen, the maximum signal intensity distribution varies from 6 to 2 ppm depending on the temperature of the thermostatic cereals. Such behavior of the lines indicates a change in the ratio of hydrogen protons that have different mobility. The different mobility of protons can be explained with varying degrees of relatedness of moisture, which is adsorbed by pattern.

A similar behavior of the proton spectra of hydrogen takes place in the study of gluten of different humidity, the results of which are shown in Fig. 10.4b.

As seen in Fig. 10.4b, the proportion of the peak corresponding to the content of weakly bound water and meeting the bandwidth of 6 ppm, increases with increasing moisture content of gluten. The increase in gluten humidity leads to an increase of only weakly bound water of gluten. Thereby, the relative proportion of strongly bounded gluten mass to weight of the total moisture is reduced accordingly. Research has revealed that the asymmetric character of the narrow spectral line corresponding to the content of moisture in the sample takes place for a significant number of cases. A similar effect was detected in the study samples of

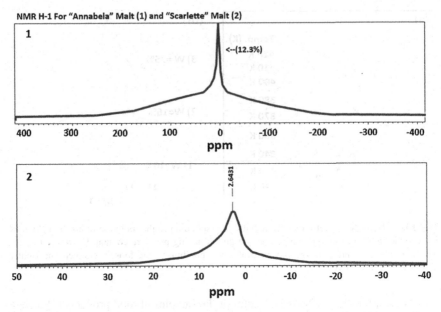

Fig. 10.5 ¹H NMR spectrum of malt: *1*—scoping spectrum, including *wide* and *narrow lines*, *2*—expanded range of the central part of the spectrum (*narrow line*) [8]

malt used in brewing. Malt must possess high extract ability, wherein the moisture content should be optimal, corresponding to the technology. Malt affinity of natural polymers to water is defined by functional groups (NH_2, OH, C = O) presented in the wort. Assessing the absorbability of malt we can be concluded about its extractable and technological efficiency.

Figure 10.5 shows the ¹H NMR—spectrum of the malt sample prepared by mixing two kind of malts, "Scarlet" and "Anabel," in 1:1 proportion [8].

As seen in Fig. 10.5 a narrow line spectrum of malt also has asymmetrical character caused by the presence of two kinds of water with different properties in the malt. The number of mobile protons corresponding to water contained in malt is 12.3 %. With an increase in the central line (curve 2) the complex nature of the curve, corresponding to the presence of protons in the sample with a different mobility, is swept. With the expansion of the center line into components (Fig. 10.6 [8]), the presence of two species of the malt moisture described by the various curves is established.

To differentiate using methods of mathematical processing of various types of water, which is presented in the samples, the narrow line was decomposed into the Lorentzian and Gaussian line shape.

Figure 10.6a shows the ¹H NMR spectrum of the narrow line of buckwheat with moisture of 10.6 % (curve 3) and its derivatives: Lorentzian curve (1) and Gaussian curve (2). The curves have the following parameters: curve 1—chemical shift—5.53 ppm, line width at its half-maximum corresponds to frequency of 789 Hz,

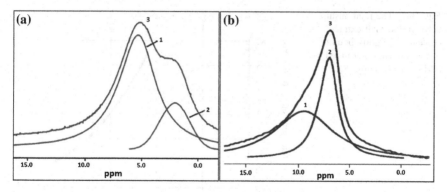

Fig. 10.6 ^{1}H NMR spectra of the narrow lines (*3*) and partial lines based on them: Lorentz curve *1*, corresponding to weakly bound water, Gaussian curve *2* (**a**) and Lorentzian curve *2* (**b**), corresponding to strongly bound water; **a** buckwheat [5]; **b** malt [8]

the integrated intensity of the curve is 82.0 % of the total intensity; curve 2—the chemical shift of 2.15 ppm, line width—624 Hz, the integrated intensity, respectively, 18.0 %.

Figure 10.6b shows the results of decomposition of the central narrow line (curve 3) in the ^{1}H NMR spectrum of malt by two Lorentzian lines: curve 1—the signal of weakly bound water (chemical shift of 6.2 ppm, $\Delta v = 2718$ Hz), curve 2—the signal of tightly bound water (chemical shift of 2.8 ppm, $\Delta v = 540$ Hz).

The presence of two types of water in the malt indicates varying degrees of water relatedness with matrix [8]. In this case, a correlation between the curve 1 and the amount of moisture removed by drying is established. Water, corresponding to the protons belonging to the curve 1 is a weakly bound, and the corresponding to protons of the curve 2—tightly bound water. Weakly bound water allows evaluating the malt humidity and tightly bound water affects the size of hydratable macromolecules, and therefore affects their rate of diffusion and, accordingly, malt extract content.

Figure 10.7 illustrates the dependence that allows estimating the malt extract content by NMR spectrum a priori, which allows the use of the method for manufacturing purposes.

10.4.5 Mathematical Model of Hygroscopicity of the Objects of Bioorganic Nature

Sorption of water on biopolymers cannot always be described by classical isotherms proposed by Henry, Langmuir, Brunauer, and BET theory [10]. Interaction of the polymer matrix with water is characterized by formation of bonds with varying strength. NMR method established presence of tightly bound water in cereal crops. Such compounds can be formed only by chemical or quasi-chemical reaction.

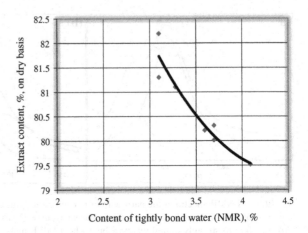

Fig. 10.7 The relationship between the malt extract content and tightly bound water content, determined by NMR spectroscopy [8]

In this connection, to describe the moisture sorption on natural biopolymers, quasi-chemical approach proposed in [11] was used.

In accordance with the law of mass action, the rate equation for quasi-chemical reaction of water molecules with reactive centers of the biopolymer [12] was obtained. The adsorption isotherm equation looks like:

$$w = a_m \frac{\alpha x}{(1 - \beta x)(1 + (\alpha - \beta)x)} \tag{10.3}$$

where α_m is the classically estimated capacity of the biopolymer, which is proportional to the concentration of reactive sites in the polymer, $x = P/p_s$, p and p_s- the water activity (partial pressure of moisture near the surface of the sorbent at equilibrium), α and β—quasi-chemical equilibrium constants:

$$\alpha = K_1 K_L \frac{1 + K_H}{1 + K_1} \tag{10.4}$$

$$\beta = K_H K_S \tag{10.5}$$

The constant β is the product of quasi-chemical equilibrium constants, reflecting the probability of reaction (K_H—the probability of a "hole" appearance near the adsorbed molecules; K_S—affinity between water molecules). For biopolymers $\beta < 1$ and decreases with the scaling index decrease, i.e., reducing the matrix availability.

The constant α is the product of the constants of quasi-chemical reactions: K_1—takes into account the likelihood of a "hole" near the reactive center; K_L—affinity between the molecule of absorbate and the active site of biopolymer. Figure 10.8 shows the isotherms of moisture adsorption by bioorganic objects with various degrees of processing obtained using the equation proposed in [13].

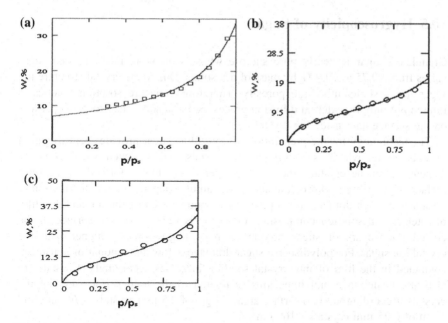

Fig. 10.8 Absorption isotherms of water vapor at 25 °C for bioorganic objects: **a** the grain of wheat; **b** rice cereal; **c** corn starch. *Dots*—experimental data, a continuous *curve*—theoretical calculation according to Eq. (10.3) [13]

Table 10.2 Parameters of the adsorption isotherms of water vapor on the caryopsis of different cultures [13]	Caryopsis	The values of the coefficients of Eq. (10.3)			
		a_m (%)	β	α	N—average number of molecules in the cluster
	Wheat	6.955	0.795	$\alpha \gg 100$	4.89
	Rye	7.739	0.711	213.67	3.46
	Triticale	7.734	0.708	$\alpha \gg 100$	3.43
	Rice	7.27	0.644	70.562	2.81
	Millet	7.106	0.669	113.373	3.02
	Oat	8.018	0.708	11.269	3.43
	Barley	8.398	0.684	23.808	3.16
	Buckwheat	8.168	0.73	56.121	3.70

Table 10.2 lists the parameters of Eq. (10.6) for individual objects of bioorganic nature.

The deviation between the theoretical and experimental data for almost all experiments was less than 5 %. The exception is wheat and triticale. When the water activity is less than 0.3, there are discrepancies that reach 9 %. However, the given humidity range is of no practical value. In general, the mathematical model developed allows evaluating the ability of a particular consignment of goods for long-term storage basing on the adsorption isotherms.

10.5 Hygroscopicity of Sugar

Granulated sugar is nearly pure sucrose whose content in manufactured sugar varies from 99.75 to 99.9 % by mass of dry solids. Due to the crystal structure the sugar has a sufficiently high porosity. Therefore, moisture sorption kinetics is limited not only by external diffusion processes, but also by the processes passing on the surface and inside the crystal.

The technology for the preparation of crystalline sugar provides centrifugation of massecuite, which is a mixture of crystals and saturated sugar solution. Depending on the crystal structure, the layer of crystals is formed on the filter surface of centrifuge, whose drainage (permeability) depends on the distribution of crystal size. High drainage takes place for monodisperse distribution and thin film of inter-crystal solution comprising nonsugars is retained on the surface of the crystal. Nonsugars of sugar production have hygroscopicity higher than the crystalline sugar. For polydisperse sugar drainage is low and amount of nonsugars contained in the film of intercrystal solution increases accordingly. Crystal size also affects drainage and impurities as related to the total surface area of the crystals. According to [5], a surface area of 1 gm of 0.5 mm crystals is 74 cm^2, and 1 gm of 0.25 mm crystals—160 cm^2.

Hygroscopicity of crystalline sugar is associated with its crystal structure through the surface area and total content of impurities. Sugar with crystals of a wide size distribution and a high content of fine fraction has high hygroscopicity.

Assuming that the sugar crystal has equal dimensions in all their optical axes, the dependences calculated for area, volume, and mass of the crystal on its characteristic size are presented in [14]. These relationships are shown in Fig. 10.9a. As can be seen from this figure, all the characteristics grow exponentially; and the size of the crystal most strongly affects its surface area.

Figure 10.9b exhibits that the size of crystals has a significant impact on the overall surface area of the crystals per unit mass of sugar. The sugar with crystal sizes of 0.5 mm or less has a special effect on the total area of the crystals.

However, large crystals with a size greater than 1.4 mm are prone to the formation of fused faces conglomerates. Figure 10.10 presents the X-ray diffractogram of major fraction of sand sugar with grain sizes of about 1.4–2.0 mm, which shows the presence of conglomerates of different configurations. The formation of such kind conglomerates also contributes to hygroscopicity of sugar. Therefore, to minimize water absorption of sugar it is necessary to optimize the distribution of its crystal size.

Figure 10.11 shows the adsorption isotherms of the crystalline sugar, from which it is clear that the absorbability depends also on the purity of the crystals. This effect can be explained by the different content of hygroscopic impurities in the film of intercrystal solution enveloping crystals.

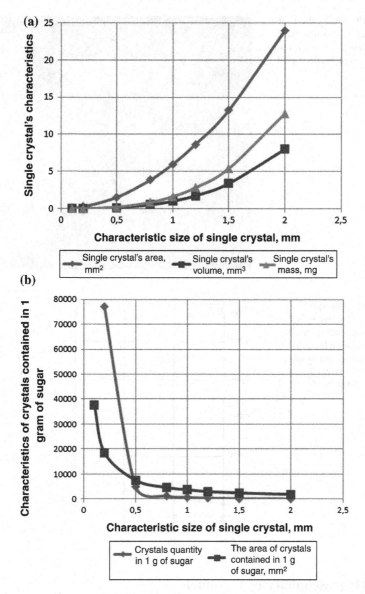

Fig. 10.9 Dependences of area, volume, and weight of a crystal on its characteristic size (**a**); dependence of the amount of crystals in 1 gm of sand sugar and their total area on the characteristic size of the crystals (**b**) [14]

Fig. 10.10 XRD pattern of
large fraction of sugar with
particle sizes of 1.4–2.0 mm
(25-fold magnification) [14]

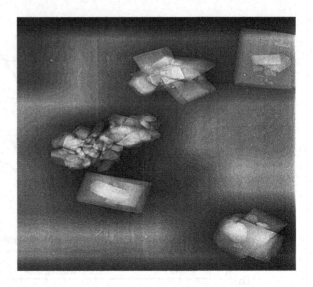

Fig. 10.11 Adsorption
isotherms of sugar (W_s, (%))
at 20 °C depending on the
ambient relative humidity (φ,
(%)) at chosen purity of
crystals, (%): 1–99.4,
2–99.75, 3–99.88 [5]

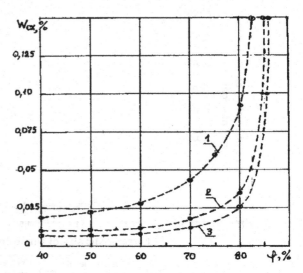

10.6 Hygroscopicity of Caramel

Caramel mass is one of the fundamental mass of confectionery production, as part
of the composition of various sugar confectioneries. From the physical and
chemical points of view, candy mass is a mixture of sugar solution subjected to
vitrification and anti-crystallizer preventing the loss of sugar crystals on cooling
mass. Starch syrup is generally used as an anti-crystallizer. Hygroscopicity of
caramel is related to the content of anti-crystallizer in it and acidity due to the need
to give caramel a flavor expression. During storage in the presence of weakly bond

moisture, the acidic hydrolysis reaction of sucrose occurs. Sucrose has a maximum chemical stability at pH = 8.5. In an acidic medium, sucrose is intensely hydrolyzed to form glucose and fructose, which are stable in the acidic environment. Monosaccharides (especially fructose) have increased hygroscopicity in comparison with sucrose due to the presence of a highly active glycosidic hydroxyl in their structure. Additional moisture sorption by caramel surface leads to further dissolution of the surface layer of caramel and further increase of its hygroscopicity. As can be seen from Fig. 10.12, to avoid autocatalytic reaction, residual moisture of caramel must not exceed 3.5—3.7 %.

Initial hygroscopicity is not connected with the autocatalytic hydrolysis reaction of sucrose in caramel and depends on its chemical composition. Due to the impossibility of achieving true equilibrium of moisture sorption by caramel, its propensity to adsorb is proposed to assess by evaluating its solubility. Solubility should be understood as the linear rate of dissolution of the caramel sample in the formalized terms.

According to the Shchukarev model [15], the rate of dissolution is proportional to the concentration difference at the surface of the solute and in the bulk of solution, and may be described by the following differential equation:

$$\frac{dM}{dt} = \beta F(C_s - C_0) \qquad (10.6)$$

where M—the sample mass, g; t—time of dissolution, s; $\frac{dM}{dt}$—rate of dissolution, g/s; F—area of the dissolving body, cm^3; C_s, and C_0—concentration at the surface of the solute and in the bulk of solution, respectively, g/cm^3; β—coefficient of proportionality. Note that the proportionality coefficient β is the rate of the substance mass transfer through the unit area under change in concentration equal to 1.

If studying the solubility, the sample is dissolved without residue in the course of time t inside the solvent volume disproportionately greater than the sample volume ($C_0 = 0$), and the area for dissolution is regarded as an average area of the sample during the time of dissolution, then Eq. (10.6) takes the form convenient for use in engineering calculations:

$$\beta = \frac{M}{tfC_s} \qquad (10.7)$$

where f—average area of the sample during the time of dissolution, $f = F/2$.

The concentration of a saturated solution of caramel mass is calculated by the equation:

$$C_s = H_0 \mathcal{K}_s \qquad (10.8)$$

where H_0—sucrose solubility at a given temperature, K_s—saturation coefficient, which is equal to weight ratio of sucrose dissolved in a saturated solution, containing nonsugars, to sucrose dissolved in a pure saturated solution [15].

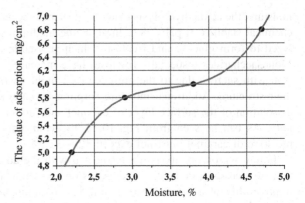

Fig. 10.12 Effect of the lollipop moisture on the amount of moisture absorption (g/g)

The coefficient β is proposed to be considered as the rate of dissolution and to be used for characterization of solubility as well as to measure water absorption of caramel. The solubility β is measured in cm/min. Solubility of caramel is determined using the apparatus shown in Fig. 10.13 [16]. A sample of caramel 6 with known surface area is placed in a calibrated cell 7, which is suspended by a thread 8 to the rocker 3 of electronic scales 4. The cell with the test sample is placed in a glass 5 mounted in the thermostat 1, equipped with a temperature sensor 2. Solubility evaluation method shows that caramel hygroscopicity is proportional to the sucrose content.

Figure 10.14 shows a diagram illustrating the effect of chemical composition and temperature on the solubility of caramel. As seen from the figure, the solubility is proportional both to the sucrose content in caramel and the temperature.

Figure 10.15 shows a typical isotherm of the moisture adsorption by lollipop caramel. As can be seen, the adsorption isotherm of caramel is close to the adsorption isotherm of sugar. Absence of drop condensation at humidity of 100 % indicates an unfavorable adsorption kinetics which cannot achieve true equilibrium [17]. Upon reaching a certain moisture content of the surface layer the process of moisture desorption begins to dominate, in which the secondary crystallization processes occur in the surface layer of sugar syrup. This process is called "making of candied caramel." Thus, candied caramel itself loses consumer properties and acquires unusual rheological characteristics.

Influence of acidity of lollipop on its consumption characteristics were studied during storage at 20 °C and relative humidity 70 % of the storage medium [18]. The results are shown in Fig. 10.16. As seen in Fig. 10.16, hygroscopicity of caramel significantly depends on its acidity. For the caramel, which has a neutral or alkaline environment, hydroscopicity rises slightly. This character of changes confirms the theory of the autocatalytic hygroscopicity of caramel.

The data show that water absorption of caramel is proportional to the concentration of hygroscopic glucose and fructose on the candy surface. Since chemical reactions occur only in the presence of water, further hydration of caramel enhances acid decomposition reaction of sucrose contained in the caramel to form new hygroscopic monosaccharides.

Fig. 10.13 Installation for evaluating solubility of caramel: *1*—thermostat *2*—thermo-contact thermometer, *3*—rocker, *4*—electronic scales, *5*—a glass, *6*—a sample of caramel, *7*—the gauging cell, *8*—a thread

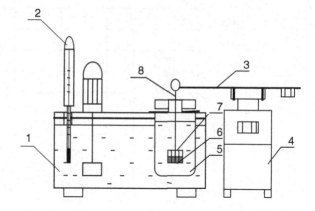

Fig. 10.14 Effect of temperature and composition of lollipop on its solubility

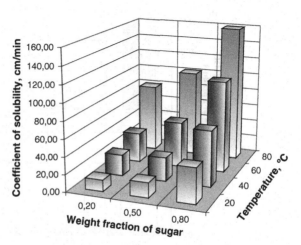

Fig. 10.15 The adsorption isotherm for lollipop

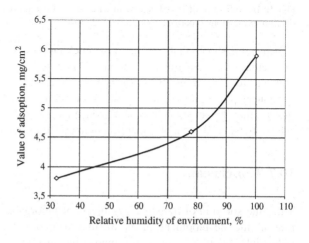

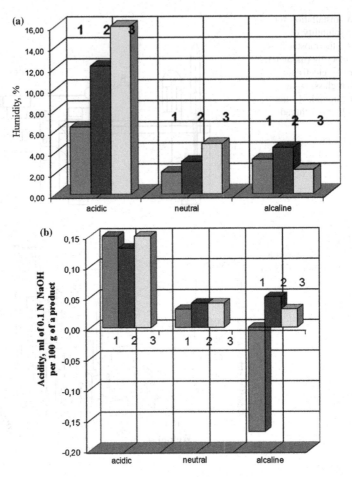

Fig. 10.16 Influence of initial acidity of caramel on its hygroscopicity (**a**) and the final acidity (**b**) during the storage period, days 1–7, 2–32, 3–120

Thus, it is shown that wetting of caramel has a place for all values of its acidity. To increase the shelf life of caramel it is necessary to take special precautions to prevent moisture access to its surface. Caramel coatings with barrier to moisture substances preventing the absorption of moisture or, at least, the decomposition of sucrose may be considered as such measures. Beeswax or glazes layer of sucrose caused by special technology are recommended as such coatings [19].

10.7 Conclusion

Thus, the sorption of moisture by objects of bioorganic nature plays a significant role in the development of production technology and food storage. In the industrial society, the need for innovative technologies in food production is increasing continuously. The problem to develop a unified design methodology for a food consumer with desired characteristics is on the agenda. Quantitative and qualitative characteristics of moisture in these foods will make the difference. In the near future there will be a technology of design and production of personalized diets, which would require taking into account not only the physicochemical, but also the physiological characteristics of water-based biological systems.

References

1. Rahman A, Stillinger FH (1973) J Am Chem Soc 95:7943
2. Rahman A, Stillinger FH (1971) J Chem Phys 55:3336
3. Franks F, Duckworth RB (eds) (1975) Water, ice and solutions of simple molecules. Water Relations of Foods, Academic Press Inc., Ltd., London
4. Rehbinder PA (1968) In: Modern Problems of Physical Chemistry, vol 3, Moscow, p.14, in Russian
5. Sidorenko YuI et al (2005) Science report: the development of sorption theory of foodstuff hygroscopicity, Moscow State University of Food Production, Moscow, 2005, in Russian
6. Egorov GA (1973) Heat and moisture influence for grain processing and storage. KolosS, Moscow in Russian
7. Egorov GA (1960) The research of food stuff water sorption isotherms. KolosS, Moscow in Russian
8. Sidorenko AYu (2008) Ph.D. Thesis, Moscow State University of Food Production, Moscow, in Russian
9. Ugrozov VV, Filippov AN, Sidorenko UY (2007) Colloid J 69(2):256
10. Egorov GA (2000) The management of technological conditions of grain. Voronezh State University, Voronezh in Russian
11. Laatikainen M, Lindstrom M (1987) Acta Polytech Scand. Chem Techn Metal Ser 178:105
12. Ugrozov VV, Filippov AN, Sidorenko UY (2007) Russ J Phys Chem A81(3):458
13. Ugrozov VV, Filippov AN, Sidorenko UY, Shebershneva NN (2009) In: Panfilov VA (ed) The fundamental theory of food production, book 1. KolosS, Moscow, 2009, p 614, in Russian
14. Beletskiy SL, Sidorenko YuI (2013) In: Abstracts of the Tovaroved-2013, Moscow State University of Food Production, Moscow, 25–26 April 2013, in Russian
15. Höhne G, Hemminger W, Flammersheim HJ (2003) Differential scanning calorimetry, 2nd edn. Springer, New York, p 176
16. Sidorenko MYu (2003) Ph.D. Thesis, Moscow State University of Food Production Moscow, in Russian
17. Skobelskaya ZG, Shebershneva NN, Shehovtsova TG (2007) Storage and Processing of Farm Products 11:14 in Russian
18. Skobelskaya ZG, Kondakova IA (2000) Storage and Processing of Farm Products 8:23 in Russian
19. Sidorenko MYu (2009) Storage and Processing of Farm Products 1:15 in Russian

Chapter 11
Soils and Plant Roots

Oksana Leonidovna Tonkha and Yuliya Sergeevna Dzyazko

Abstract The interaction of soil and root system of plants is considered in this chapter. Particularly, the hierarchical structure of soils and soil agents, which stabilize it at different organization levels, pore classifications, depending on their size, origin, and functionality are presented. Regarding the plants, soil pores were shown to act as a medium, in which the root system is formed. The pores provide a transport of water and nutrients on the one hand and gas exchange on the other hand. The roots are involved to aggregation of soil particles, and the waste products of plant stabilize their shape. During growth and development of the root system, the waste products are the medium for the expansion of symbiotic and nonsymbiotic nitrogen-fixing bacteria. The structure of root and its constituent tissues has been considered. Special attention was paid to the porous structure of cell membranes, which carry out, in particular, accumulating and transporting functions. Mass transport in the soil–root system has been also analyzed. It is shown, that higher hydraulic conductivity of the roots in a comparison with the soil is the necessary condition to provide water maturation by the plants.

11.1 Soil Functions

Soil (pedosphere) is an independent complex specific biological shell of the earth, which covers the land of continents and the shallow of seas and lakes. Soil always interacts with other shells of the planet, it is also involved in the complex processes of exchange and transformation of energy and matter.

Soil cover carry out global and socioeconomic functions. Global functions comprise: (i) providing the life on the Earth—the soil cover is the habitat for plants, animals, and microorganisms; (ii) providing a continuous interaction of large geological and small-scale biological cycles of substances on the earth's surface; (iii) regulation of chemical composition of the atmosphere and hydrosphere, as a result, oxygen, hydrogen, nitrogen, carbon in various forms are involved in the synthesis of organic matters by plants, (iv) regulation of

Y. M. Volfkovich et al., *Structural Properties of Porous Materials and Powders Used in Different Fields of Science and Technology*, Engineering Materials and Processes, DOI: 10.1007/978-1-4471-6377-0_11, © Springer-Verlag London 2014

Table 11.1 Soil physical properties and processes, which affect agricultural, engineering, and environmental soil functions (adapted from [1])

Process	Properties	Soil functions
Biomass productivity (agricultural functions)		
Compaction	Bulk density, porosity, particle size distribution	Root growth, water and nutrient uptake by plants
Erosion	Structural stability, erodibility, particle size, infiltration and hydraulic conductivity, transportability, rillability	Root growth, water and nutrient uptake, aeration
Water movement	Hydraulic conductivity, pore size distribution, tortuosity	Water availability to plants, chemical transport
Aeration	Porosity, pore size distribution, soil structure, concentration gradient, diffusion coefficient	Root growth and development, soil and plant respiration
Heat transfer	Thermal conductivity, soil moisture content	Root growth, water and nutrient uptake, microbial activity
Engineering functions		
Sedimentation	Particle size distribution, dispersibility	Filtration, water quality
Subsidence	Soil strength, soil water content, porosity	Bearing capacity, trafficability
Water movement	Hydraulic conductivity, porosity, pore size distribution	Seepage, waste disposal, drainage
Compaction	Soil strength, compactability, texture	Foundation strength
Environmental functions		
Absorption/ adsorption	Particle size distribution, surface area, charge density	Filtration, water quality regulation, waste disposal
Diffusion/ aeration	Total and aeration porosity, tortuosity, concentration gradient	Gaseous emission from soil to the atmosphere

biospherical processes by dynamically restoring soil fertility; (v) accumulation of organic substances and related chemical energy at the earth's surface.

Socioeconomic function involves, primarily, agriculture (Table 11.1) [1]. The soil carries out technical functions, if buildings and constructions are built on the earth's surface and communications are routed. Ecological functions of soil are supporting of human health, they are associated with soil biota in the destruction of organic residues, sterilization of toxic substances, and oppression of pathogenic microorganisms.

Functional properties of the soils depend not only on their chemical composition, but also on the physical properties and parameters, such as particle size, porosity, volumetric and specific weight, hardness, ductility, and so on. Physical and functional relationship of soil properties is the subject of young and dynamically developed science—physical edaphology, which is the part of Soil Science (pedology).

Such parameter as bulk density of a dry soil, which is associated with porosity and the particle density, is most commonly used to characterize the pore structure of the soil. Total porosity involves pores both between the primary particles and their aggregates, it is subdivided by capillary and noncapillary (macropores) porosity.

Table 11.2 Physical parameters of soils, which are necessary for plant development (adapted from [1])

Parameter	Value
Particle density	2.6–2.8 g cm^{-3}
Bulk density (of absolutely dry soil)	0.7–1.8 g cm^{-3}
Volumetric soil moisture content (ratio of volumes of water and soil)	0–0.7
Void ratio (ratio of volumes of pores and solid)	0.4–2.2
Air ratio (ratio of volumes of gas and solid)	0–1
Wet bulk density	1–2 g cm^{-3}
Total porosity	30–70 %
Saturation degree (ratio of volumes of water and total porosity)	0–1
Gravimetric soil moisture content (relation of mass of water and solid)	0–0.3
Dry specific volume (for absolutely dry soil)	0.5–1 cm^3 g^{-1}
Liquid ratio (ratio of volumes of water and solid)	0–1
Aeration degree (ration of volumes of gas and water)	0–1

Fig. 11.1 Non-capillary porosity, which provides the best crop capacity of potatoes (*1*) and sugar beets (*2*), as a function of total porosity (adapted from [2])

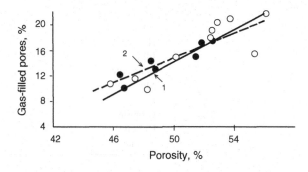

Capillaries are filled with water containing dissolved matters, mainly K^+, NH_4^+, Ca^{2+}, Mg^{2+}, Cl^-, NO_3^-, PO_4^{3-}, SO_4^{2-}, Na^+ ions. Such gases as N_2, O_2, CO_2, CH_4, C_2H_6, H_2S, N_2O are inside macropores (gas pores). Agricultural activities of people and the associated function of the soil lead to a change in parameters of solid, liquid, and gaseous phases of the soil (Table 11.2).

Figure 11.1 illustrates the contribution of non-capillary porosity to the total porosity, at which the maximum harvest of potatoes and sugar beet is reached: non-capillary porosity is ≈ 25–50 % of total porosity [2]. In general, from the point of view of the physical properties, the optimal soil for the cultivation of crops has the following parameters: solid phase of 40–60 %, non-capillary porosity of 20–30 %, capillary porosity of 20–30 %.

Non-capillary porosity provides gas exchange and capillary porosity is responsible for moisture exchange. If non-capillary porosity of the soil is less than ≈ 50 % of the total porosity, anaerobic conditions are formed, which leads to reduced crop capacity.

Regarding soils, which perform an engineering function, the porosity should be 10–20 %: such low values is a result of compaction.

Table 11.3 Particle density of some soil constituents (systematized [3] data)

Substance	Particle density (g cm^{-3})
Peat	0.50–0.80
Decomposed peat	1.00–1.20
Humus	1.30–1.40
Montmorillonite	2.00–3.00
Kaolinite	2.61–2.68
Serpentine	2.55
Talc	2.58–2.83
Tourmaline	3.03–3.25
Vermiculite	2.30
Hornblende	3.02–3.45
Dolomite	2.86
Geothite	3.30–4.30
Biotite	2.70–3.30
Quartz	2.60–2.65
Gypsum	2.30–2.47
Gallit	4.2
Orthoclase	2.55–2.63
Anorthite	2.72–2.75
Hematite	5.26
Illite	2.60–2.90
Magnetite	5.17
Muscovite	2.77–2.88
Pyrite	5.02
Gibbsite	2.38–2.42
Chlorite	2.60–3.30
Brucite	2.38–3.40

11.2 The Composition and Structure of Soils

The important parameter, which determines the soil functions, is the bulk density. It depends on the content of inorganic components (2.0–3.5 g cm^{-3}) as well as organic substances (0.5–1.4 g cm^{-3}) (Table 11.3 [3]). The soil is a polydisperse system, it consists of particles of different sizes, which are formed by the weathering of dense rocks and soil formation. Soil mass may contain large rock fragments, the smaller sand particles, very fine clay, and silt units. Depending on the texture, the inorganic components can be divided into three classes: sand (0.02–2 mm), silt (0.002–0.02 mm), and clay (<0.002 mm). Suspended particles of sand and silt form a rigid skeleton of the soil matrix.

Sand granules, which are the primary particles, are formed mostly from quartz with inclusions of trace minerals like zircon, tourmaline, and hornblende. Sand does not swell in water, ion-exchange properties of this material are practically absent.

The mineralogical composition of silt, which relates to the intermediate soil fraction, is similar to that of the sand. These particles, which are dense in a dry state, have an irregular shape. Silt demonstrates slight tendency to aggregation and swelling, low water permeability, and high capillary rise.

The composition of clay, which is the finest mineral fraction, comprises soil minerals like aluminosilicates containing Fe_2O_3, Al_2O_3, $CaCO_3$ as well as other oxides and salts. Clays swell in water, they are characterized by high specific surface area (5–20 m^2 g^{-1} for kaolinite and 700–800 m^2 g^{-1} for smectite) and considerable surface charge density (75 and 19 mC m^{-2} for kaolinite and smectite, respectively). Clay soil retains water and provides adsorption of ionic impurities. Clay particles, which have lamellar or cylindrical shape, are admirable to aggregation and practically waterproof.

The content of organic component in most mineral soils do not exceed 5–7 % in the upper horizons; this component may predominate in organic soils. The soil organic matters include both plant and animal debris, which have not lost their anatomical features, as well as specific chemical acidic compounds called humus. Humus comprises the substances of known structure (lipids, carbohydrates, lignin, flavonoids, pigments, wax, pitches, and so on), their content in humus is up to 10–15 %. These substances form specific humic acids in soil.

Humic acids have no defined formula and represent a class of high molecular compounds, which can be subdivided into (i) insoluble humic acid, (ii) soluble salts of humic and fulvic acids (alkali metal salts). Humus is characterized by high specific surface, ion-exchange capacity, and buffer capacity. It also demonstrates significant hydrophilicity and tendency to swelling, which provide agricultural functions of soils. The components of humus form stable complexes with Cu^{+2}, Mn^{+2}, Zn^{+2}, Al^{+3}, Fe^{+3} cations, they can accumulate toxic ionic contaminants, such as Pb^{2+} and Cd^{2+}. The ability of humus to adsorb toxic species attaches environmental function to soil. However, the accumulation of heavy metals make the soils unsuitable for crops growing.

According to the classification of International Society of Soil Science (ISSS), particles, which are larger than 2 mm, are considered as nonsoil inclusions. New growths based on compounds of iron and manganese are widespread. These inclusions are tubes along roots, cortical formations, pseudomicelles, etc. Sometimes, the soil mass is even cemented by ferruginous material. New growths of readily soluble salts, calcareous, silica, and clay minerals are fairly common. Archaeological finds, bones, the shells of mollusks and protozoa, rock fragments, and garbage are also attributed to the inclusions. They have a negative impact on agricultural and environmental soil functions. Only biogenic new growths formed under the influence of soil fauna, have the beneficial effect on crop capacity.

Soil functions are determined not only by its composition but also the structure, which involves quantitative ratio, the nature of the relationship and the location of both the native particles and their aggregates. The size of the aggregates can influence the harvest considerably. For growth of plants, the optimum structure has at least 65–70 % of predominant units with a size of 1–5 mm as well as macro-aggregates (up to 93 %). However, the optimum particle size is 10 mm for humid conditions and 2 mm for drier areas. In the first case, the large structural

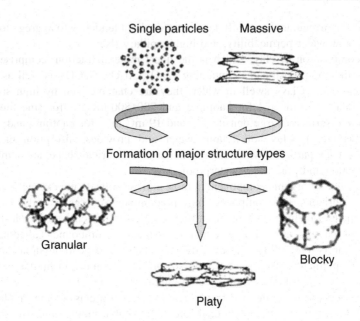

Fig. 11.2 Soil particles of different shape (adapted from [4])

assemblies provide better water–air regime, since small units with sufficient soil aeration retain moisture.

The predominant form of aggregates is an important diagnostic feature of the soil. There are granular, platy, and blocky (prismatic, columnar) structures (Fig. 11.2), and a number of transitional forms and gradations in size [4]. The first type is typical for the upper humus horizons and determines the large porosity. The second one is attributed for elluvial horizons, which are characterized by the removal of organic and/or mineral components. The block structure is typical for the illuvial horizons, which accumulate substances learned from elluvial horizons.

Native particles of humus and clay minerals tend to the aggregates formation. Factors for aggregate formation are as follows: swelling, shrinkage, and crumbling of soil during cycles of wetting–drying and freezing–thawing, activity of soil flora and fauna also affects soil structure [1, 4–6]. Swelling of clay aggregates sometimes leads to their destruction; however, in some cases, the swollen particles become stronger in a comparison with dry ones under loading. This phenomenon, which is due to formation of hydrogen bond between the primary particles, is called "tixotropy".

At the lowest level, the aggregate stabilization can be achieved by hydrogen bonds, the difference between the surface energy of physically bound water and free water, formation of bridges like clay—Cat–COO–R–OOC–Cat–COO–R–OOC–Cat–clay, where Cat is Ca, Mg or Fe, R is the organic compound (e.g., a polysaccharide). The aggregates can be also stabilized by means of electrostatic

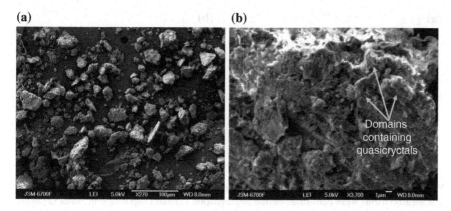

Fig. 11.3   SEM images of clay particles: **a**—microaggregates, **b**—a microaggregate consisting of domains, which include quasi-crystals

interaction between the charges: positive (edge surface of the clay particles) and negative (face surface) ones [7, 8].

The domain model has been developed to describe hierarchical structure of soil. In accordance with the model, the clay particles are in a form of domains, a size of which is several microns in a diameter. The domains are separated from each other by "bonding pores" [9]. Domains are clustered together in micro-aggregates (5 μm–1 mm), the micro-aggregates form aggregates (1–5 mm). The aggregates also include particles of sand and silt. A vehicle function is performed by poorly soluble inorganic compounds, organic substances, oriented layers of clay or colonies of microorganisms.

Afterward, the domain theory was improved. In particular, a new model, which provides the domain formation by quasi-crystals (0.01–1.3 microns), has been proposed. The quasi-crystals are the fields, which are characterized by a parallel orientation of alumino-silicate layers [10]. Micro-aggregates, domains, and quasi-crystals of the clay are visible on SEM images obtained by authors of this chapter (Fig. 11.3).

Hierarchical models suggest different mechanisms of stabilization of the structure at each level: at the level of quasi-crystals, the aggregation occurs due to the deposition of insoluble inorganic compounds. Stabilization at higher levels is reached due to organic matter and microorganisms. Such stabilizers as the particles of organic matters (humins, hyphae of fungal microflora, and roots) are necessary to maintain soil fertility [11]. The period of the stabilizing effect of the hyphae is short: after the death of fungal microorganisms the macro-aggregates are broken down [12]. According to [13], carbonates, which are formed by the decomposition of organic matters, stabilize aggregates at higher levels. The authors of [14] took the opposite point of view: the calcium compounds do not contribute to the stability of macro-aggregates.

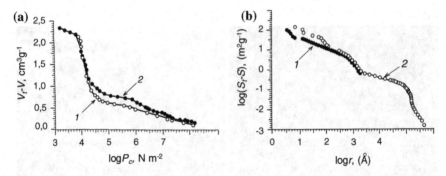

Fig. 11.4 Integral curves for peat: **a** distribution of pore volume plotted vs capillary pressure. The measurements were performed during impregnation (*1*) and evaporation of water (*2*) using the method of standard contact porosimetry; **b** distribution of specific surface area on obtained by means of method of standard contact porosimetry under impregnation with water (*1*) and by the adsorption method under water condensation (*2*) (adapted from [20])

To estimate fractal (self-similar) structure of soils at the nanoscale, the following methods are used: small-angle X-ray scattering [15], water vapor adsorption [16], sieve analysis [17], and computed tomography by means of synchrotron radiation (this method allows us to obtain a three-dimensional image) [18]. More universal techniques are a combination of scanning and transmission electron microscopy [19] as well as the method of contact porosimetry [20]. They allow us to investigate both the nano and micro levels of the structure.

Standard contact porosimetry was used to study peat (Sphagnum) in the range of pore radius r from 1 nm to 50 microns. Figure 11.4a shows the integral pore volume V distributions, which are plotted as $V_t - V$ (where V_t is the total pore volume) *vs* the capillary pressure values P_c. The capillary pressure determines the force, which holds water in the pores. Hysteresis between the curves, which were obtained by impregnation with water and its evaporation, indicates corrugated pores in the peat. The fractal dimension (d_f) is determined from the integral distribution of specific surface area (S), which is plotted as $\log(S_t - S)$ (where S_t is the total specific surface area) vs $\log r$ (Fig. 11.4b) according to [21]:

$$\log(S_t - S) = \text{const} + (2 - d_f) \log r. \tag{11.1}$$

The d_f values can be estimated from the slope of extended linear fields of the plots. Regarding the peat, $d_f \approx 2.55$ ($r = 1.5$–80 nm) and ≈ 2.42 ($r = 0.25$–9 μm). Fractal structure on the microlevels is observed in Fig. 11.5.

For aggregates, whose size is in the range of 25 μm–2 mm, the bulk density decreases with a reduction of the fractal dimension [22]. Moreover, the d_f value depends on the type of plants growing in the soil [22]. In the case of ryegrass and bromegrass, the values of fractal dimension have been estimated as 2.51 and 2.12, respectively.

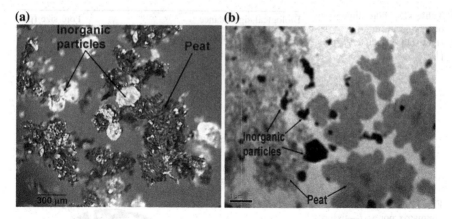

Fig. 11.5 Images of peat, obtained with optical (**a**) and transmission electron (**b**) microscopy. Peat—dark (**a**) and light (**b**) particles, inorganic impurities—light (**a**) and dark (**b**) particles (TEM image has been obtained by Dr. N. N. Scherbatyuk (M. G. Kholodny Institute of Botany of the National Academy of Science of Ukraine))

Soil texture and stability of the aggregate determine steadiness of soils against erosion, which involves destruction and demolition of the upper horizons of the most fertile soil due to affect of water, wind, marine breaker, glaciers, gravitation, etc. The soils containing about 40 % of fine sand and silt particles (0.02–0.2 mm), which demonstrate a weak tendency to aggregation, are the most susceptible to erosion [23]. Considerable content of well-swelling clay minerals, for instance, montmorillonite, as well as sodium salts contained in the pore solution also promote erosion. At the same time, bad swelling clay minerals, in particular kaolin or chlorite in a combination with humus, form a hydrophobic layer on the aggregate surface. This layer protects soil against erosion.

11.3 The Pores of Soil

In edaphological terminology pores are divided into several types depending on their size, geometry, origin, functions, etc. For example, according to the origin, there are matrix pores (between mechanical particles), interstructural (between aggregates) and nonmatrix (formed by roots, soil fauna, air, and water flows). Biogenic factors, which cause nonmatrix pores, are shown in Table 11.4. Nonmatrix pores are also formed due to anthropogenic factors, however, this factor does not influence matrix and interstructural pores.

It should be emphasized, that the terms of "macro-", "meso-", and "micropores" should be used for soils very cautiously. For example, according to the IUPAC classification, which is widely used in chemistry, the pores, a diameter of which is lower than 2 nm, are attributed to micropores. A size of mesopores is 2–50 nm, larger

Table 11.4 Pore dimensions of biological origin or significance (adapted from [1])

Biological significance	Pore size (μm)
Ant hills and channels	1,500–50,000
Worm channels	500–11,000
Tap roots	300–10,000
Nodal roots	500–10,000
Seminal roots	100–1,000
Root hairs	1,000
Lateral roots	50–100
1st- and 2nd-order laterals	20–50
Fungal hyphae	0.5–2
Colonies of microorganisms	0.2–2

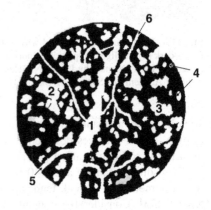

Fig. 11.6 Shape of nonmatrix pores: *1*—fissures, *2*—irregular pores, *3*—chamber (restricted) pores, *4*—bubble pores, *5*—simple tubular pores, *6*—dendritic tubular pores (adapted from [5])

pores are related to macropores. There are a number of classifications in soil science, in accordance with some of them the pores up to 30 microns are considered as micropores.

According to geometric characteristics, the nonmatrix cavities are subdivided into irregular, restricted (chamber), bubble, tubular, pores, and fissures (Fig. 11.6) [5]. Irregular pores are extended or compact shapeless voids with jagged, angular, or rounded edges. Restricted ones are roundish relatively large pores with rough walls. Bubble cavities are rather small pores, their shape is close to spherical. Tubular cylindrical pores are more or less elongated in one direction, their cross-section is round or oval, they can be simple (one non branching tube) or dendritic (branching tube). All the listed types of pores can be opened or closed. Fissures are the cavities with rather parallel walls, these pores can be oriented vertically or horizontally, they are also able to form a network.

Sometimes the pores of soil are systematized taking into account their functions. For instance, according to the Luxmoore classification, the pores, a diameter of which is larger than 1 mm, are related to channels [24]. The pores of 10 μm– 1 mm are considered as gravitational. A pressure gradient is realized in pores, which are smaller than 10 μm.

More complicated classification proposed by Greenland includes 5 classes of pores: (i) fissures, where a root system develops (>500 μm), (ii) pores providing

Table 11.5 Bulk density and porosity for different types of soils

Soil	Porosity (%)	Bulk density (g sm^{-3})
Sandy	35–60	1.60
Argillaceous	30–70	1.10
Loamy	30–60	1.10
Peat	80–85	0.25

ventilation and drainage of excess water (50–500 μm), (iii) accumulating pores, which hold water (500 nm–50 μm), (iv) residual pores, where ions are transported through soil solution (5–500 nm), and (v) bonding pores, where the interaction between soil particles is realized (<5 nm) [25].

Porosity and bulk density of common types of soil are shown in Table 11.5: sandy soils are the densest, peat soils are the most friable. When the abovementioned optimal ratio of the components is realized, the soil porosity of 50–60 % (bulk density of 1.1–1.3 g cm^{-3}) is reached. This porosity is required for normal development of the most cultivated plants.

The increased bulk density negatively affects the soil moisture regime, gas exchange, the number and activity of soil microorganisms. The gas exchange is broken or completely stops at the density of more than 1.45 g cm^{-3}. As a result, the amount of oxygen in the air-filled soil pores decreases. Simultaneously, the content of ethylene increases due to slow decomposition of organic matters. The reduced bulk density (<1.1 g cm^{-3}) also unfavorable for the development of cultivated plants. This is caused by low reserve of water in this soil on the one hand and difficult access of nutrients from large aggregates on the other hand.

The destruction of soil aggregates, which results in soil compaction, can be caused by mechanical impacts, such as tillage and traffic [1, 4]. Cultivated lands are also characterized by a high content of microorganisms, which destroy humates [26]. Graizing livestock has a less negative impact on soil fertility [27]. As compared with virgin soils, cultivated ones are characterized by a high content of micropores and a low content of macropores (Fig. 11.7). Thus, water and air regimes of these soils are different.

Compaction caused by mechanical stress is more typical for sandy soils than for clayey ones, whose aggregates tend to swelling. Adverse changes in the structure of topsoil, which is treated periodically, develop for 3–6 years. The increase in the density of topsoil (the thickness of this layer is about 75 cm) by 0.1 g cm^{-3} (up to 1.55–1.6 g cm^{-3}) leads to 10–15 % loss of harvest of cereals. A variety of ways to minimize mechanical treatment of soil (deep tillage, application of subsoil ploughs, simultaneous use of different methods of mechanical treatment, etc.) is applied to prevent its compaction.

The bulk density increases along the direction from surface to depth, however the profile of this parameter is sharper for virgin lands than for cultivated soils (see Fig. 11.7). The high microporosity is characteristic for soils, which are treated regularly with agricultural machinery. The treatment prevents damage and loss of plants due to the root protrusion. The protrusion is caused by the beginning of

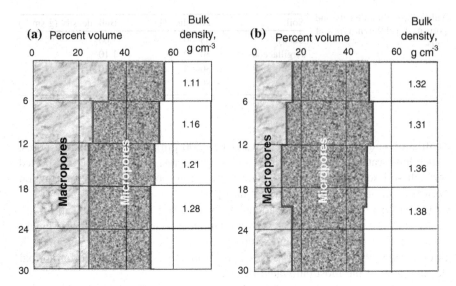

Fig. 11.7 Porosity and bulk density of the soils: **a** the virgin and **b** cultivated for 90 years (Houston Black Clay soil) (adapted from [4])

water freezing at a certain depth (in macroporous layers), when frost comes. Simultaneously water near the ground surface, which is characterized by high micro- and mesoporosity, remains unfrosted. Water begins to freeze at a depth, when the fall frost occurs under an absence of a snow cover or in the case of low water content at the surface layer of soil (arid autumn). This is also possible after thawing, if snow water is soaked into the soil. The ice interlayer, which is formed at a depth, bulges the upper layers of the soil together with the plants. This leads to breakage of plant roots, which has penetrated to a significant depth. Furthermore, nonsoil inclusions like stones are imposed to the surface.

Thus, the structure of the soil has a significant impact on the growth and development of plants. On the other hand, plants play an important role in soil formation: particularly, in the formation of its organic constituent. Consequently, the soil and the root are in continuous interaction, changing each other.

11.4 Functions and Anatomy of the Root System

Functions of the root are to fix plants in soil, provide absorption, and transport of water, which contains dissolved minerals, toward the stem and leaves. In some cases, the roots accumulate nutrients and synthesize organic substances. A root never form a leaf, it lacks chlorophyll. It goes deep into the soil, branches, forming lateral roots. A root never form leaves, it lacks chlorophyll. It deepens into the soil and arborizes forming lateral roots. Adventitious roots occur on the stems and leaves of many plants, in particular, of most monocotyledons and cereals.

Fig. 11.8 Longitudial section of a root (adapted from [28])

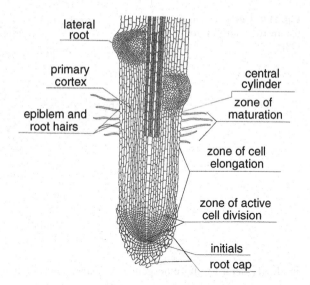

Different parts of the root (zones) perform different functions (Fig. 11.8) [28–30]. The tip of the root is covered with a root cap consisting of living cells, which renew constantly. The root cap is formed by the fundamental tissues including thin-walled parenchyma cells. Sometimes the roots are deprived root caps (aquatic and some parasitic plants).

The root cap is a source of slime (highly hydrated polysaccharide), which covers the root tip and provides adherence of soil particles to the roots. Slimy coating protects the root tip against harmful substances and drying. The slime may also perform ion-exchange functions: it forms chelate complexes with certain nutrients. Root cap controls geotropism of the root (usually nodal one), which grows down. Giving horizontal orientation to the root causes its subsequent growth along the curve until downward orientation of the tip is reached.

The root cap also prevents damage of apical cells, sometimes it protects only one apical cell, which is called "initial." The apical meristem is tissue consisting of thin-walled cells of different shapes (rectangular, multifaceted), which fit snugly to each other. Apical cells function throughout the life of the plant. Actively dividing meristematic cells always appear from them, these cells are also located under the root cap. Meristematic cells as well as initials form zone of active cell division with a length of ≈1 mm.

The meristematic cells lose their ability to divide during the root growth, the tissue structure becomes looser due to formation of intercellular spaces. These cells grow in length pushing the root end deep into the soil. They form the zone of cell elongation, which is located behind the zone of active cell division. The zone of cell elongation is also small in extent (a few millimeters).

Just above the elongation field, the zone of maturation is located. Differentiation of tissue cells (dermal, ground, conductive, etc.) begins here. A length of this zone is from several millimeters to several centimeters. The beginning of this zone

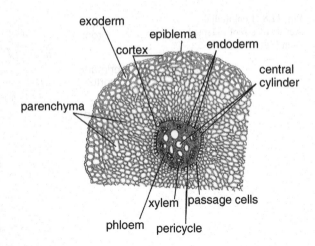

Fig. 11.9 Cross-section of the iris root (adapted from [28])

is clearly seen by the emergence of numerous root hairs (trichomes), which are outgrowths of epithelial tissue (epiblema). In contrast to the zone of cell elongation, the maturation field is not displaced relatively to the soil particles.

Further the transition zone is located, where lateral roots are formed. Epiblema as well as root hairs die and flake off along this field, which is responsible for transport of water and solutions of mineral salts toward the overlying plant organs.

A cross-section of the root in the maturation zone is subdivided into three areas: epiblema with root hairs, the primary cortex, and the central cylinder. Epiblema (dermal tissue), which consists of a series of serried cells, performs a function of absorption of soil water and nutrients. Root hairs, whose lifetime is up to 10–20 days, increase the suction surface of the root. Their total surface for a plant, such as winter wheat, is 4.2 m^2 g^{-1}, and total length of 10,000 m. Length of a single hair in different plants varies from 0.05 to 10 mm, its diameter is from 5 to 20 μm. Hairs also secrete metabolism products into the soil, particularly carbonic acid or formic acid, which dissolve the soil particles. The acids are not formed in a very dry soil, at the same time their formation becomes slower with the increase of soil moisture [28].

The bulk of the primary cortex, which has a complex structure, contains living parenchyma cells (ground tissue), the morphology of which varies in the direction from the epiblema to the central cylinder (Fig. 11.9). The outer part of the primary cortex (exoderm), which is formed by closely spaced living cells with thickened walls, is located under epiblema and performs protective and suction functions. Exoderm is absent for some plants, for example, aqueous ones. In the transition zone, the exoderm becomes a peripheral cell layer instead of epiblema, which is peeled off.

Mesoderm cells, which are located under exoderm, are very small on the periphery and arranged closely to each other. Toward the middle, a size of the mesoderm cells increases, they become relatively loosely arranged. Closer to the center, a size of the mesoderm cells decreases again, the cells tend to pack closely. Slack arrangement of a bulk of the mesoderm parenchyma is essential for aeration.

Fig. 11.10 Cross-section of the reed root (adapted from [28])

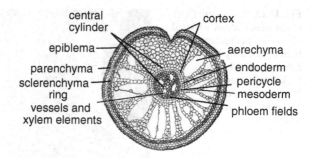

central cylinder
cortex
epiblema
aerechyma
parenchyma
endoderm
sclerenchyma ring
pericycle
mesoderm
vessels and xylem elements
phloem fields

Fig. 11.11 A cross-section of the stem of horsetail (*Equisétum arvénse*). Here *Phl* is the phloem, *Xyl* is the xylem. (SEM image has been obtained by Dr. N. N. Scherbatyuk (M. G. Kholodny Institute of Botany of the National Academy of Science of Ukraine))

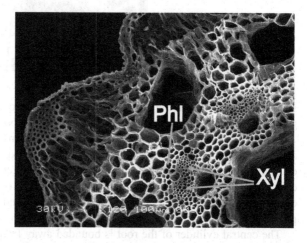

Large cavities filled with air are developed in the mesoderm of the plants of aquatic and waterlogged lands. These cavities are intercellular spaces, the tissue containing them is called "aerenchyma" (Fig. 11.10). Squeezing of aerenchyma in the transverse direction is warned by annular layer of sclerenchyma, which is a mechanical tissue with very thick walls. This tissue is located directly below the epiblema (for the Carex limosa sedge). Damage of the air-containing cavities can be also prevented by several peripheral layers of suberized cells (for the Carex uralensis sedge).

The inner layer of the primary cortex consisting of 1–2 layers of thick-walled cells, which are alternated with thin-walled passage cells, is called endoderm. Its function is not only aeration, but also transportation of aqueous solution of mineral salts from epiblema to the central cylinder of the root toward the horizontal direction. The solution flow is regulated by the endoderm.

Alternating conducting vascular tissues are placed in the central cylinder of the root. One of these tissues is phloem, where organic substances, which are synthesized by the plant itself, are transported from the leaves to other organs (Fig. 11.11,[1] see also Figs. 11.9 and 11.10). Phloem consists of fibers as well as

---

[1] SEM images have been obtained by Dr. N. N. Scherbatyuk (M. G. Kholodny Institute of Botany of the National Academy of Science of Ukraine).

Fig. 11.12 The xylem (*Xyl*) of the stem node of horsetail (*E. arvénse*) (TEM image has been obtained by Dr. N. N. Scherbatyuk (M. G. Kholodny Institute of Botany of the National Academy of Science of Ukraine))

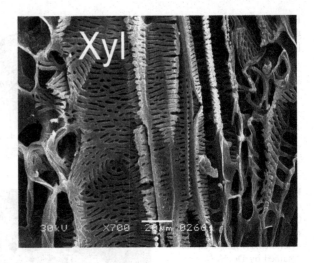

sieve elements, such as sieve plates, sieve tubes and companion cells. The fibers give mechanical strength to root, the sieve elements provide basic transport.

Another conducting tissue is xylem, which is responsible for transport of water and nutrients from bottom to top [28–31]. Xylem includes tracheids (dead lignified cells with holes) as well as vessels formed by the merger of fibers and parenchyma cells (Fig. 11.12; see also Figs. 11.9–11.11). Xylem cells form conducting vascular-fibrous bundles. Phloem and xylem are alternated with parenchyma, which provides short-range radial transport in the maturation zone.

The central cylinder of the root is bounded away from the primary cortex with one layer of meristematic cells. This layer is called pericycle, it forms the lateral and additional roots.

The structure of the root, which has been described above, is related to the primary root structure. Monocotyledons and filices persist the primary structure of the root throughout their life. During the growth of gymnosperms and dicotyledons, their roots thickens and their primary structure is replaced by secondary one due to expansion of the cambium. Cambium arises from the parenchyma cells located in the central cylinder, as well as from the pericycle.

11.5   The Porous Structure of Plant Tissues

The "roots porosity" term means often a porosity of aerenchyma, which is determined by pycnometric method without removal of the living content of a cell [32]. However, this term in the broad sense of the word involves a porosity due to not only intercellular voids, but also the cell membranes and the cell wall. Known porosimetry methods involve preliminary removal of liquid from the living tissues, including the cell membranes. Thereby intercellular voids, cavities insulated by

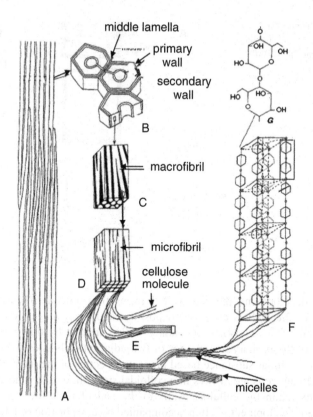

Fig. 11.13 Structure of cell wall. *A*—strand of fiber cells. *B*—transverse section of fiber cells showing layering; a layer of primary wall and three layers of secondary walls. *C*—fragment from the middle layer of the secondary wall shoeing microfibrils (*white*) of cellulose and interfibrillar spaces (*black*), which are filled with noncellulosic material. *D*—fragment of a macrofibril showing microfibril (*white*). The spaces among macrofibrils (*black*) are filled with noncellulosic material. *E*—structure of microfibrils: chain-like molecules of cellulose, which in some parts of microfibrils are orderly arranged. These parts are the micelles. *F*—fragment of a micelle showing parts of chain-like cellulose molecules arranged in a space lattice. *G*—fragment of a cellulose molecule (adapted from [29])

walls of dead cells, as well as pores of the walls of living cells contribute to the porosity. Hierarchical structure of the cell walls is of special interest.

The cell wall performs protective and supporting function, it is involved in the uptake and the transport of substances. Chemical basis of plant cell wall is polysaccharide, namely cellulose $(C_6H_{10}O_5)_n$. The macromolecules of cellulose are combined into bundles (micelles) with maximum diameter of 10 nm (Fig. 11.13). The micelles form the microfibrils (25 nm), which are oriented parallel or perpendicular or randomly to the cell axis. Porosity caused by microfibrils is 20–50 % of the total porosity of the cell walls [29]. In turn, the microfibrils form macrofibrils (400 nm).

Fig. 11.14 Scheme of a
field, where two *A* and *B* cells
are connected (adapted from
[29])

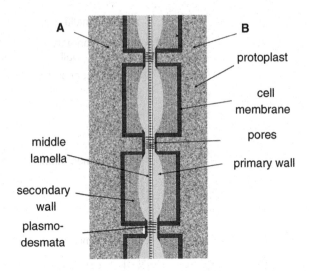

With the development, the cell wall thickens from outside to inside, so the oldest layer of the wall is an outward position and the youngest one occupies internal position (Fig. 11.14). A secondary wall is formed by this manner. Connection area of the primary membranes of two adjacent cells is called the middle lamella or intercellular substance based on pectin. Pectins of the intercellulars and walls of the living cells are always in a colloidal state, they soak water and swell.

Many cells retain the cellulose walls to the end of their life. However, the process of cell development is often accompanied by a deposition of new layers of another substance on the wall. As a result, the cell wall gets new chemical and physical properties. Lignin deposition leads to lignifications, formation of waxy-like substance like suberin caused suberization. Suberin is deposited mainly on the secondary wall as one or more lamellas. This causes special thickenings (Casparian bands) on radial walls of the cell. These formations block the solution motion along the cell walls.

Deposition of cutin, the composition of which is similar to the suberin, results in cutinization. Formation of layers of silica and calcium carbonate results in mineralization of the cell wall.

The secondary cell walls have characteristic deepening, which are considered as pores. In the case of adjacent cells, the pores are located opposite to each other (see Fig. 11.14). The pores are the result of nonuniform deposition of secondary wall material: no substance is deposited over the pore membrane. Therefore the pores are the fields, where the secondary wall is interrupted. The primary walls also form pores, where the wall is essentially thin and riddled with plasmodesmata, which are filament-like strands of the cytoplasm. The plasmodesmata are connected directly with the neighboring cells. Thus, the primary wall is not interrupted throughout these fields, except only those fields, where plasmodesmatic channels are located. The pores in the secondary wall are positioned above the primary cell.

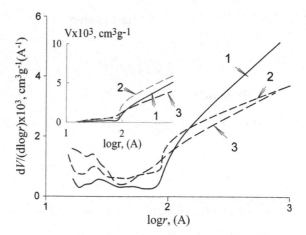

Fig. 11.15 The differential pore volume distribution for cell walls. Plants: spruce (*Picea abies*) (*1*), drooping birch (*Betula pendula*) (*2*), obovate magnolia (*Magnolia obovata*) (*3*). The *inset* shows the integral porogramms

The pores, which are formed by sudden interruption of the secondary wall, are related to simple ones. They are characteristic mainly for the ground and mechanical tissues. Conducting tissues involve bordered pores, in these cases, the secondary wall hangs over them. The bordering is supposed to perform the function of the valves for mass transfer.

Porosity, which is caused by certain elements of hierarchical structure of cell membranes, varies considerably. We investigated porous structure of apical fields of the branches of some trees using a method of low temperature nitrogen adsorption (living tissue contents had been previously removed). The integral and differential distribution of the pores, which are formed by the elements of the lowest structural levels, are shown in Fig. 11.15. Characteristic peaks at 2.5 nm correspond obviously to the pores between the micelles of cellulose macromolecules. Pores, which are formed by macromolecules (r < 1.9 nm), provide water storage. The larger pores, a size of which is 10–100 nm, are related to microfibrils, they are responsible for transport of water and nutrients. The maxima of pore distributions, which is located in this region, is probably superimposed on the peak occurring due to the presence of pores formed by macrofibrils.

The porosity of living tissues is also provided by the cell membranes (plasmolemmas), which are located both directly under the cell wall, ensuring the integrity of the cells, and within cells, isolating specialized closed cell compartments, where certain conditions of the intracellular environment are supported.

Vital activity of the cells is caused by the pores of the cell membranes. The microporous membranes are lipid bimolecular layer with a thickness of 3.5 nm, which is coated on each side with a protein layer, whose thickness is 2.5 nm (Fig. 11.16) [33]. In the bimolecular lipid layer, the lipid molecules are directed by their hydrophilic or polar ends (marked by circles) to the layers of proteins. Inside areas of hydrophilic pores, the polar tails of lipid molecules are facing each other. Protein layers are composed of individual subunits (tonofilaments), some of them (the tunnel proteins) cross the entire membrane. A special role in the system of

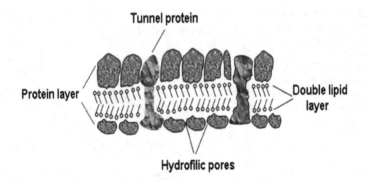

Fig. 11.16 Three-layer model of the cell wall (protein–lipid–protein) of biomembrane. Polar tails of the lipid molecules are indicated by circles (adapted from [33])

water transport in plants is played by aquaporins. These substances are the membrane proteins, which form the "water channels" in the cell membranes of plants. The channels permit water molecules selectively, enabling them to enter and leave the cell but at the same time preventing the transport of ions and other solutes. This property provides the electrochemical membrane potential. Other proteins form micropores, which are responsible for transport of certain ions.

11.6 Mass Transfer in Soil-Plant System

Because the plant receives water and nutrients through the root, so it is important to know the forces, which govern the process. In order to characterize the forces of various nature, which affect soil moisture, the concept of water potential (Ψ_w) is usually used. It is defined as useful work to make a reversible isothermal transfer of the unit of free pure water from a predetermined level to a certain point of the soil system or plants. The ratio of the amount of work to the mass is the unit of potential (J kg^{-1}), while the ratio of the work amount to the volume of transferred water is the unit of pressure (Pa). Isobaric-isothermal potential and the Gibbs free energy are applied to evaluation the energy status of water. Since this energy depends on mass (volume) of each constituent of the multicomponent system, it is typically used in the form of a specific magnitude called chemical potential. The chemical potential characterizes the state of the components in the absence of external forces, however soil moisture is always under their influence. Regarding soil moisture, the chemical potential has been agreed to call water potential [30].

Water potential, whose value is negative, approaches to zero in a wet soil. As the soil dries, its water-retaining forces increase, and the water potential decreases. The average value of water potential (1.5×10^6 Pa) corresponds to the moisture of sustainable wilting of plants. Water is supplied to the roots, if their water potential is below than that of the soil. In other words, a gradient of the water potential affects the transport of water in the soil-root system. The magnitude of this gradient is determined by the inflow rate of water to the roots and the rate of its uptake by the roots. The field of the most rapid absorption of water is usually near the root tips. Therefore, both the water content and water potential of the soil vary strongly in different parts of the root location.

The value of water potential involves three components: gravitation, osmotic (Π) potentials, and potential of tensometric pressure or pressure potential (P):

$$\Psi_w = \rho g h - \Pi + P \qquad (11.2)$$

where g is the acceleration of free fall, ρ is the density of water, h is the height. The gravitation potential is expressed by the difference of heights between two levels of the soil profile or the plant organ. Osmotic potential is caused by the salt content of the soil solution and can be expressed in terms of experimentally measured osmotic pressure. The pressure potential is the measured pressure of the liquid phase of the soil. In the case of plants, the pressure potential corresponds to intracellular pressure. Regarding soils, the pressure potential is associated with the curvature of the surface of the liquid phase and the geometry of the soil particles. This potential also depends on pressure in the gas phase. In turn, the pressure potential is divided into two components: pneumatic and matrix potential. The first one is caused by an excess of pressure, the second one is a result of the influence of the soil solid phase to the decrease in free energy of water. In swellable soils, the matrix potential depends on the moisture content and the mechanical pressure of the mass surrounding certain volume. The matrix potential plays a very significant role to estimation of moisture availability of crop. It is measured using tensiometer and capillarometer, i.e., under the conditions, at which capillary forces dominate.

The osmotic potential, which is a constituent of the water potential, is determined by the composition of the substances of the soil solution. The osmotic pressure of this solution increases rapidly with increment of concentration of dissolved compounds. The value of osmotic pressure is not the same for different soils, it is in a diapason from 0.1–0.3 to 1.2 MPa [28]. The plants are able to suck water and nutrients, when the osmotic pressure of the soil solution is below the osmotic pressure of the cell sap of a root system. The osmotic pressure of the cell sap is about 0.1 MPa for most cultivated plants and 0.3 MPa for many plants in drylands. It reaches 10 MPa or higher value for the plants of salted soils. If the osmotic pressure of the soil solution is higher than that of the cell sap, the water transport into the plant is stopped. Therefore, cultivation of crops is impossible in salted soils without previous recultivation of the land.

Water potential gradient is also characteristic for the roots. It varies from higher value inside the cells, which form the root hairs, to lower magnitude inside cells adjacent to the xylem. This gradient is maintained by lower osmotic pressure of the root hairs in a comparison with that for the xylem liquid [30].

Since the cell membranes are semipermeable, the cell roots suck sufficient amount of water from the soil. Water is taken by the cell sap; the diluted sap stretches elastic wall of the cell. As a result, the cell strains maximally. However, the cell wall is elastic, thus it tends to shrink and pressure on the internal contents of the cell. This state of tension of the cell and the plant organ, which is caused by water uptake, is called turgor. The plant organs and tissues are tense and relatively mechanically stable due to water absorption. Turgor pressure helps growing roots to break through hard soil and even rock. When the cell is in the media with increased salt content, the cell loses water (plasmolysis).

Physiological activity of roots, related to the supply of the plant with water and nutrients, is enhanced by its symbiotic association with a specific fungus. This association is called mycorhiza. The fungi are introduced into the primary cortex, but the root cells do show no pathological symptoms and retain their characteristic features. The functions of fungi are assumed to transform minerals of the soil and to decompose organic debris down to the forms, which are available for the host plant. The fungi probably reduce the hydrodynamic resistance of the root. Roots can be associated with bacteria (for example, nitrogen-fixing bacteria called *Rhizobium*). The transport system, in which the nutrients between bacteria and host plant are exchanged, is formed by this manner [29].

Mass transport in the direction of the soil–root can be described mathematically. Two main types of models are known. The models of the first type relate the rate of water absorption by the roots with their hydraulic conductivity in the radial direction and the pressure drop, the geometrical factor is taken into consideration. The hydraulic conductivity, the value of which is about 10^{-8}–10^{-7} m^2 s^{-1} MPa^{-1}, is inverse to hydraulic resistance. The second group of models joins the experimental and theoretical values of the water absorption rate.

The difference in matrix potentials of the soil pressure (the third term of Eq. (11.2)) at the point corresponding to the half distance between two roots (ψ_s) and at the soil–root interface (ψ_a), is defined by the Gardner equation [34]:

$$\psi_s - \psi_a = (I_w/4\pi K_s)\ln(b^2/a^2). \tag{11.3}$$

Here I_w is the rate of water uptake by the root, a length of which is 1 m (m^3 m^{-1} s^{-1}), K_s is the hydraulic conductivity of the soil, which is unsaturated with water (m^2 s^{-1} MPa^{-1}), b is the half distance between roots (m), a is the root radius (m). Van den Honert and Newman proposed the following equation to describe the transport of water in the soil and roots [35]:

$$W = \frac{\psi_s - \psi_x}{(1/K_s + 1/K_x)}LV, \tag{11.4}$$

where W is the volume of water absorbed per unit time by the root (m^3 s^{-1}), ψ_s and ψ_x correspond to the xylem magnitude at the base of the stem, L is the root length per 1 m^3 of the soil (m m^{-3}), V is the soil volume (m^3).

The values of Eqs. (11.3) and (11.4) can be obtained experimentally, in particular, the water potential of the roots can be found by psychrometric measurements [36]. The rate of water absorption by roots is calculated using the found values of water potential and hydraulic conductivity. This value is compared with the experimental one. Then the coefficients, which join experimental and

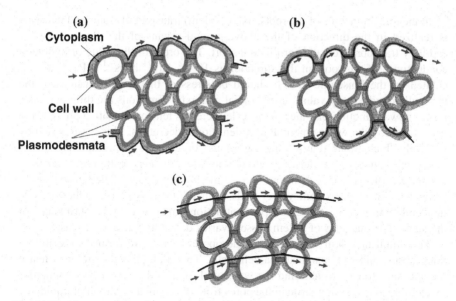

Fig. 11.17 Mass transport in the root cells: **a**—apoplastic, **b**—symplastic, **c**—transcellular (adapted from [29])

theoretical values of S or I_w, is found. At last, the dependence of this coefficient on the varied parameter (for example, the salt concentration in the soil solution) is plotted and the corresponding function is approximated.

Combined models, which join two abovementioned types, as well as the models developed to describe simultaneous absorption of water and nutrients are also known [37].

The radial transport of water and nutrients in the direction from epyblema to the xylem is implemented in the maturation zone of the roots. The transport is realized in three directions: (i) apoplastic path through cell walls, (ii) symplastic path through the cytoplasm and plasmodesmatas, and (iii) transcellular path through the cell membrane (Fig. 11.17) [30].

As seen from Eq. (11.4), one of the main parameters, which determines the transport of water from the soil to the roots, is the hydraulic conductivity of the roots. A death of the cells causes a release of the space formerly occupied by protoplast, as a result, the hydraulic conductivity in the apoplastic path increases. Simultaneously, this reduces symplastic and transcellular transport due to violation of the continuum. A thickening of the cell walls improves the apoplastic transport on the one hand and deteriorate the transport in the symplastic way due to increase of plasmodesmata length. At the same time, the transcellular transport is promoted by aquaporins. Casparian bands, layers of suberin, cutin, lignin, or minerals also reduce hydraulic conductivity of the roots in the radial direction. Thus, the aging plants are able to control their water consumption. Radial transport is reduced also by aerenchyma for plants of waterlogged lands.

In the transition zone of the roots, axial (xylem) transport of water and nutrients is realized in the direction of the above-ground organs of the plants. Laws of capillary transport are valid for this transport. The order of the value of hydraulic conductivity of the roots in this direction is 10^{-12}–10^{-11} m^4 s^{-1} MPa^{-1}, as calculated by the Poiseuille law. It should be stressed, that the dimensions of the hydraulic conductivity values in the radial and axial directions are different. As a result, upward motion of water in the xylem creates tension (negative pressure) in it, and thereby the water potential of xylem liquid decreases. This is suggested by the daily fluctuations (day-to-day variations) of the tree stems diameters. The minimum diameter is reached in the daytime, when the evaporation of water from the surface of the leaves (transpiration) is the most intensive. The tension of the water column in the xylem vessel pulls a bit of the wall due to adhesion, and the combination of these microscopic contractions gives an overall "shrinkage" of the stem. The diameter of xylem vessels varies depending also on a season.

The liquid in xylem vessel fills its mesopores, which are formed by cellulose microfibrils, and wets parenchyma cell walls. This multiplies the area and hence the adhesion forces, which bind the water molecule to the pore walls. Numerous "water threads," passing through the pores from the capillary column of liquid, are bounded between each other by the cohesion forces and by the forces of adhesion with the walls of the parenchyma cells. Together with parenchyma cells, this gives a possibility to suspend a large mass of water, which is able to rise to a great height from the roots to the top of the tree stem in spite of a small diameter of vasculars (0.01–0.2 mm).

The most complicated mechanism is the transport of nutrients synthesized in the leaves to other organs of the plant, including the roots (the phloem transport) [31, 37]. According to the original hypothesis, the driving force of this transport is the turgor pressure. The cells leaves where saccharas are formed (donors) have a high concentration of the cell sap and high turgor pressure, and cells in which the saccharas are consumed (acceptors) have a low turgor pressure. If these cells are interconnected, the fluid must flow from high-pressure cells into cells with low pressure. However, the organic substances is not always transported along a gradient of turgor pressure. An alternative is the hypothesis, according to which their transport is accompanied by energy consumptions. A source of the energy can be adenosine triphosphoric acid. There are some ideas, that the periodic reduction of protein strands of sieve tubes can facilitate the motion of substances in a certain direction. These protein strands are capable for peristaltic contractions, which push solutions. According to another hypothesis, the phloem transport is a result of electroosmosis: the gradient of electric potential through each sieve plate is caused by circulation of K$^+$ ions. The transport through the phloem is assumed to realize due to several mechanisms listed above.

Thus, the radial transport in the maturation zone of roots is a result of a gradient of water potential. The transport through the xylem is realized in the transition zone due to capillary forces, while the phloem transport is evidently a combination of processes of different nature.

11.7   Influence of Soil Porosity on the Formation of the Root System of Plants

Conditions of root growth, whose mass varies from 20–30 to 90 % relative to the total phytomass, are significantly different from those of shoot growth, since growing root breaks of solid soil particles. Regarding sandy soils, the root is well-branched and strongly deepens toward ground water. Alternately, in the case of clayey soils, the root growth in depth is difficult, horizontal roots are mainly developed. Some plants, such as juniper, figs, beech, and pine-tree are able to grow and develop on the rocks. This is due to the features of their root system and the turgor pressure. During the growth period, the roots crack the rocks. Moreover, acids which are secreted by them, dissolve even the rocky substrate. Regarding the plants of deserts and semi-deserts, the height of the above-ground part of camel-thorn (alhagi) is 50–60 cm, however, its roots grow to a depths of 15–20 m. The absolute record (68 m) was fixed for the roots of boscia albitrunka (the Shepherd tree), which was found in the Kalahari desert [28].

For many plants of droughty localities, the limiting length of the roots is determined by the depth of subsoil water. The tree roots extend to deep of 10–15 m, and the radius of the lateral roots spread is much higher than the radius of projection of the tree crown. The roots extend to a depth of 2–3 m under cultivation conditions, while their length in the horizontal direction is much larger (sometimes it reaches 10–18 m) than the circle of the crown projection. For many cereals, the average area, occupied by the roots, is 40–50 cm^2. The roots of maize or cabbage penetrate to a depth of 1.5–2 m and up to 1.5 m, respectively. The rate of development of the roots of most plants exceeds the rate of growth of side shoots. In the case of the alfalfa acrospire, the root is branched and being deepened to 1 m during the phase of the second or third leaf.

Bulk density of the soil affects the deepening of the roots. Development of the root system of the plants is stopped completely at the density of more than 1.55–1.6 g cm^{-3}. However, the roots of some cosmos plants, such as ryegrass Italian, are able to penetrate into the pores, which that are smaller than their thickness [30]. Moreover, the roots of this plant compact the soil [22].

Penetration of the roots in the compacted horizons with the packed density of 1.4–1.55 g cm^{-3} is very slow. In this case, the root growth predominantly in the horizontal direction. This negatively affects the plant development. The growth in the horizontal direction is caused by a violation of geotropism of the nodal root, which is controlled by the root cap, as mentioned above. Vertically arranged nonmatrix pores (fissures) are favorable for the formation of the root systems. In this case the nodal roots occupy the largest pores, which form the angle of 75–90° with a surface (Fig. 11.18). The lateral roots, the geotropism of which is very weak, form smaller angle. They as well as the roots of the 2nd, 3rd, etc. orders grow in all directions.

Fig. 11.18 The content of
the additional roots of wheat
in the largest nonmatrix pores
(fissures), which are located
at a different angle to the
surface (adapted from [6])

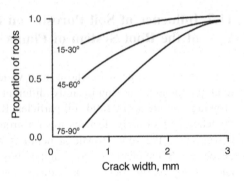

Fig. 11.19 Depth of the root
system of winter barley as a
function of a degree of soil
compactness in the first (*1, 2*)
and fourth (*3, 4*) year of
cultivation. The soil
components: loamy silt (*1, 3*),
loamy sand (*2, 4*) (adapted
from [6])

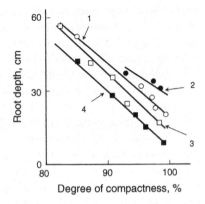

The thickness of the roots of cultural annual plants is within the interval of
0.06 μm (root hairs) to several millimeters (nodal roots), thereby the root penetration depth in the soil depends on the size and location of macropores. When the
compactness degree (ratio of the soil bulk density to the maximum possible
density) is about 75–80 %, the root system is concentrated throughout the topsoil.
Maximum compaction provides the root location only in its upper part of the
topsoil (Fig. 11.19).

The roots, from which the living substrate has been removed, are characterized
by the larger porosity in a comparison with substrates, where the roots are
developed. This porosity is caused by the pores with a radius of 1.5–100 nm
(Fig. 11.20). The pore size distributions for the roots of plants, such as dandelion
(teraxacum) and aloe (aloe suprafoliata), which grow in different climate zones,
show higher capillary porosity for hardy plant as compared with the inhabitant of
hot regions. High capillary porosity of the hardy plants obviously prevents them
from freezing, since liquid freezes in these pores under lower temperatures than in
macropores. It is also possible to assume, that high porosity, which is caused by
pores with a diameter of less than 2.5 nm, is due to necessity to accumulate water
and nutrients before the hibernation period.

Fig. 11.20 Differential pore
size distributions for the roots
(*1*, *2*) and soil substrates (*3*,
4), in which these roots were
developed. Plants: dandelion
(*1*, *3*), aloe vera (*2*, *4*)

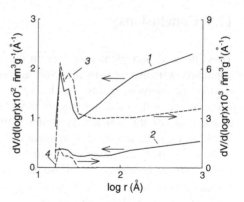

Table 11.6 Aid voids of roots for some plants (adapted from [30])

Plant	Root type	Relation of summary volume of intercellular air voids and aerenchyma to total tissue volume	
		Deaired/aerated soils	O_2-deficient
Wetland plants			
Oryza sativa	Additional roots	15–30	32–45
Typha domingensis		10–13	28–34
Phragmites australis or Phragmites commúnis		43	52
Juncus effusus		31–40	36–45
Carex acuta		10	22
Rumex palustris		15–30	32–45
Plantago maritima	Entire root system	8	22
Ranunculus fammula		9–11	30–37
Non-wetland plants			
Triticum aestivum	Additional roots	3–6	13–22
Hordeum vulgare		7	16
Zea mays		4	13
Festuca rubra	Entire root system	1	2
Vicia faba		2	4
Pisum sativum		1	4
Brassica napus		3	3
Trifolium tomentosum		7	11

The droughtier region is, the smaller pores are in the soil and plant roots.
Indeed, the contribution of intercellular and air voids into the total porosity of
tissues of the root plants, which grow on wet soils, is much lower than that for the
plants of waterlogged lands (Table 11.6). Moreover, the lack of oxygen leads to
increase of the volume of gaseous pores of roots.

11.8 Conclusions

Thus, the soil and plant roots are in constant interaction: the roots cause the nonmatrix porosity and stabilize soil aggregates at the highest level. Fungi and colonies of microorganisms, which live on the roots, interacts with micro-aggregates of the soil. Vital activity of the plants affects changes in the chemical composition of the soil and promotes stabilization of aggregates at the lowest levels. In turn, the normal development of root systems of the most crops requires a certain porosity of the soil, in particular, a balance between pores of different sizes. This is necessary to ensure the transport of the soil solution and gas exchange. The porosity of the roots depends on the properties of the soil: air-containing tissue is developed for the plants of wetlands, the cell walls of roots of hardy plants show high microporosity. Regarding the cell walls, higher mesoporosity has been found in a comparison with soil, where the plants are developed. This provides improvement of hydraulic conductivity of the roots and intensification of the mass transport in the soil-root system.

References

1. Lal R, Shukla MK (2004) Principles of soil physics. Marsel Dekker Inc., New York
2. Brussaard L, Ferrera-Cerrato R (ed) Soil ecology in sustainable agricultural systems. Advances in agroecology. Lewis Publishing, Boca Raton
3. Anthony JW, Bideaux RA, Bladh KW, Nichols MC (1995) Handbook of mineralogy, vol II. Silica, Silicates. Mineral Data Publishing, Tucson
4. Foth HD (1990) Fundamentals of soil science. Wiley, New York
5. Rozanov BG (2004) Soil morphology, M. V. Lomonosov Moscow State University, Academy project, Moscow (in Russian)
6. Chesworth W (ed) (2008) Encyclopedia of soil science. Springer, New York
7. Tisdall JM, Oades JM (1982) J Soil Sci 33(2):141
8. Oades JM, Waters AG (1991) Austr J Soil Res 29(6):815
9. Williams BG, Greenland DJ, Quirk JP (1967) Austr J Soil Res 5(1):77
10. Aylmore LAG, Sills ID, Quirk JP (1970) Clays Clay Miner 18(2):91
11. Daynes CN, Field DJ, Saleeba JA, Cole MA, McGee PA (2013) Soil Biol Biochem 57:683
12. Daynes CN, Zhang N, Saleeba JA, McGee PA (2012) Soil Biol Biochem 48:151
13. Fernandez-Ugalde O, Virto I, Barre P, Gartzia-Bengoetxea N, Enrique A, Imaz MJ, Bescansa P (2011) Geoderma 164(3–4):203
14. Gargiulo L, Mele G, Terribile F (2013) Geoderma 197–198:151
15. Mumby PJ, Harborne AR, Rice JA, Tombacz E, Malekani K (1999) Geoderma 88(3):251
16. Sokołowska Z, Warchulska P, Sokołowski S (2009) Ecol Complex 6(3):254
17. Millan H, Gonzalez-Posada M, Benito RM (2002) Geoderma 109(1–2):75
18. Zhou H, Peng X, Peth S, Xiao TQ (2012) Soil Tillage Res 124:17
19. Dathe A, Thullner M (2005) Geoderma 129(3–4):279
20. Neimark AV, Robens E, Unger KK, Volfkovich YM (1994) Fractals 2(1):45
21. Neimark AV, Unger KK (1993) J Colloid Interface Sci 158(2):412
22. Parent LE, Almeida CX, Hernandes A, Egozcue JJ, Gulser C, Bolinder MA, Katterer T, Andren O, Parent SE, Anctil F, Centurion JF, Natale W (2012) Geoderma 179–180:123
23. Morgan RPC (2005) Soil erosion and conservation. Blackwell, Malden

24. Luxmoore R (1981) Soil Sci Soc Amer J 45(3):671
25. Greenland DJ, Pereira HC (1977) Phil Trans Royal Soc B 281:193
26. Tonkha OL, Bykova OY, Ievtushenko TV (2013) Annal Agrar Sci 11:1
27. Bell LW, Kirkegaard JA, Swan A, Hunt JR, Huth NI, Fettell NA (2011) Soil Tillage Res 113(1):19
28. Kh V (1980) Tutayuk *Anatomiya I Morphologiya Rastenii*. Vysshaya Shkola, Moscow
29. Evert RF (2006) Esau's plant anatomy: meristems, cells, and tissues of the plant body: their structure, function, and development. Whiley, Hoboken
30. Gregory PJ (2006) Plant roots. Growth, activity and interaction with soil. Blackwell, Oxford
31. Lucas WJ, Groover A, Lichtenberger R, Furuta K, Yadav S-R, Helariutta Y, He X-Q, Kachroo P, Integrat J (2013) Plant Biol 55(4):294
32. Soyka RE (1988) Env Exp Bot 28(4):275
33. Gennis RB (1989) Biomembranes, Molecular structure and functions. Springer, Berlin
34. Green SR, Kirkham MB, Closier BE (2006) Agricult Water Manag 86(1–2):165
35. Smith DM, Inman-Bamber NG, Thorburn PJ (2005) Field Crops Res 92(2–3):169
36. Adeoye KB, Rawlins SL (1981) Plant Physiol 68(1):44
37. Turgeon R, Wolf Sh (2009) Ann Rev Plant Biol 60:207

Chapter 12
Hide and Skin of Mammals

Olena Romanovna Mokrousova and Yury Mironovich Volfkovich

Abstract The section summarizes information about the structure of collagen—the basis of natural leather, which is characterized by a hierarchical organization of the main structural elements and chemical multifunctional nature. It is shown that the native collagen comprises the elements and pores of various sizes in the range from 1 nm to 200 μ forming micro-, meso-, and macro levels. In leather manufacture, collagen undergoes structural transformations because of an influence of a variety of chemical materials. This phenomenon contributes to the leather structure formation with the necessary complex of functional properties and due to porosity.

12.1 Features of the Structure of Leather

Natural leather (NL) is a product of successive changes and transformations of the fibrous structure of collagen (the main protein of animal's hides and skins) under the action of chemical, physicochemical, and mechanical influences, with a view to the application of the NL in the leather and footwear industry. Features of the structure of collagen can specifically affect its fiber structure during processing treatments, while significantly altering its porosity, the colloid-chemical, physico-mechanical and deformation properties with the ultimate purpose of finished leather. Collagen is characterized by a multilevel organization and has a hierarchical structure. Thus, lower levels promote the formation of the upper levels. The lowest level is the consistent position of amino acid residues in the polypeptide chains, and the top level—the dermis of an animal hide and skin [1]. Between the mentioned levels, a number of other structures exist (Table 12.1). Thus, the spatial association of polypeptide chains in a triple helix is a macromolecule. Groups of adjacent and located parallel to five macromolecules form microfibrils, and their combination—a fibril, whose diameter does not exceed tenths of a micron. Fibrils respectively form the next level of the primary structure of fibers with diameters of about 5 μ in flooded condition. Further, combining the primary fibers in bundles corresponds to the aggregates with dimensions of 200 μ. The spatial organization

Y. M. Volfkovich et al., *Structural Properties of Porous Materials and Powders Used in Different Fields of Science and Technology*, Engineering Materials and Processes, DOI: 10.1007/978-1-4471-6377-0_12, © Springer-Verlag London 2014

Table 12.1 Structural levels of collagen

Structure	Structural level	Structural elements	Average size of elements, nm
Molecular	Molecule	Polypeptide chains with molecular weight of 300,000	
	Macromolecule	Spatial association of polypeptide chains in a triple helix	$\sim$1.5
Supramolecular	Microfibril	Conformation of 5 macromolecules	$\sim$3–5
	Fibril	Combination of 900–2,000 microfibrils	$\sim$50–200
Supra fibrillar	Primary fiber	Formation of 900–1,000 fibrils	$\sim$5,000
	Secondary fiber (fiber bundles)	Aggregates of 30–300 primary fibers	$\sim$200,000
	Dermis	Interweaving bundles of fibers with their orientation in different directions	

of the fiber bundles shows the structure of the dermis. With the transition from the intermediate to the final formation of collagen (dermis), the parallel packing of the individual structural elements (polypeptide chains, macromolecules, microfibrils, fibrils, etc.) is lost substantially.

Every structural level forms a certain number of capillaries and pores. Taking this into consideration, the dermis as the porous material comprises pores of different sizes that vary in the range from 1 nm to 200 μ and form a micro- ($r_{pore} \leq 1$ nm), meso- (1 nm $\leq r_{pore} \leq 200$ nm) and macroporous ($r_{pore} \geq 200$ nm) [2] structure. The percentage of pores of different diameters varies within wide limits and depends on the physicochemical treatment and the effect of the materials [3].

From the standpoint of the action of chemical materials in structural changes of the protein, reactivity of collagen is important. It is caused by the presence of different functional groups in lateral chains: carboxyl, amino, hydroxyl, etc. Therefore, bonds of different nature and location can be formed in the collagen structure: longitudinal intrachain and intramolecular, transverse interchain and intramolecular, transverse interchain and intermolecular [4].

The structure of collagen has relatively ordered hydrophobic and disordered hydrophilic zones. Hydrophobic or crystal blocks contain nonpolar amino acid residues with hydrocarbon lateral chains. Hydrophilic or amorphous zones contain amino acid residues with polar lateral chains. Besides the hydrophilic–hydrophobic surface of the structural elements of collagen is coated with water-soluble globular proteins that should be removed during the conversion of animal hides and skin into leather. The presence of collagen in the structure of such zones significantly affects the nature of the interaction of the protein with chemical materials and their diffusion accounting for compatibility of the chemical nature of the last with collagen and matching the size of their particles with pore sizes of the dermis. Such concepts are important in the understanding of the derma formation of and fixing its structure and porosity.

Hides and skin of all mammals are characterized by an identical hierarchical organization of the collagen structure of the dermis, and virtually the same chemical composition. However, hides and skins of different mammals may vary the porosity, the average angle of inclination of collagen bundles to the surface of the dermis and packing density of the structural elements. This is due to the histological structure of animal hides and skins, namely:

- The nature and extent of hair,
- Thickness of the epidermis,
- The relation between the papillary and reticular layers of the dermis, etc.

So, the sheep skin is characterized by a loose structure of the dermis with the large number of macropores. The pig skin is characterized by the presence of bristles, which permeate the entire thickness of the dermis. This leads to the absence of boundary between the papillary and reticular layers of dermis, dense epidermis with a higher thickness and a horizontal packing of collagen fibers and their bundles. Most spatial arrangement of collagen bundles of the dermis and their even weave is a characteristic of cattle hides. With regard to age, degree of hair and skin mass the following skins are segregated: calf, kip, cow, bull-calf, cattle hide and oxhide. More objective findings of formation of capillary-porous structure of the dermis can be obtained by using skins of kip and bull-calf.

12.2 Formation of Porous Structure and Properties of the Leather

A difference between the properties of fresh hides and finished leather is determined by the porosity, total proteins per unit volume, the average angle of inclination of the bundles to the surface of the dermis, the number of intermolecular bridges, and bonding strength of adjacent elements of the microstructure. All these factors affect the mechanical properties of the finished leather. Porosity is determined by the degree of separation of the structural elements and the formation of collagen of the dermis. During the processing of animal hides and skins, dynamic fracture of bonds between the various structural-determining elements of the protein and the occurrence of new bonds by the action of chemical reagents are originated. As a result, the leather obtained from animal hides and skins is characterized by a complex structure, high porosity, and a large specific surface area, which provides it with a number of necessary functional, hygienic, and consumer properties [5].

Technological processes of three main groups: preparation, tanning, and finishing have a significant influence in obtaining leathers of the necessary quality and specific purpose. Preparatory processes are aimed mainly at the division of the collagen structure and removal of noncollagenous proteins that is achieved by the action of alkalis, acids, and enzyme preparations (liming, bating, and pickling). The resulting collagen structure is called a pelt (untanned hide) [6].

Tanning promotes fixation and stabilization of the resultant structure of the dermis. For this purpose, such types of compounds as: the inorganic compounds (basic salts of chromium, aluminum, iron, titanium, and others), fatty organic compounds (formaldehyde, glutaraldehyde, highly unsaturated fats) and organic aromatic compounds (benzoquinone, multiaromatic sulphonic acids, tannins of oak, spruce, quebracho, mimosa, and others) [5]. Tanning radically alters the properties of pelts. The most common manifestations of the effect of tanning are: the reduction of the deformability of the watered collagen, reducing shrinkage and increase (and then maintenance) of porosity in the drying process, reducing the adhesion of the surface of microstructure elements (bundles, fibers, fibrils), increasing the strength of the dermis, reducing the overall ordering of the fine structure collagen.

The final formation of the leather is achieved in a subsequent wet and coat finishing (retannage, filling, fat-liquoring, dyeing, staking, etc.). For high quality leather, it is necessary to achieve the optimal state of micro-and macrostructure of the protein at each stage of the process.

Native dermis of fresh hide and skin consists of collagen, 25 %, and water, 70 %, indicating about significant moisture capacity of the protein. Collagen interaction with water is the result of hydration, diffusion swelling, wetting and capillary condensation [4–6]. Structure elements of the dermis increase in volume when swollen in water, and the intervals between them are compressed. During dried-air drying of untreated dermis, it is homogenized by bonding of structural elements, which leads to the disappearance of the pores. This is confirmed by the fact that the specific weight of collagen in this case, measured by mercury porosimetry, corresponds to its true specific weight [6]. Using of the dewatering organic liquids allows to eliminate homogenization of the dermis, that prevents the pore restriction to a greater or lesser extent, but can not avoid convergence of the collagen molecules, located in the interstices between pores, in the process of drying [5].

According to [7] after hydration of the fresh hide or skin with ethyl alcohol the total pore volume equaled 43.7 % was found using the volumetric method. Moreover, the authors point out that the total porosity of the hide and skin depends on the volume of open pores.

Recent researches in the field of porous structures were fulfilled using MSCP [8, 9]. The main advantages of MSCP in comparison with mercury porosimetry method used previously are no deformation of the structure under high pressure of mercury intrusion (up to hundreds of MPa), a wide range of pore sizes measured and no toxicity. MSCP allowed for fresh skin to diagnose an extremely wide range of pore sizes from tenths of a nanometer to about of 100 μ. This range is more than five orders of magnitude.

Integral and differential pore volume distributions according to their pore radii for the fresh calf-skin (Figs. 12.1 and 12.2), obtained using octane for porosimetric measurements, show portions that correspond to certain levels of collagen structure: macromolecules (log $r = 0–0.4$ nm), microfibrils (log $r = 0.4–0.6$ nm), fibrils (log $r = 1.9–2.5$ nm) and the primary fibers (log $r > 3.2$ nm). Based on these relationships, it is clear that the total porosity of the fresh skin is 52 %.

Fig. 12.1 Integral curves of the pore volume distribution on radii of the porous structure of fresh hide and skin, as measured by octane (*1*) and water (*2*)

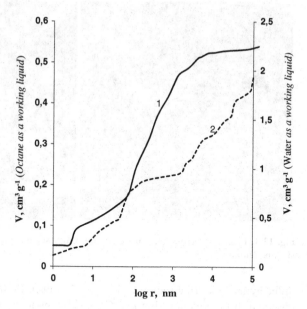

Fig. 12.2 Differential curves of the pore volume distribution on radii of the porous structure of fresh hide and skin, as measured by octane (*1*) and water (*2*)

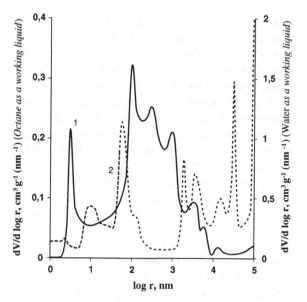

When swelling in water, the total porosity of 73 %, as well as porosity due to micropores, increases as compared with similar characteristics determined using octane. There is a clear division of pores (Figs. 12.1 and 12.2 [9]), corresponding to those or other elements of the hierarchical structure: the positions of the peaks, corresponding to the pores formed by the fibrils and the thinnest primary fibers,

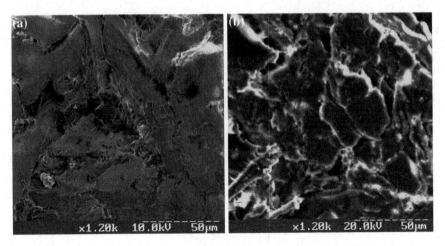

Fig. 12.3 Electron-microscopic images of the cross section of fresh calf skin (**a**) and after alkaline plumpness (**b**)

differ by an order (the region of log $r = 2$–3 nm). Furthermore, the maximum is observed in the region of log $r \geq 5$ nm corresponding to the secondary fibers. In the transition from macromolecules to microfibrils and onwards to regularly arranged fibrils these picks are narrowed and their intensity increases, indicating about the structural ordering under the transition from lower to higher levels.

Modern microscopic studies using scanning microscope allowed visually to fix state of the porous structure of the native dermis, which includes both the individual fibers and bundles of fibers, and voids between them (Fig. 12.3a) [10].

Under the action of alkaline materials (the liming process), swelling of the structure occurs, in which, first of all, the largest pores are "closed", and then more and more small pores. Visual examination of the pelts in the state of plumpness (excessive watering of the dermis) reveals an almost complete homogenization of the dermis. Thus, there is a seal structure of the dermis (Fig. 12.3b) [10].

Of great importance is the state of the alkali plumpness in changing the microstructure and porosity of the dermis. Swelling, which is accompanied by an increase in the thickness of the untanned hide, shortening its filaments, leads to loosening and opening-up of the fine structure of collagen. Furthermore, interfibrillar proteins dissolved under the action of alkali [6]. These structural changes contribute to increase of specific surface area of the dermis from 47 m^2/g (for fresh calf skin) to 119 m^2/g (for the hide after liming). Herewith, a total amount of the capillary moisture during irrigation of raw material is 81 %, and increases to 93 % in the case of hide after liming [10]. It is positive that change of the relative position of the structural elements of the dermis is typical only for the pelts, without significant changes in the chemical composition of collagen [6].

The subsequent action of the enzyme preparations (the bating process) enhances the degree of loosening of the dermis, mainly due to the removal of products of collagen degradation and interfibrillar proteins. This increases the permeability of

Fig. 12.4 Integral and differential curves of the pore volume distribution on radii of the porous structure of fresh hide and skin, as measured by octane

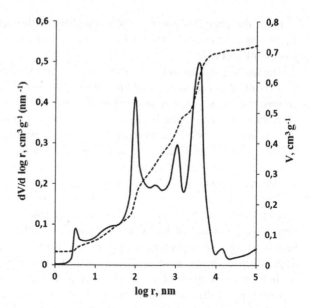

the dermis, and accordingly, increasing the porosity. The ratio of meso- and micropores increases after bating (5.5:1) as compared to the native dermis (4.0:1) [11]. Thus, the specific surface area is equal to 134 m^2/g [10]. The bating increases the softness of leather, their malleability and smoothness. The process of bating is particularly relevant in the production of flexible, soft, fancy and garment leather.

A significant increase in the permeability and porosity of the pelts takes place under the action of acid and neutral salts (pickle solution). Ya. Frenkel showed that especially strong broadening of the dermis micropores is a result of its processing by sulfates or other salts of the third group [5]. Mikhailov's work indicates a decrease in the compressibility of the pelts and an increase in porosity as a result of dewatering activities of sulfates. The maximum pore diameter calculated according to the filtration rate of the equilibrium solution is 1–10 μ in these pelts [5].

More recent work of V.I. Chursin to study the porosity of the collagen structure by mercury porosimetry [12] confirms the improvement of disjoining ability of the structure of untanned pelt after acid-salt treatment. Along with the increase of the pore volume in the range of 3–30, 0.1–2, and 0.02–0.06 μ, the pores with size of 30–200 μ are formed additionally.

In recent studies using the MSCP, an increase of the pelts pore volume (Fig. 12.4 [9]) for log $r > 1$ (nm) is found, indicating that there is a removal of noncollagenous impurities when processing of fresh skins, far except the pores, which correspond to the maximum under log $r = 2.5$ (nm). This locks the ordering of the porous structure on the level of fibrils. In addition, the transformation of two ill-defined peaks in the log $r = 3$–4 (nm) (for the fresh skin) in a fairly narrow and intense (for the pelts) indicates loosening effect of primary fibers under the influence of chemical treatment.

From the chemical point of view, pickling of the pelts creates the right conditions for fixation and stabilization of the resultant structure of the dermis in the process of tanning.

The tanning treatment leads to irreversible changes in the properties of collagen as a result of appearance of additional bridges and bonds between its functional groups and tanning agent that guides to the fixation of the microstructure [5]. Additional intermolecular bridges that occur in collagen as a result of tanning fix the relative position of the molecules and reduce the compressibility of the dermis flooded by external loads. The resulting structure is characterized by a greater resistance to shrinkage (reduction in area), reduced deformability, hydrothermal resistance, the action of chemicals and water.

The porosity of the structure in the tanning process is fixed and is the result of reducing the adhesion of the microstructure elements of the dermis. Even when you move the walls of interstructural capillaries of a tanned semi-finished item before drying by compaction, the porosity is easily restored by mechanical staking. This effect is widely used in the practice of the finishing operations in the production of soft leather with a view to increasing their porosity.

Changing of the leather porosity is always associated with a change in its total specific weight. If the values of the porosity and total specific weight for hide in the state of plumpness are 5.8 and 1.21 %, respectively, then for chrome-tanned leather these numbers are −40.0 % and 0.87, and for vegetable-tanned leather − 53.6 % and 0.67 [5].

The indicator of fixing for porosity is often used to develop the best options of the tanning treatment. Using tanning compounds which have the ability to be actively adsorbed on the surface of structural elements of the dermis, it is possible to regulate the degree of fixation of the dermis, and correct the shrinkage of a semi-finished and structural rigidity [13].

According to [7], the total porosity of semi-finished items increases in the organic and mineral tanning as compared with fresh hide. Moreover, this increase is associated with the formation of small pores and with the appearance of closed pores as well. The latter fact, as the authors note, is caused by the action of tannins used. Besides, when using chrome tanning an increase in porosity is associated with formation of a wider range of pore sizes. Probably smaller structural elements of the leather are available for chromium compounds and also the structuring effect from the side of chromium salts reduces the number of large pores.

Significant impact on the character of the porosity of the tanned semi-finished item material has not only type but also consumed amount of tannin. It is shown in the works of Chursin [3, 12] that the base of the porous structure of chromium leather is formed at low consumption of tannin −0.5 % (based on chromium oxide). Thus, there are pores with a diameter of 10–20 μ as well as there are a large number of pores with a diameter of 0.1–5 μ. Finally, the porous structure of chromium leather is formed at the tanning agent amount of 2 % per weight of a semi-finished item.

Fig. 12.5 Integral curves of
the pore volume distribution
on radii for the dermis
after chrome tanning (*1*),
retanning (*2*), and filling with
bentonite (*3*)

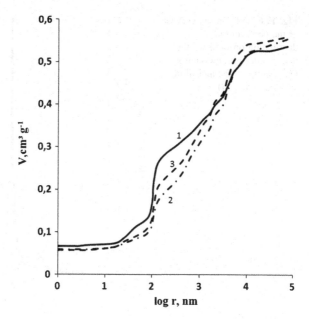

The integral and differential porosimetric curves measured by the MSCP
(Figs. 12.5 and 12.6) allowed to analyze changes in the porous structure of the
dermis after chrome tanning, retanning, and filling with bentonite.

In [9], it was found that all samples of dermis have very high values of volume
porosity (54–60 %) and specific surface area (193–216 m^2/g). In this case,
according to the authors, the pores formed by various elements of the structure
cause certain functional properties of leather. Meso- and micropores have a significant impact on the physical and mechanical properties of the finished leather, in
which the capillary condensation of vapor is going on. This ensures the presence of
adsorbed moisture and protects the dermis from cracking and violations of elasticity. The regularity of the structure of semi-finished product at the fibrils level,
lack of tortuosity and dead-end pores, indicate a speed transport of moisture in
these pores. For pores, which are formed by smaller elements of the structure,
regularity is not typical. Such pores provide for accumulation of moisture. The role
of macro pores is reduced to the heat and mass transfer with the environment,
which are intensified by increasing the porosity at the level of primary fibers. Due
to the elasticity of the dermis, the macropore volume is increased at high temperatures and decreased at low temperatures. For living organisms, such self-
regulation provides protection against adverse weather conditions.

Chrome-free tanning methods, at a rate of aldehyde compounds 9 % by the pelt
weight, demonstrate complete formation of the porous structure of the dermis having
major pore diameters of 5–20 and 0.5–2 μ. Increased consumption of tanning agent
up to 12 % is not accompanied by significant changes in the pore structure, with
the exception of reducing the fraction of pores with a diameter of 20–100 μ.

Fig. 12.6 Differential curves
of the pore volume
distribution on radii for the
dermis after chrome tanning
(*1*), retanning (*2*), and filling
with bentonite (*3*)

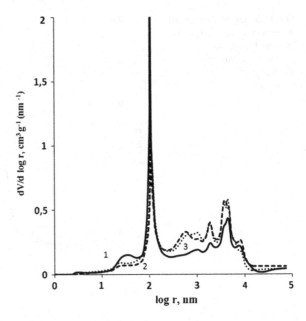

A semi-finished item of chrome-free tanning can be regarded as a coarse-grained. The authors attribute the appearance of the defect of openness of such leather with this. Keeping in mind the characteristics of the porous structure of the chromium-free tanning leather, the authors also point out the need to adjust the following methods of semi-finished processing to produce elastic and soft leather [12]. Besides that, the aldehyde-tanned leather is less thick. According to the physical and mechanical properties of the glutaraldehyde-tanning leather is characterized by high perspiration-proof, fastness to lime, softness, and elasticity.

The vegetable-tanned leather, as found by Y.L. Kavkazov, has a higher number of narrow pores, in which capillary condensation of vapor is possible in comparison with the chrome-tanned leather [5]. The presence of a large amount of the condensed moisture inside capillaries greatly affects the mechanical properties of the dermis. This is because the water, filling the pores of the respective structural elements, serves as a flexibilizer [9]. Not only specific chemical interactions with the active groups of the collagen are characterized for tannins, but also the physical deposition in the form of filler inside structural interstices of the porous structure of the dermis. This phenomenon gives a number of valuable properties to the finished leather: increased thickness, flexibility, but at the present time purely vegetable-tanning is not applied due to the low hydrothermal resistance and fastness to wear.

Taking the advantages of each individual type of tanning into account, it is expedient to combine several ways of tanning in order to obtain high-quality leather in accordance with the standards. The use of chromium tanning provides leathers with a high fastness to wear, hydrothermal resistance; the use of tannins gives flexibility, thickness; zirconium tannins—density, fastness; aldehydes—softness and perspiration-proof. In this regard, in practice of the leather manufacturing, chromium–vegetable, chromium–syntan–vegetable, and chromium-titanium-syntan tanning, etc., were extended.

As a result of the tanning, the average inclination angle of bundles of dry leather to its surface increases. Studies of [13] show the relationship between the angle of the fiber bundles inclination to the surface and the leather shrinking over surface area. If the fibers of a semi-finished item are uniformly intertwined, then the increasing of the inclination angle of 4° gives the shrinkage of about 8 % [14]. Reducing the fiber length of 5.9… 6.4 % at the average angle 38° of inclination diminishes the area of semi-finished products by 10 %. Changes in the microstructure cause the formation of the dermis volume, whose quantitative characteristic is the change in porosity or in the ratio of the pore volume to the total volume of the leather.

An important role in forming of the leather properties belongs to finishing, during which the final porous structure of finished leather is produced and fixed [15]. Significant decrease in the number of large pores is observed after the finishing process (retanning, filling, fat-liquoring, dyeing, etc.). According to [14], if during the tanning process smaller pores with a diameter of 0.6–0.8 μ were fixed, then, while the filling, larger pores with the pore diameter more than one micron are filled. The distribution function of the pore diameter moves toward the micropore side that equalizes the porosity of leather along topographic areas. Also the pore distribution of the belly and butt zones, is observed for the leather of combined-tanning with the use of mineral and vegetable-tanning agents.

Vegetable retanning reduces a porosity of the microstructure [9]. This is due to crosslinking of macromolecules and a corresponding decrease in the thickness of microfibrils, which, in general, causes a disordering of the structure of collagen on the level of fibrils. Thus, there is an increase in the thickness of the primary fibers and macroporosity, thereby increasing the thickness of dermis, and an increase of volumetric yield. Retannage also leads to a reduction in the volume of mesopores, which determines a moisture transport, and a simultaneous increase in the volume of pores that provide the required moisture content. Besides that, the volume of macropores increases, providing a heat and mass interchange with the environment. These structural changes have an impact on the deformation and mechanical properties of finished leathers.

Mineral content promotes adsorption of filler, advantageously in the macropores. The increase in the level of macro-porosity of the primary fibers causes the ordering of the structure of the dermis, promotes a quality formation of the dermis and improves the relaxation properties of the leather. Complex effect of filling significantly reduces the stiffness of the structure and the resulting leathers are characterized by softness and elasticity.

Subsequent fat-liquoring, dyeing of semi-finished item reduces the porosity of collagen structure, which negatively emerges hygienic properties of the finished leathers [10]. Blockage of the through pores with a covering film in the finishing of leather grain leads to a reduction of air permeability.

The open porosity of the structure of collagen, which mainly determines the specific surface area of the leather, depends on the method of processing and tanning manner [3].

Indicators of open porosity determine a complex of hygienic properties of the finished leather [16]. A type of leather, which provides contact with the human body, such as foot, should supply sorption of moisture (to 8 g/h), which is allocated by foot legs, and it's desorption. Closed porosity determines, to a greater extent, thermal characteristics of the leather.

Open and closed porosity totally form such technological characteristics of the finished leather as the volume yield. A wide range of this indicator (150... 400 cm^3 per 100 g of collagen [15]) shows the influence of the level of the dermis structure, where the interaction of tannins with collagen occurs. The volume yield is also determined by the method of tanning and sizes of tanning particles. In this, the level of fixation of the dermis structure by tanning agent is determined, to a greater extent, by the index of the yield of leather area.

Porosity of the finished leather determines uniformity of its thickness and mechanical properties [17]. A large porosity of macrostructure is typical for a belly as compared to a butt zone. Large porosity of a belly indicates less packing density of the structural elements of collagen, which appears to reduce the objective tensile strength and increased elongation in comparison with the butt portion of the leather. Relation between the tensile strength and elongation in both longitudinal and transverse directions is often unequal values that indicates and confirms the level of orientation of interlocking elements of fibrous structure of the dermis. This causes a smaller increase in thickness in the peripheral areas of the leather. Uneven thickness of the skin increases with its area.

Fat-liquoring of semi-finished leather and filling it with polymer dispersions increases the uniformity of the thickness on the area and improve the physical and mechanical properties of finished leather. At the same time, orientation and fixation of fibrous collagen structure are aligned [14].

The final formation of the porous structure determines the consumer properties of the leather, including indicators of comfort, softness or hardness, weight, hygiene, and thermal insulation properties (thermal conductivity and total thermal effect) [15]. Thermal conductivity of dry chrome leather for shoe upper is 0.051–0.156 W/m × °C, and increases after fat-liquoring and filling.

In general, changes in the porous structure of the dermis can serve as a measure of purposeful formation of the functional properties of finished leather of a wide assortment during processing treatments and getting the objective laws of transformations between the capillary-porous structure and properties of the dermis of the leather.

12.3 Conclusion

The features of organization and the principles of formation of the porous structure of the dermis were investigated during main technological stages of the leather manufacturing. It is shown that the native collagen includes elements and pores of different sizes that range from 1 nm to 200 μ forming micro-, meso- and macrolevel structures. Hides and skins of all mammals are characterized by identical hierarchical organization of collagen dermis and almost the same chemical composition. However, hides and skins of various mammals have different porosity, average angle of inclination of collagen bundles to the derma surface and packing density of the structural elements. In the process of the leather production, the hierarchical structure of collagen undergoes structural transformation, which is due to the influence of various chemical materials. As a result, the finished leather is characterized by a complex structure, high porosity and a large surface area. Special role of non-collagenous, hydrophilic, and hydrophobic inclusions, which not only accumulate moisture and provide uniformity of its distribution, but also clog the space between the dermis structural elements, was established in the porosity formation of native hide and skin by the method of standard contact porosimetry (MSCP).

After the chemical treatment, there are significant changes in porous structure of pelts. At that, changes of the dermis macroporosity are associated mainly with pulping structure under the influence of chemicals at the level of primary fibrils and fibers. Changes in the dermis microporosity of pelts at macromolecular and microfibrillar levels are mostly related to the removal of noncollagenous inclusions. Tanning processes lead to changes at the micro- and meso-levels, providing the necessary moisture. Finishing processes help to order the porous structure at the level of fibrils. Accordingly, transport of moisture is intensified trough these pores and, as a consequence, it improves elasticity. As a result of the above changes, the redistribution of porosity and ordering the dermis structure at all levels of the hierarchical organization of collagen occur.

This promotes formation of the structure of the dermis with certain porosity and provides the ready-made leather with a number of necessary operational, hygiene, and consumer properties. The results of our investigation demonstrate the ability to control changes in the porous structure of the dermis, and accordingly, consumer properties of a wide assortment of leathers during processing treatments.

References

1. Heidermann E (1995) Das Leder 6:149
2. Dubinin MM (1982) Russ Chem Rev 7(51):1065 (in Russian)
3. Chursin VI (1994) Leather Shoes Ind 5–8:28 (in Russian)
4. Mikhailov AN (ed) (1980) Chemistry and physics of the collagen of skin integument. Light Industry, Moscow, p 232 (in Russian)
5. Mikhailov AN (ed) (1953) Chemistry of tanning agent and tanning process. Gizlegprom, Moscow, p 794 (in Russian)
6. Reich G (ed) (1969) Collagen. Light industry, Moscow, p 327 (in Russian)
7. Zeldina AE, Kondratikov EF, Zurabian KM, Kytiin VA (1974) Technologies of the light industry In: Proceedings of higher education, vol 3, p 60
8. Volfkovich YuM, Bagotzky VS (1994) J Power Sources 48:327
9. Mokrousova OR, Volfkovich YuM, Nikolskaya NF (2010) Leather Shoes Manufact 6:19 (in Russian)

10. Lishchuk VI, Danylkovich AG, Zhugodskiy AG (2005) Light Ind 4:51 (in Ukrainian)
11. Lishchuk VI, Danylkovich AG (2005) Light Ind 2:51 (in Ukrainian)
12. Chursin VI (1995) Leather Shoes Ind 2:36 (in Russian)
13. Dumnov VS, Truhin OY (1977) Leather Shoes Ind 9:48 (in Russian)
14. Zakharenko VA, Pavlin VA (1973) Leather Shoes Ind 11:53 (in Russian)
15. Strakhov IP (ed) (1985) Chemistry and technology of leather and fur. Legprombutizdat, Moscow, p. 496 (in Russian)
16. Zubin YP, Avilov, Gvozdev YM, Chernov NV (1968) Materials science of leather. Light Industry, Moscow, p 383 (in Russian)
17. Kuprianov MP, Elen BL, Morekhodov GA (1975) Leather Heterogeneity. Light Industry, Moscow, p 166 (in Russian)

Part IV
Mathematical Modeling of Filtration Processes in Porous Media

Part IV
Mathematical Modeling of Filtration
Processes in Porous Media

Chapter 13
Mathematical Modeling of Filtration Processes in Porous Media

Anatoly Nikolaevich Filippov

Abstract The current variety of filtration theories suggests that the porous medium and the fluid filling it form some continuous medium. That is, the elements of the porous medium-fluid system, though considered physically infinitesimal, however, are large enough compared to the size of pores and particles (grains, fibers) forming a porous medium. Averaged characteristics of the porous medium, which are introduced for mathematical description, may be sufficiently substantiated only for the volume with a large number of enclosed pores and particles. In terms of the elementary theory of filtration, the meaning of the solid skeleton of the porous medium is, above all, geometric—the skeleton limits the region of space in which the fluid moves. In more complex cases, we have seen strong interaction between the skeleton and the adjacent layers of the fluid. Therefore, the properties of the porous medium in the theory of filtration are usually described by a set of geometric averages. In this chapter, several cell models are considered for calculation of hydrodynamic permeability of porous media. It is assumed that porous media in general may consist of partially porous spherical or cylindrical particles. Different limiting cases are investigated and theoretical results are compared with experimental data.

13.1 Darcy's Filtration Law (1856)

In the theory of filtration the chosen element of the porous medium is characterized by the *average porosity* ε, which is equal to the ratio of $\tilde{V}_p$—the volume occupied by pores to the total volume $\tilde{V}$ of this element:

$$\varepsilon = \tilde{V}_p/\tilde{V}, \tag{13.1}$$

where tilde $\sim$ over a symbol marks dimensional values. The limit of (13.1) is regarded as local or point porosity in a selected physical point under the contraction of the allocated volume at this point. Note, that in this limiting transfer case,

Y. M. Volfkovich et al., *Structural Properties of Porous Materials and Powders Used in Different Fields of Science and Technology*, Engineering Materials and Processes, DOI: 10.1007/978-1-4471-6377-0_13, © Springer-Verlag London 2014

linear dimensions of the element must be large (macroscopic), compared to a microscopic scale of the porous medium (pore size or constituent grains). The described situation has analogs in other branches of continuum mechanics. For example, in determining the local density of the gas, the characteristic linear dimension of the volume is always chosen much larger than the average free path of the gas molecules.

Usually one distinguishes between the *total porosity*, i.e., taking into account all the pores and the *active or true porosity*, taking into account only the pores, which are connected together in a single system, i.e., can be filled by an external fluid. For the purpose of filtering only the active porosity is significant, so in what follows we shall consider only the active porosity. Note, however, that in a classic work by Happel and Brenner [1] the volume fraction of voids is termed *fenestration*, related only to the percentage of voids between the particles forming the porous medium through which the fluid moves (equivalent to the active porosity). Meanwhile, the same authors define the porosity as the volume fraction of the pore space within the particles of the porous material (equivalent to the total porosity).

Along with the porosity ε, sometimes the notion of *translucence*, as the ratio of the area of active pores in any section, passing through a given point, to the entire cross-sectional area is introduced. Under the assumptions above, translucence at this point does not depend on the direction of the cross-section and is the same as the porosity ε [2].

The porosity value is the same for geometrically similar media and does not characterize the pore size. Therefore to describe the porous medium, we also have to specify the characteristic size $\tilde{d}$ of the pore space. There are many equivalent ways to define this size. For example, the characteristic dimension is taken as some average value of the radius of the pore channels or individual grains of the porous skeleton (understood as the average values of the corresponding random variables).

It is clear that the curves of pore or grain size distribution contain much more information about the microstructure of the porous medium than the average of its geometrical parameters. Therefore, there are numerous attempts to determine all of the geometric and hydrodynamic characteristics of the porous medium on the basis of the distribution curves. This is especially widespread for different types of membranes—thin porous structures which are able, for various reasons, to pass through it some substances and reject others when an external field (pressure, concentration, electric potential drop, etc.) is applied. However, the dependence of the hydrodynamic characteristics of the porous medium on the parameters of the distribution curves cannot be universal, that is, the same for different species. Indeed, introducing thin impermeable film can radically change the hydrodynamic characteristics of the medium, without changing or only slightly changing the form of the distribution curves. At the same time, for dissimilar processes different statistical characteristics of pore sizes and grains are important. Thus, for the transport processes in porous media, the degree of heterogeneity of the porous medium components (like pores and grains or fibers) is essential. In this case, along with an average size, its variance, which characterizes the degree of deviation from the average value, is also significant.

The main characteristic of the filtration flow is the filtration velocity vector $\tilde{\mathbf{U}}$, which is defined as follows. Let the fluid mass $\Delta\tilde{m}$ run at a unit of time through an arbitrary area element $\Delta\tilde{S}$ with the unit normal vector $\mathbf{n}$, being at a certain point of the porous medium. Then the projection of the vector $\tilde{\mathbf{U}}$ onto the vector $\mathbf{n}$ is the limit of the following ratio,

$$\tilde{\mathbf{U}} \cdot \mathbf{n} = \lim_{\Delta\tilde{S} \to 0} \frac{\Delta\tilde{m}}{\tilde{\rho}\Delta\tilde{S}}, \tag{13.2}$$

where $\tilde{\rho}$—density of the liquid. We emphasize that the limit is understood in the sense described above, and that the mass of the fluid is divided by the total area $\Delta\tilde{S}$ (not only by the part occupied by the pores).

The basic relation of filtration (the filtration law) relates the filtration velocity vector and the pressure field, which causes flow of a fluid. Here and below, the pressure means the difference between the total and the hydrostatic pressure. When the fluid does not move, pressure in the pores is distributed according to the hydrostatic law. As soon as motion of the liquid starts, the excess (above hydrostatic) pressure becomes variable in space, occupied by a porous medium and the liquid. Fluid flow in a porous media is different from the flows discussed in ordinary hydrodynamics, so that in any macroscopic volume there is a fixed solid phase, on the boundary of which the liquid is also immobile, due to the standard conditions of adhesion. Therefore, the system of pore channels in the elementary macroscopic volume is hydrodynamically equivalent to the system of closely related pipes or a tube of variable section. Filtration rate describes the flow rate through the system. On the other hand, the flow is determined by the pressure on the ins and outs of the pore channels. Since the flow rate is an integrable value over many pore channels, it is determined by the averaged pressure gradient of the liquid. That is why, in contrast to the usual equations of hydrodynamics, there is a local relationship between the pressure gradient and the vector velocity in the theory of filtration. Some of the information on the form of filtration law relating the filtration velocity and the pressure gradient can be obtained on the basis of the most common representations of the dimension theory. The porous medium is described by geometrical parameters: the characteristic size $\tilde{d}$ and some dimensionless quantities, e.g., porosity ε, parameters of the distribution curve, and others. The law of the fluid filtration should follow from the equations of the fluid motion in the pore space, so the system of governing quantities includes the characteristics of liquid that are included in those equations: density $\tilde{\rho}$ and viscosity $\tilde{\mu}^o$ of a clear liquid. Thus, we are looking for a form of dependence of the pressure gradient, $\mathbf{grad}\tilde{p}$, on the filtering velocity $\tilde{\mathbf{U}}$, the geometric characteristics of the porous medium ε, $\tilde{d}$, etc. and the physical characteristics of the liquid, i.e., $\tilde{\rho}$ and $\tilde{\mu}^o$. Among the variables that affect $\mathbf{grad}\tilde{p}$, only the filtration velocity $\tilde{\mathbf{U}}$ is a vector.

Therefore, if the porous medium is isotropic, the vector $\mathbf{grad}\tilde{p}$ should be collinear with the velocity vector $\tilde{\mathbf{U}}$, so that we have

$$\mathbf{grad}\tilde{p} = -\tilde{C}\tilde{\mathbf{U}} \tag{13.3}$$

where $\tilde{C}$ is a scalar quantity that depends on the modulus of the velocity $\tilde{\mathbf{U}}$, and the quantities $\tilde{d}$, ε, $\tilde{\rho}$, $\tilde{\mu}^o$. Let us consider the filtration flows when the inertial forces can be neglected. Most of the seepage flows encountered in practice belong to noninertial motions, because they are slow. The density $\tilde{\rho}$, which characterizes the inertial properties of the liquid, is excluded from the governing parameters. Thus, when we have noninertial motion, the value of $\tilde{C}$ depends only on $\tilde{\mathbf{U}}$, $\tilde{d}$, ε and $\tilde{\mu}^o$. It is easy to see that only three $(\tilde{U}, \tilde{d}, \tilde{\mu}^o)$ from the five governing parameters $(\tilde{C}, \tilde{U}, \tilde{d}, \varepsilon, \tilde{\mu}^o)$ have independent dimensions. Then, according to the theory of dimensions, the dimensionless combination $\tilde{C}\tilde{d}^2/\tilde{\mu}^o$ can depend solitary on porosity ε [2] which is the only dimensionless parameter among the governing parameters:

$$\frac{\tilde{C}\tilde{d}^2}{\tilde{\mu}^o} = f(\varepsilon). \tag{13.4}$$

Using (13.4), Eq. (13.3) can be written as Darcy filtration law (named after the French engineer Henri Darcy, who established it experimentally in 1856 in the study of water filtration through layers of fine sand):

$$\tilde{\mathbf{U}} = -\frac{\tilde{k}_D}{\tilde{\mu}^o}\mathbf{grad}\tilde{p}, \tag{13.5}$$

where the value $\tilde{k}_D = \tilde{d}^2/f(\varepsilon)$, called *specific permeability*, has the dimension of area (m^2), does not depend on the fluid properties and is a purely geometric characteristic of the porous medium. Darcy's law is often used in the study of all types of flow through porous media: inflow to ground wells, the flow of water in irrigated soil, filtering through the dam, the penetration of the solvent through the artificial membrane in reverse osmosis, nano, ultra, microfiltration, etc. In addition, the flow of oil in underground reservoirs is also subject to Darcy's law, and now the unit of permeability, named darcy (D), is commonly used in the oil industry. A porous medium (rock) with a permeability of 1 darcy permits a flow of 1 cm³/s of a fluid with viscosity 1 cP (centipoise) under a pressure gradient of 1 atm/cm acting across an area of 1 cm² (1D $\sim$ 1 μm²). Typical values of permeability range from as high as 100,000 darcys for gravel, to less than 0.01 microdarcy for granite. Sand has a permeability of approximately 1 darcy.

Note, that if instead of $\tilde{p}$, the true pressure is considered in the fluid $\tilde{P} = \tilde{p} - \tilde{\rho}\tilde{g}\tilde{z}$, where $\tilde{g}$—acceleration of gravity, $\tilde{z}$—height of the point over a settlement level, (13.5) can be written as

$$\tilde{\mathbf{U}} = -\frac{\tilde{k}_D}{\tilde{\mu}^o}\mathbf{grad}\left(\tilde{P} + \tilde{\rho}\tilde{g}\tilde{z}\right). \tag{13.6}$$

In calculation of hydraulic pressure the magnitude $\tilde{H} = \tilde{p}/\tilde{\rho}\tilde{g}$ is commonly used, so from (13.5) we have,

$$\tilde{\mathbf{U}} = -\tilde{c}\cdot\mathbf{grad}\tilde{H}, \quad \tilde{c} = \frac{\tilde{k}_D\tilde{\rho}\tilde{g}}{\tilde{\mu}^o}, \tag{13.7}$$

where $\tilde{c}$—filtration coefficient, which has the dimension of velocity. As seen from the derivation here, Darcy's law is a consequence of the assumption that fluid motion is noninertial. Filtration flow obeying the Darcy law is a special case of a "creeping" flow, which is characterized by the predominance of viscous forces over inertial forces (very low Reynolds numbers, Re $\ll$ 1). Therefore, attempts to derive the Darcy law by averaging the hydrodynamic equations are reduced to the calculation of permeability given by the geometrical structure of the porous medium. The paper [3] should be, therefore, mentioned in which Darcy's equation has been theoretically derived based on the method of volume averaging.

13.2 Brinkman Equation (1947)

Despite its simplicity and ease of use for calculations, Darcy's equation has a drawback. In solving problems when the porous medium is in contact with the pure liquid, a difficulty to correctly formulate the boundary conditions arises. Since Darcy's equation is of the first order, and the Stokes equation—of the second, then the pairing of these equations at the interface requires formulation of special conditions, which in itself is a difficult task. There are two ways to avoid these difficulties:

(i) Taking into account that the velocity of the fluid in the porous medium is much less than the velocity of the free (pure) fluid, we believe that the former does not affect the second. Therefore, the velocity and the pressure fields in the free fluid may be obtained from the solution of the Stokes equation with the so-called *adherence condition* [4] on a porous surface, which, in essence, is a no-slip condition (zero velocity at the boundary). Then the velocity and the pressure in a porous medium can be found from the equations of Darcy and the mass balance simultaneously with the pressure on its boundary obtained from the solution of the Stokes equation.

(ii) Application of the Brinkman equation, which also has the second order, for the fluid flow in a porous media [5] using standard conditions of continuity of the velocity vector and the stress tensor on the boundary of a porous medium with a clear liquid.

The Brinkman equation in the vector form is given in [5]:

$$\tilde{\mu}^i \Delta \tilde{\mathbf{U}} - \frac{\tilde{\mu}^o}{\tilde{k}} \tilde{\mathbf{U}} = \mathbf{grad}\tilde{p}, \qquad (13.8)$$

where $\tilde{k}$—permeability of the porous Brinkman medium, $\tilde{\mu}^o$—pure liquid viscosity again, $\tilde{\mu}^i$—viscosity of the liquid in the pores (the effective viscosity of the Brinkman medium). As can be seen, the Brinkman Eq. (13.8) differs from Darcy's Eq. (13.6) by the presence of additional viscous term, which takes into account the effect of the porous skeleton to the fluid flow. In Ref. [6] it was shown that the effective viscosity $\tilde{\mu}^i$ of the Brinkman medium, filled with the same liquid as the medium itself, is always higher than the viscosity $\tilde{\mu}^o$ of the pure liquid. However, if the mixture of vapor/air and liquid to be filtered are located near the pore walls, the effective viscosity of the Brinkman medium will be reduced compared to the viscosity of the pure liquid [7]. Thus, in general, the viscosities $\tilde{\mu}^o$ and $\tilde{\mu}^i$ are different.

It should be noted that the Brinkman equation is often used for modeling flow in porous grains which form a porous medium. Moreover, hydrodynamic permeability $\tilde{L}_{11}$, representing the first diagonal element of the Onsager kinetic coefficients matrix, and associated in the case of a porous isotropic medium with a specific permeability by the formula

$$\tilde{L}_{11} = \frac{\tilde{k}_D}{\tilde{\mu}^o}, \qquad (13.9)$$

can be calculated for such a medium by the cell method for example. If the medium is not isotropic for viscous flow in it, instead of (13.5), the generalized Darcy's law, to find i-th component of the filtration velocity, is used

$$\tilde{U}_i = -\frac{\tilde{k}_{ij}}{\tilde{\mu}^o} \cdot \frac{\partial \tilde{p}}{\partial \tilde{x}_j}, \qquad (13.10)$$

where $\tilde{k}_{ij}$ is a tensor of permeability coefficients, $\tilde{x}_j$—generalized coordinates. The following sections of this chapter presents various cell methods for calculating $\tilde{k}_D$—the specific permeability coefficient (in the case of an isotropic medium consisting of partially or completely porous spherical particles or partially or fully porous cylinders randomly distributed in the space), as well as the principal values of the tensor components $\tilde{k}_{ij}$ (in the case of transversal isotropic porous medium composed of a partially or completely porous cylinders arranged in parallel to their axes).

13.3 Modeling Hydrodynamic Permeability of Porous Media by the Cell Method

Most particles in the nature do not have a smooth homogeneous surface but rather have a rough surface or a surface covered with a porous shell. The roughness of the surface can be modeled by a thin porous shell. Investigations of flow in concentrated disperse systems, built up by porous particles or particles covered with a porous shell, is important for both natural and industrial processes. The most important examples of such processes are flows through a layer built up by porous particles, flows in sand beds, flows through porous membranes under applied pressure difference, those in petroleum reservoir rocks, in soils and sands, slurries, underground flows of oil and water, sedimentation, etc. [8, 9]. Such flows can be effectively modeled using the cell model suggested by Happel and Brenner [1] and afterward developed by others. The cell model is one of the most effective ones for investigating the behavior of concentrated disperse systems and porous media. The cell method has been successfully used for investigating electrokinetic phenomena in concentrated dispersions [10]. According to the cell method, investigation of a flow in a system of chaotically distributed in space dispersed particles of a concentrated dispersion is replaced by a consideration of a flow inside an array of identical spherical (or cylindrical) cells with spherical (or cylindrical) particles in the centre. The size of each cell is selected in such a way that the porosity inside each cell is equal to the porosity of the initial dispersion. The boundary conditions on the cell surfaces are to mimic the conditions in the real dispersion. Spherical cells are suitable for modeling porous layers (membranes) having globular-cellular structure while cylindrical cells can be used for modeling porous layers with a fibrous-capillary frame.

Boundary conditions on cell surface, which have to take into account the influence of the neighboring particles, have to be discussed in more detail according to the cell models. It is assumed that each particle is surrounded by a fluid cell and all the disturbances due to each particle are transmitted to the surrounding particles only through the boundary conditions on the surface of the cell. The fluid part of each cell is assumed to be the same volumetric proportion of fluid to solid as the one that exists in the entire dispersed media. In the case of a spherical particle and a cell, the cell radius is calculated in the following way: the volume fraction of the particles in the cell, γ^3, should be equal to the volume fraction in the real porous medium (membrane) (Fig. 13.1):

$$1 - \varepsilon = \gamma^3 = \left(\frac{\tilde{a}}{\tilde{b}}\right)^3,$$
(13.11a)

where ε is an outer porosity of the membrane, and $\tilde{a} = \tilde{R} + \tilde{\delta}$ is the total radius of the particle. In the case of a cylindrical particle inside a cylindrical cell the relation (13.1) has to be changed as follows (Fig. 13.1 is the same for spherical and cylindrical particles):

$$1 - \varepsilon = \gamma^2 = \left(\frac{\tilde{a}}{\tilde{b}}\right)^2. \qquad (13.11b)$$

Equations (13.11a) and (13.11b) determine the unknown radius of the spherical and cylindrical cell as $\tilde{b} = \tilde{a}\big/(1 - \varepsilon)^{1/3}$ and $\tilde{b} = \tilde{a}\big/(1 - \varepsilon)^{1/2}$ correspondingly.

The presence of porous shells on solid surfaces substantially modifies both hydrodynamic [11–13] and electrokinetic [14–18] phenomena. Hydrodynamic interaction of two particles covered by a porous shell is substantially different from that of nonporous particles [16–18]. From now on, we refer to any dispersed or porous medium as "a membrane" for abbreviation. The presence of a porous shell on the particle surface introduces a new internal degree of freedom in the membrane performance. In the course of filtration of a liquid solution through the membrane the structure of the membrane can undergo a substantial change. The latter can be caused by either partial dissolution of the particle or fiber surfaces [14], which build the membrane, or by adsorption of polymers on the same surfaces ("poisoning") [19]. As a result, a porous shell (or gel-layer) which is usually difficult to remove [19] is formed on the particle's surface. In the case of partial dissolution of the particle surfaces, the membrane hydrodynamic permeability increases because of the increase of the total porosity, however, the selectivity decreases. In the case of polymer adsorption, just the opposite process occurs: the total porosity decreases and the hydrodynamic permeability decreases, while the selectivity increases as a rule.

The presence of porous shells on the particle surface always alters a hydrodynamic drag force exerted on the particle by the flowing liquid [11–13, 20–23]. The presence of a porous shell results in a modification of a diffusion coefficient inside the membrane [23, 24] and the overall hydrodynamic permeability [25]. Hydrodynamic effects caused by the presence of adsorbed polymers on the particle surface were considered in [19, 24, 26], where the thickness of polymer shells was assumed much smaller compared to the particle radius.

Special cases of the problem under consideration were investigated in [27–31]. The flow around and inside a completely porous particle was considered in [27, 28] using the cell model. The exact analytical expression was deduced for the hydrodynamic permeability of a membrane built up by such particles and various limiting and special cases were considered [27, 28]. Hydrodynamic permeability of particles covered by a porous shell was investigated in [29]. In Refs. [27–29, 31] the Mehta–Morse boundary condition was used. In Ref. [30] the problem was solved on a motion of an isolated particle covered by a porous shell in an unbounded liquid, viscosities of outer and inner liquid were assumed to be different. Note, that the cell method is now frequently used as the first step in a new version of the mean field approximation [6]. The paper [32] presents a review of applications of the cell method for investigations of hydrodynamic permeability of porous/dispersed media and membranes. Based on the cell method, hydrodynamic permeability of a porous layer/membrane built up by solid particles with a porous shell and nonporous

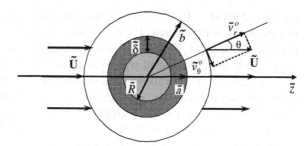

Fig. 13.1 A cell with a solid particle covered by a porous layer: $\tilde{U}, \tilde{a}, \tilde{b}, \tilde{\delta}$ are the velocity outside the cell, radius of the particle, radius of the cell, and the thickness of the porous layer, respectively [38, 39]

impermeable interior is calculated. Four known boundary conditions on the outer cell boundary are considered and compared: Happel's, Kuwabara, Kvashnin's, and Cunningham's (usually referred to as Mehta–Morse's condition). For description of a flow inside the porous shell Brinkman's equations are used.

The aim of this chapter is to generalize, in the framework of different cell models, the results concerning a flow around a spherical or cylindrical particle with a porous shell and to give exact formulas for hydrodynamic permeability of the porous media built up by such particles. Different boundary conditions on the outer surface of spherical or cylindrical cells are considered.

Let us introduce a spherical $(\tilde{r}, \theta, \varphi)$ or cylindrical $(\tilde{r}, \theta, z)$ coordinate system with the origin located in the particle centre (at the cylinder axis) and (additional) axis $\tilde{z}$, directed along a uniform flow with velocity $\tilde{\mathbf{U}}$ $(|\tilde{\mathbf{U}}| = \tilde{U})$ on the boundary of the cell (Fig. 13.1). The liquid flows at a low Reynolds number inside the cell but outside the porous shell $(\tilde{a} \leq \tilde{r} \leq \tilde{b})$ the flow is described by Stokes equations with the continuity condition:

$$\begin{cases} \tilde{\nabla}\tilde{p}^o = \tilde{\mu}^o \tilde{\Delta}\tilde{\mathbf{v}}^o \\ \tilde{\nabla} \cdot \tilde{\mathbf{v}}^o = 0, \end{cases} \tag{13.12}$$

Inside the porous shell $(\tilde{R} \leq \tilde{r} \leq \tilde{a})$ the flow is described by Brinkman's equations and the continuity condition [1]:

$$\begin{cases} \tilde{\nabla}\tilde{p}^i = \tilde{\mu}^i \tilde{\Delta}\tilde{\mathbf{v}}^i - \frac{\tilde{\mu}^o}{\tilde{k}} \tilde{\mathbf{v}}^i, \\ \tilde{\nabla} \cdot \tilde{\mathbf{v}}^i = 0, \end{cases} \tag{13.13}$$

where o, i are superscripts, which mark flow in the cell $(\tilde{a} \leq \tilde{r} \leq \tilde{b})$ and in the porous shell $(\tilde{R} \leq \tilde{r} \leq \tilde{a})$, respectively; $\tilde{\mu}^o$, $\tilde{\mu}^i$ are viscosities in the corresponding zones; $\tilde{p}^o, \tilde{p}^i$ and $\tilde{\mathbf{v}}^o, \tilde{\mathbf{v}}^i$ are pressures and velocity vectors in the corresponding zones; $\tilde{k}$ is the specific permeability of the porous shell. Brinkman's Eq. (13.13) means that the flow in a real porous shell is replaced by the flow of an effective liquid with the effective viscosity $\tilde{\mu}^i$ and the friction between the liquid and the porous skeleton is effectively represented by the friction force with the friction coefficient, $\frac{\tilde{\mu}^o}{\tilde{k}}$. The dependency of $\tilde{\mu}^i$ and $\frac{\tilde{\mu}^o}{\tilde{k}}$ on the local porosity of the shell were considered in [6]. In Ref. [22] the problem of a flow around a particle covered by a porous shell was considered. The flow inside the porous shell was described using Brinkman's equation. However, it was assumed that the effective viscosity of the liquid inside the porous shell was equal to the liquid viscosity outside the shell, $\tilde{\mu}^o = \tilde{\mu}^i$. This assumption allows for a substantial simplification of all calculations. It has been shown, however, that these viscosities are always different [6] and the inner effective viscosity is always higher than the real liquid viscosity. However, if the porosity inside the porous shell is very low the deviation of the effective viscosity from the viscosity of the real liquid tends to zero, hence, $\tilde{\mu}^i \to \tilde{\mu}^o$. That is, the solution presented in [22] though valid from the mathematical point of view, can be applied only in the case of low porosity of the shell. It is the reason why in [27–32] different viscosities inside the porous shell and in the surrounding liquid were used.

Note, Einstein equation results in

$$\tilde{\mu}^i/\tilde{\mu}^o = 1 + 5c/2, \tag{13.14}$$

where c is the volume fraction of solid material inside the shell. In Ref. [6], the following dependency for the effective viscosity was deduced:

$$\tilde{\mu}^i/\tilde{\mu}^o = \frac{1}{(1 - c/c^i)^{2.5A^i}}, \tag{13.15}$$

where A^i and c^i are determined by internal porosity and structure of the porous shell. The latter equation shows that the internal effective viscosity is always greater than the real liquid viscosity and tends to the real liquid viscosity, if the volume fraction of solid material inside the shell tends to zero. Taking that into account we consider below the case when viscosities inside the porous shell and in the liquid differ. We will discuss the consequences of this assumption. In order to finalize the boundary value problem for Eqs. (13.12)–(13.13), we should require appropriate boundary conditions. No-slip boundary conditions are imposed on the surface of the solid core of the particle:

$$\tilde{\mathbf{v}}^i = 0, \ \tilde{r} = \tilde{R}. \tag{13.16}$$

On the boundary between the porous shell and the liquid, $\tilde{r} = \tilde{a}$, we assume continuity of velocity, tangential $\tilde{\sigma}_{r\theta}$ and normal $\tilde{\sigma}_{rr}$ components of the viscous stress [17, 18, 33]:

$$\begin{cases} \tilde{\mathbf{v}}^o = \tilde{\mathbf{v}}^i \\ \tilde{\sigma}_{rr}^o = \tilde{\sigma}_{rr}^i \\ \tilde{\sigma}_{r\theta}^o = \tilde{\sigma}_{r\theta}^i \end{cases} \tag{13.17}$$

The physical background behind these boundary conditions (13.7) is discussed in [17, 18, 33].

Special attention is usually paid to the boundary conditions on the outer cell boundary, $\tilde{r} = \tilde{b}$. Four different types of boundary conditions on the cell surface have been suggested in the literature [1], which are referred to below as Happel's, Kuwabara's, Kvashnin's, and Mehta–Morse's (Cunningham's) models. All four models assume continuity of the radial component of the liquid velocity on the outer cell surface ($\tilde{r} = \tilde{b}$):

$$\tilde{v}_r^o = \tilde{U} \cos \theta. \tag{13.18}$$

Let us consider an additional condition used in each of the aforementioned models. According to Happel's model [1], the tangential viscous stress vanishes on the cell boundary ($\tilde{r} = \tilde{b}$):

$$\tilde{\sigma}_{r\theta}^o = 0. \tag{13.19a}$$

According to Kuwabara's model [34] the curl vanishes on the cell boundary ($\tilde{r} = \tilde{b}$), that is the flow is assumed to be a potential one:

$$\mathrm{rot}(\tilde{\mathbf{v}}^o) = 0. \tag{13.19b}$$

According to Kvashnin's model [35], a symmetry condition is introduced as follows:

$$\frac{\partial \tilde{v}_\theta^o}{\partial \tilde{r}} = 0, \tilde{r} = \tilde{b}. \tag{13.19c}$$

Mehta–Morse's model [36] assumes homogeneity of the flow at the cell boundary ($\tilde{r} = \tilde{b}$):

$$\tilde{v}_\theta^o = -\tilde{U} \sin \theta. \tag{13.19d}$$

The condition (13.19d) was proposed for the first time by Cunningham [37] in 1910. Later, Mehta and Morse [36] used Cunningham's boundary condition; however, they made a mistake in calculations and did not obtain the correct limiting case for permeability of the membrane composed of a set of completely impenetrable particles (see formula (13.52) below, which was first derived by Cunningham [37]). However, due to the known Happel and Brenner monograph [1], the names of Mehta and Morse were also affixed to the Cunningham condition. Here, we will use both these titles. There are no decisive arguments in the literature in favor of any of the four models. Even worse than that: in the case of a flow in a flat chamber (which is the limiting case of a cell of an infinite radius) all four mentioned boundary conditions are satisfied [1] in the centre of the chamber (which corresponds to the boundary of the cell). That is the reason why we consider and compare below all four models.

By using the following dimensionless variables and parameters,

$$
r = \frac{\tilde{r}}{\tilde{a}}, \quad z = \frac{\tilde{z}}{\tilde{a}}, \quad \nabla = \tilde{\nabla} \cdot \tilde{a}, \quad \Delta = \tilde{\Delta} \cdot \tilde{a}^2, \quad \delta = \frac{\tilde{\delta}}{\tilde{a}}, \quad R = \frac{\tilde{R}}{\tilde{a}} = 1 - \delta,
$$

$$
\frac{1}{\gamma} = \frac{\tilde{b}}{\tilde{a}}, \quad \mathbf{v} = \frac{\tilde{\mathbf{v}}}{\tilde{U}}; \quad p = \frac{\tilde{p}}{\tilde{p}_0}, \quad \tilde{p}_0 = \frac{\tilde{U} \cdot \tilde{\mu}^o}{\tilde{a}}, \quad m = \frac{\tilde{\mu}^i}{\tilde{\mu}^o}, \quad s_0 = \frac{\tilde{a}}{\tilde{R}_B}, \quad s = \frac{s_0}{\sqrt{m}},
$$

(13.20)

where $\tilde{R}_B = \sqrt{\tilde{k}}$ is the Brinkman length, which is a characteristic depth of penetration of the flow inside the porous shell, the system of governing equations (Eqs. (13.12) and (13.13)) in dimensionless form becomes:

$$
\begin{cases} \nabla p^o = \Delta \mathbf{v}^o, \\ \nabla \cdot \mathbf{v}^o = 0, \end{cases} \quad (1 \le r \le \frac{1}{\gamma}),
$$

(13.21)

$$
\begin{cases} \nabla p^i = m \Delta \mathbf{v}^i - s_0^2 \mathbf{v}^i, \\ \nabla \cdot \mathbf{v}^i = 0, \end{cases} \quad (R \le r \le 1).
$$

(13.22)

The liquid flow is axisymmetric in both (spherical and cylindrical) cases. Taking this into account the solution of the problem (13.21)–(13.22) was found as a first order spherical [32, 38],

$$
v_r^o = \left(\frac{b_1}{r^3} + \frac{b_2}{r} + b_3 + b_4 r^2 \right) \cos \theta,
$$

(13.23)

$$
v_\theta^o = \left(\frac{b_1}{2r^3} - \frac{b_2}{2r} - b_3 - 2b_4 r^2 \right) \sin \theta,
$$

(13.24)

$$
p^o = \left(\frac{b_2}{r^2} + 10 b_4 r \right) \cos \theta,
$$

(13.25)

$$
v_r^i = \left\{ \begin{array}{l} c_1 \left(\frac{\cosh(sr)}{r^2 s^2} - \frac{\sinh(sr)}{r^3 s^3} \right) + c_2 \left(\frac{\sinh(sr)}{r^2 s^2} - \frac{\cosh(sr)}{r^3 s^3} \right) \\ + \frac{c_3}{r^3} + c_4 \end{array} \right\} \cos \theta,
$$

(13.26)

$$
v_\theta^i = \left\{ \begin{array}{l} c_1 \left(\frac{\cosh(sr)}{2r^2 s^2} - \frac{\sinh(sr)}{2r^3 s^3} - \frac{\sinh(sr)}{2rs} \right) + \\ c_2 \left(\frac{\sinh(sr)}{2r^2 s^2} - \frac{\cosh(sr)}{2r^3 s^3} - \frac{\cosh(sr)}{2rs} \right) + \frac{c_3}{2r^3} - c_4 \end{array} \right\} \sin \theta,
$$

(13.27)

$$
p^i = ms^2 \left(\frac{c_3}{2r^2} - c_4 r \right) \cos \theta,
$$

(13.28)

or cylindrical harmonics [39]:

$$v_r^o = \left(\frac{b_1'}{r^2} + b_2' \ln r + b_3' r^2 + b_4' \right) \cos \theta, \tag{13.29}$$

$$v_\theta^o = \left(\frac{b_1'}{r^2} - b_2'(1 + \ln r) - 3 b_3' r^2 - b_4' \right) \sin \theta, \tag{13.30}$$

$$p^o = \left(-\frac{2 b_2'}{r} + 8 b_3' r \right) \cos \theta, \tag{13.31}$$

$$v_r^i = \left\{ c_1' - c_2' \frac{rs - 2 I_1(rs)}{rs^2} - \frac{c_3'}{r^2} - c_4' r \mathrm{Meij} \left(\frac{rs}{2} \right) \right\} \cos \theta, \tag{13.32}$$

$$v_\theta^i = \left\{ \begin{array}{l} -c_1' + c_2' \left(\frac{1 - I_0(rs) - I_2(rs)}{s} \right) - \frac{c_3'}{r^2} + \\ c_4' \left(\frac{8 K_0(rs)}{s} - rs \, \mathrm{Meij} \left(\frac{rs}{2} \right) \right) \end{array} \right\} \sin \theta \tag{13.33}$$

$$p^i = m \left\{ \begin{array}{l} -c_1' rs^2 + c_2' rs - c_3' \frac{s^2}{r} - \\ c_4' \left(8 K_1(rs) + r^2 s^2 \mathrm{Meij} \left(\frac{rs}{2} \right) \right) \end{array} \right\} \cos \theta, \tag{13.34}$$

where $\mathrm{Meij} \left(\frac{rs}{2} \right) = \frac{1}{4\pi i} \int \frac{\Gamma(1 + (1 - \xi)/2) \cdot \Gamma^2((-1 + \xi)/2)}{\Gamma(1 + (3 - \xi)/2)} \left(\frac{rs}{2} \right)^{-\xi} d\xi$ is the Major G-function, $\Gamma(x)$ is the Euler gamma-function, $I_i(rs)$, $K_i(rs)$ are the i-th order modified Bessel functions of the first and the second rank, respectively.

Upon substitution of expressions (13.23)–(13.28) and (13.29)–(13.24) into boundary conditions (13.16)–(13.19a, 13.19b, 13.19c, 13.19d) rewritten in dimensionless forms, we obtain a system of algebraic equations for determining the unknown constants b_j, c_j and b_j', c_j'. Depending on the model chosen, one of the conditions (13.19a, 13.19b, 13.19c, 13.19d) is used. In general, the solution of algebraic system is cumbersome and is not reported here.

The main value in which we are interested in, is the hydrodynamic drag force, $\tilde{F}$, exerted onto a spherical or cylindrical particle per unit length by the flowing liquid [1]. After substitution of the general expressions for the velocity components (13.23)–(13.28) or (13.29)–(13.24) we arrive, after integration, at the following expression:

$$\tilde{F} = \oiint (\tilde{\sigma}_{rr}^o \cos \theta - \tilde{\sigma}_{r\theta}^o \sin \theta) \mathrm{d}s = 4\pi \left\{ \begin{array}{c} -b_2 \\ b_2' \end{array} \right\} \tilde{a} \tilde{\mu}^o \tilde{U}$$

$$\equiv F \tilde{a} \tilde{\mu}^o \tilde{U} \tag{13.35}$$

where the integration is performed with respect to the external surface of the porous layer; F is the dimensionless force. Note that constants b_2 and b_2' have opposite signs. The hydrodynamic permeability of a membrane, $\tilde{L}_{11}$, which represents one of the elements of Onsager's matrix [40], is determined as the ratio of the cell flux of liquid to the cell pressure gradient [1] as shown below:

$$\tilde{L}_{11} = \frac{\tilde{U}}{\tilde{F}/\tilde{V}} \tag{13.36}$$

where $\tilde{V} = \frac{4}{3}\pi\tilde{b}^3$ is the cell volume for a spherical cell and $\tilde{V} = \pi\tilde{b}^2$—the cell volume for a cylindrical one. Substitution of expressions for the hydrodynamic drag force (13.35) into Eq. (13.36) allows to determine the hydrodynamic permeability as:

$$\tilde{L}_{11} = -\frac{1}{3b_2\gamma^3}\frac{\tilde{a}^2}{\tilde{\mu}^o} \equiv L_{11}\frac{\tilde{a}^2}{\tilde{\mu}^o} \quad \text{(for spheres)}, \tag{13.37}$$

$$\tilde{L}'_{11} = \frac{1}{4b'_2\gamma^2}\frac{\tilde{a}^2}{\tilde{\mu}^o} \equiv L'_{11}\frac{\tilde{a}^2}{\tilde{\mu}^o} \quad \text{(for cylinders)}, \tag{13.37'}$$

where

$$L_{11} = \frac{-1}{3\gamma^3 b_2}, \tag{13.38}$$

$$L'_{11} = \frac{1}{4\gamma^2 b'_2} \tag{13.38'}$$

are dimensionless hydrodynamic permeability of the membrane, consisting of a swarm of spheres and cylinders, respectively.

The hydrodynamic permeability, $L_{11}(\delta, \gamma, m, s_0)$, is a function of four parameters. Parameters δ and γ are geometrical characteristics of the particles and shells which characterize the fraction of a porous phase in the particle and in the cell; m and $s_0 = s\sqrt{m}$ are characteristics of the internal structure of the porous shell. In the general form, the expressions for the hydrodynamic permeability, L_{11}, are too lengthy and for this reason we do not present them here.

13.3.1 Hydrodynamic Permeability of a Porous Medium Modeled as a Swarm of Partially Porous Spherical Particles with Impermeable Core

The dependence of the dimensionless hydrodynamic permeability, L_{11}, of a membrane on γ is presented in Fig. 13.2 for all four models at $\delta = 0.5$; $s_0 = 8$; $m = 4$. The maximum value of $\gamma = 0.905$ is reached in the case of hexagonal packing of spheres, while for a simple cubic packing $\gamma = 0.806$.

Fig. 13.2 Variation of the dimensionless hydrodynamic permeability, L_{11}, of a membrane built up by solid spherical particles covered by a porous shell, with the parameter γ, that is, (volume fraction of particles)$^{1/3}$, at $m = 4$, $s_0 = 8$, $\delta = 0.5$ for the following models: *1* Happel, *2* Kuwabara, *3* Kvashnin, *4* Cunningham (Mehta–Morse) [38]

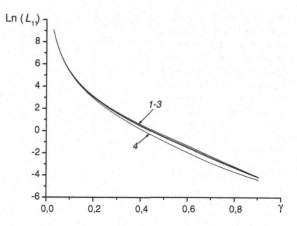

Figure 13.2 shows that as γ increases, that is, as the volume fraction of a solid phase increases, the hydrodynamic permeability of the membrane decreases. The rate of a decrease of L_{11} is higher at lower γ values (that is, low volume fraction of particles). As $\gamma \to 0$ the membrane hydrodynamic permeability increases unboundedly, however, as $\gamma \to 1$ the membrane hydrodynamic permeability tends to zero.

Figure 13.2 shows that the hydrodynamic permeabilities calculated for three of the models (Happel's, Kuwabara's, and Kvashnin's) almost coincide. Calculations based on Mehta–Morse's model result in a slightly lower hydrodynamic permeability at higher volume fractions when compared with the other three models. From a physical point of view, the Cunningham (Mehta–Morse) condition is the joining of a flow in the cell with a uniform flow at its boundary, which is the most "hard" condition out of the four considered types of boundary conditions. This condition should lead to an increased dissipation of energy in the system and, hence, to a more noticeable decrease in the hydrodynamic permeability of a membrane.

Figure 13.3 shows the variation of the dimensionless hydrodynamic permeability with the dimensionless thickness of the porous shell, δ for all four models at $\gamma = 0.3$; $s_0 = 5$; $m = 1$. All the calculated values are increasing functions of the thickness of the shell (as expected), as the more porous the particles the higher their hydrodynamic permeability. The calculated hydrodynamic permeability increases in the following direction: from the lowest for Mehta–Morse's model, through Kuwabara's, Kvashnin's model to the highest possible value for Happel's model. The parameter s_0 characterizes the depth of penetration of the flow inside the porous shell: the higher s_0 the lower the depth of penetration of the flow inside the porous shell. At high values of s_0, the flow only takes place in a thin shell (with the thickness roughly equal to Brinkman's length) rather than in the whole porous shell. That is, when the thickness of the shell becomes greater than Brinkman's length, the hydrodynamic permeability ceases to depend on the thickness of the porous shell. This is shown by the leveling off of the curves in Fig. 13.3.

Fig. 13.3 Variation of the dimensionless hydrodynamic permeability, L_{11}, of a membrane built up by solid spherical particles covered by a porous shell, with the dimensionless thickness of the porous shell, δ, at $m = 1$, $s_0 = 5$, $\gamma = 0.3$ for the following models: *1* Happel, *2* Kuwabara, *3* Kvashnin, *4* Cunningham (Mehta–Morse) [38]

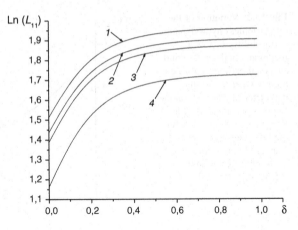

Fig. 13.4 Variation of the dimensionless hydrodynamic permeability, L_{11}, of a membrane built up by solid spherical particles covered by a porous shell, with the viscosity ratio m at $\gamma = 0.8$, $s_0 = 5$, $\delta = 0.5$ for the following models: *1* Happel, *2* Kuwabara, *3* Kvashnin, *4* Cunningham (Mehta–Morse) [38]

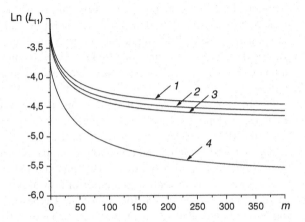

The influence of the viscosity ratio, m, on the dimensionless hydrodynamic permeability, L_{11}, is presented in Fig. 13.4 for all four models at $\gamma = 0.8$; $s_0 = 5$; $\delta = 0.5$. Figure 13.4 shows that an increase of the inner viscosity results in a decrease of the membrane hydrodynamic permeability, as expected. Figure 13.4 also shows that the hydrodynamic permeability decreases sharply in the beginning and then levels off when the inner viscosity inside the porous shell becomes sufficiently high, that is, when the flow inside the porous shell practically vanishes.

In Fig. 13.5 the dependence of the dimensionless hydrodynamic permeability, L_{11}, on the parameter s_0 is presented for all four models at $\gamma = 0.8$; $m = 1$; $\delta = 0.5$. As pointed out before, the dimensionless parameter s_0 characterizes the depth of penetration of the flow inside a porous shell. The latter depends on the porosity of the porous shell: the lower the porosity, the higher is the value of the parameter s_0.

Fig. 13.5 Variation of the dimensionless hydrodynamic permeability, L_{11}, of a membrane built up by solid spherical particles covered by a porous shell, with the parameter s_0 at $m = 1$, $\gamma = 0.8$, $\delta = 0.5$ for the following models: *1* Happel, *2* Kuwabara, *3* Kvashnin, *4* Cunningham (Mehta–Morse) [38]

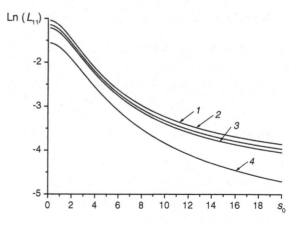

As $s_0 \to \infty$ the porous shell becomes completely impermeable. Figure 13.4 shows that a limiting value of the hydrodynamic permeability is reached as m increases; however, Fig. 13.5 does not show that a limiting value is reached as s_0 increases because at higher s_0 calculations become unstable. As in Fig. 13.4, the calculations, based on Mehta–Morse's model, result in lower values of the hydrodynamic permeability when compared to the other models.

Note, that the presented conditions of porous media (membranes) built up by spherical particles with a porous shell on their surfaces provide two additional degrees of freedom (namely, internal rigidity R is equal to the ratio of the radius of the rigid core to the radius of the entire particle and the viscosity ratio m inside the porous layer) to control the membrane performance as compared with the membranes built up by completely porous particles in [27–29], and three additional degrees of freedom (such as internal rigidity R, the resistance of porous layer to filtration, s_0, and the dimensionless viscosity of the liquid inside the porous layer, m) as compared with the consideration in [1, 34–37], where the hydrodynamic permeability of a membrane built up by nonporous spherical particles was considered. The new degree of freedom introduced here allows to describe a wider range of phenomena including the process of internal poisoning and/or dissolution of membranes on their hydrodynamic permeability in the course of filtration.

In the next paragraph, important limiting cases of the above theory are considered.

13.3.1.1 Completely Porous Spherical Particles

Completely porous particles correspond to the condition of $\delta = 1$. Expressions for the hydrodynamic permeability, $L_{11}(\delta, m, s)$, for the different models take the following forms:

Happel's model:

$$L_{11} = \{(2\gamma^6 - 3\gamma^5 + 3\gamma - 2)m^2s^2\omega_3 + 18(\gamma^5 - 1)(\omega_1 - 2\omega_2) - 3m[2s^2\gamma^6(\omega_1 - 2\omega_2)$$
$$+ \gamma^5(-\omega_1s^2 + 4\omega_2s^2 + 4\omega_1 - 8\omega_2 + 4\omega_3 + 10) - 2s^2\gamma(\omega_1 - 2\omega_2) + \omega_1s^2 - 4\omega_1 - 4\omega_2s^2$$
$$+ 8\omega_2 + \omega_3 - 10)]\}/\{-3\gamma^3ms^2[(-6\omega_1 + 12\omega_2)(\gamma^5 - 1) + m\omega_3(2\gamma^5 + 3)]\}.$$

$$(13.39)$$

Kuwabara's model:

$$L_{11} = \{2(\gamma^6 - 5\gamma^5 + 9\gamma - 5)m^2s^2\omega_3 - 90(\omega_1 - 2\omega_2) - 3m[2s^2\gamma^6(\omega_1 - 2\omega_2)$$
$$+ 5\gamma^3(\omega_1s^2 + 4\omega_1 - 8\omega_2 + 2\omega_3 + 10) - 12s^2\gamma(\omega_1 - 2\omega_2) + 5(\omega_1s^2 + 4\omega_2s^2 - 4\omega_1 + 8\omega_2$$
$$+ \omega_3 - 10)]\}/\{-45\gamma^3ms^2(2\omega_1 - 4\omega_2 + m\omega_3)\}.$$

$$(13.40)$$

Kvashnin's model:

$$L_{11} = \left\{m^2s^2\omega_3(\gamma - 1)^3(8\gamma^3 + 15\gamma^2 + 21\gamma + 16) + 18(3\gamma^5 - 8)(\omega_1 - 2\omega_2)\right.$$
$$- 3m[8s^2\gamma^6(\omega_1 - 2\omega_2) - 3\gamma^5(\omega_1s^2 - 4\omega_2s^2 - 4\omega_1 + 8\omega_2 - 4\omega_3 - 10)$$
$$+ 5\gamma^3(\omega_1s^2 + 4\omega_1 - 8\omega_2 + 2\omega_3 + 10) - 18s^2\gamma(\omega_1 - 2\omega_2) + 8(\omega_1s^2 - 4\omega_2s^2 - 4\omega_1$$
$$\left. + 8\omega_2 + \omega_3 - 10)]\}/\{-18\gamma^3ms^2[(8 - 3\gamma^5)\omega_1 + (6\gamma^5 - 16)\omega_2 + m\omega_3(\gamma^5 + 4)]\}.$$

$$(13.41)$$

and for Mehta–Morse's model:

$$L_{11} = \left\{m^2s^2\omega_3(\gamma - 1)^4(4\gamma^2 + 7\gamma + 4) + 18(3\gamma^5 + 2)(\omega_1 - 2\omega_2)\right.$$
$$- 3m[4s^2\gamma^6(\omega_1 - 2\omega_2) - 3\gamma^5(\omega_1s^2 - 4\omega_2s^2 - 4\omega_1 + 8\omega_2 - 4\omega_3 - 10)$$
$$- 5\gamma^3(\omega_1s^2 + 4\omega_1 - 8\omega_2 + 2\omega_3 + 10) + 6s^2\gamma(\omega_1 - 2\omega_2) - 2(\omega_1s^2 - 4\omega_2s^2 - 4\omega_1$$
$$\left. + 8\omega_2 + \omega_3 - 10)]\}/\{-18\gamma^3ms^2[-(3\gamma^5 + 2)\omega_1 + (6\gamma^5 + 4)\omega_2 + m\omega_3(\gamma^5 - 1)]\},$$

$$(13.42)$$

where ω_1, ω_2, ω_3 are

$$\omega_1 = 30\left(\frac{\cosh(s)}{s^4} - \frac{\sinh(s)}{s^5} - \frac{1}{3s^2}\right),$$

$$\omega_2 = -\frac{15}{2}\left(\frac{\cosh(s)}{s^4} - \frac{\sinh(s)}{s^5}(1 + s^2) + \frac{2}{3s^2}\right),$$

$$\omega_3 = -90\left(\frac{\cosh(s)}{s^4}\left(1 + \frac{s^2}{6}\right) - \frac{\sinh(s)}{s^5}\left(1 + \frac{s^2}{2}\right)\right).$$

$$(13.43)$$

Note, that in the limiting case when $s \to 0$, which corresponds to the absence of the solid phase in the particle, Eq. (13.43) result in

$$\lim_{s \to 0} \omega_1 = \lim_{s \to 0} \omega_2 = 1; \quad \lim_{s \to 0} \omega_3 - 3 \tag{13.44}$$

and the solution of the problem is transformed into the Hadamard-Rybczynski general solution for a liquid sphere [1]. In this case the hydrodynamic drag force disappears, which is as it should be in the case of complete mixing of liquids.

As we already noted before, the case $\tilde{\mu}^i < \tilde{\mu}^o$ is not realistic from the physical point of view [6]. However the case $\tilde{\mu}^i \ll \tilde{\mu}^o$ ($m \ll 1$, or $m \to 0$) which corresponds to a strongly reduced viscosity of a liquid inside the porous layer can be easily investigated from the mathematical point of view, using Eqs. (13.39)–(13.42). According to [7], this situation may arise in systems with superhydrophobic surfaces, systems containing air (vapor) between microfibers, or in presence of an air interlayer between solid and liquid surfaces. The latter limit gives the following expressions for hydrodynamic permeability:

Happel's model:

$$L_{11} = \frac{1}{3\gamma^3} - \frac{1}{3\gamma^2} + \frac{1}{\gamma^3 s_0^2}. \tag{13.45}$$

Kuwabara's model:

$$L_{11} = \frac{1}{3\gamma^3} - \frac{2}{5\gamma^2} + \frac{\gamma^3}{15} + \frac{1}{\gamma^3 s_0^2}. \tag{13.46}$$

Kvashnin's model:

$$L_{11} = \frac{1}{3\gamma^3} - \frac{4}{9\gamma^2} + \frac{5}{9\gamma^2(8 - 3\gamma^5)} + \frac{1}{\gamma^3 s_0^2}. \tag{13.47}$$

Cunningham's (Mehta–Morse's) model:

$$L_{11} = \frac{1}{3\gamma^3} - \frac{3 + 2\gamma^5}{\gamma^2(6 + 9\gamma^5)} + \frac{1}{\gamma^3 s_0^2}. \tag{13.48}$$

The variation of the dimensionless hydrodynamic permeability, L_{11}, with the parameter γ calculated according to Eqs. (13.39)–(13.42) (that is, for a completely porous particles) is presented in Fig. 13.6 at $m = 1$; $s_0 = 5$. For all models, the hydrodynamic permeability decreases along with volume fraction of particles. Note that the difference between different models in this special case of completely porous particles is less pronounced than in the general case of rigid spherical particles covered by a porous shell (Fig. 13.2).

Fig. 13.6 Variation of the dimensionless hydrodynamic permeability, L_{11}, of a membrane built up by completely porous spherical particles with the parameter γ, that is, (volume fraction of particles)$^{1/3}$ at $m = 1$, $s_0 = 5$ for the following models: *1* Happel, *2* Kuwabara, *3* Kvashnin, *4* Cunningham (Mehta–Morse) [38]

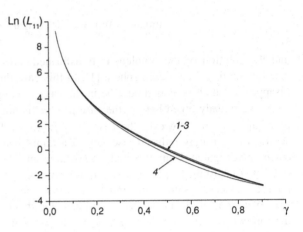

13.3.1.2 Solid Nonporous Spherical Particles

When $\delta = 0$, or $s \to \infty$, we arrive at a membrane built up by solid nonporous particles. Expressions for the dimensionless hydrodynamic permeability, $L_{11}(\gamma)$, for the different models take the following forms:

Happel's model:

$$L_{11} = \frac{2 - 3\gamma + 3\gamma^5 - 2\gamma^6}{3\gamma^3(2\gamma^5 + 3)}, \tag{13.49}$$

which coincides with the corresponding expression in [1];

Kuwabara's model:

$$L_{11} = \frac{2(5 - 9\gamma + 5\gamma^3 - \gamma^6)}{45\gamma^3}, \tag{13.50}$$

which coincides with the expression deduced in [34];

Kvashnin's model:

$$L_{11} = \frac{(1 - \gamma)^3(8\gamma^3 + 15\gamma^2 + 21\gamma + 16)}{18\gamma^3(\gamma^5 + 4)}, \tag{13.51}$$

which coincides with the expression derived in [35];

Cunningham's (Mehta–Morse's) model:

$$L_{11} = \frac{(1 - \gamma)^3(4\gamma^2 + 7\gamma + 4)}{18\gamma^3(\gamma^4 + \gamma^3 + \gamma^2 + \gamma + 1)}, \tag{13.52}$$

which was deduced by Cunningham [37] and then repeated in [27–29, 31]. It is necessary to mention here that the boundary condition (13.19d), which we, following Happel and Brenner [1], usually refer to as the Mehta and Morse condition, was proposed for the first time by Cunningham in 1910 [37]. Mehta and Morse only used Cunningham's condition in their analysis [36], which has an error. We corrected that error and it is the reason why Eq. (13.52) differs from that of Mehta and Morse [36].

The semi-empirical Kozeny–Carman equation [1] gives the following expression for the hydrodynamic permeability of a swarm of solid particles:

$$L_{11} = \frac{(1 - \gamma^3)^3}{45\gamma^6}.$$

(13.53)

The variation of the hydrodynamic permeability for the different models with the parameter γ is presented in Fig. 13.7. These results are calculated according to Eqs. (13.49)–(13.53). This figure shows that the dependencies calculated for all four cell models are very close to each other (although the calculations according to the Mehta–Morse model give slightly lower hydrodynamic permeability values at high volume fractions when compared with the other models).

The calculations based on the semi-empirical Kozeny–Carman model give a higher hydrodynamic permeability when compared with the cell models. The latter deviation is very substantial at low volume fractions but decreases and almost disappears at high volume fractions.

13.3.2 Hydrodynamic Permeability of a Porous Medium Modeled as a Swarm of Partially Porous Cylindrical Particles with Impermeable Core (the Flow is Normal to the Axes of Cylinders)

Figure 13.8 presents dependences of the natural logarithm of dimensionless hydrodynamic permeability $L_{11}^{\perp}$ of a system composed of cylindrical particles covered with a porous layer on parameter γ at $m = 2$, $s_0 = 4$, and $\delta = 0.5$ for the following different models: (curve 1) Happel, (2) Kuwabara, (3) Kvashnin, and (4) Cunningham (the flow is perpendicular to the axes of cylinders). Curves 1–4 are plotted for cylindrical particles according to Eq. (13.38).

The largest possible γ value for the packing of identical cylinders is equal to $\sqrt{\pi/(2\sqrt{3})} = 0.968$. Permeability of the membrane decreases with an increase in γ, i.e., with an increase in the fraction of the solid phase. In this case, the rate of permeability reduction is higher at low values of γ, which is typical for all models considered. Note that, at $\gamma \to 0$, permeability of the membrane increases indefinitely, while, at $\gamma \to 1$, porosity tends to zero and the flow proceeds only in the

Fig. 13.7 Variation of the
dimensionless hydrodynamic
permeability, L_{11}, of a
membrane built up by
nonporous solid spherical
particles with the parameter
γ, that is, (volume fraction
of particles)$^{1/3}$, for the
following models: *1* Happel,
2 Kuwabara, *3* Kvashnin, *4*
Cunningham (Mehta–Morse)
and *5* Kozeny–Carman [38]

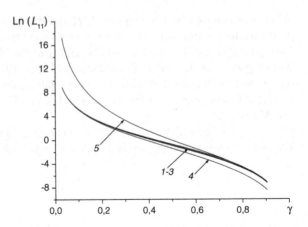

Fig. 13.8 Dependences of
the natural logarithm of
dimensionless hydrodynamic
permeability $L_{11}^{\perp}$ of a system
composed of cylindrical
particles covered with porous
layer on parameter γ at $m = 2$,
$s_0 = 4$, and $\delta = 0.5$ for
different models: (*1*) Happel,
(*2*) Kvashnin, (*3*) Kuwabara,
and (*4*) Cunningham [39]

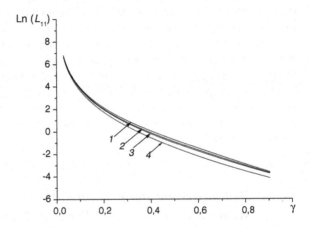

Brinkman medium. In this case, permeability tends to the limiting nonzero value. As can be seen from Fig. 13.8, permeabilities for the Happel, Kuwabara, and Kvashnin models nearly coincide with one another. Deviation (decrease) of permeability from the aforementioned models is observed for the Cunningham model in the case of concentrated media. From a mathematical viewpoint, this fact can be explained as follows. The same terms used in different combinations participate in the Happel, Kuwabara, and Kvashnin boundary conditions, whereas the Cunningham condition differs significantly from other conditions. From a physical point of view, the Cunningham condition, i.e., joining with the uniform flow at the cell boundary, is the most rigorous condition of the four considered types of boundary conditions. This condition must result in the increased dissipation of energy in a system and, hence, in a more noticeable reduction in the hydrodynamic permeability of a membrane.

The presented models are compared in the logarithmic coordinate (for the force) in Fig. 13.9 with the data reported in [41], where systems of cylindrical

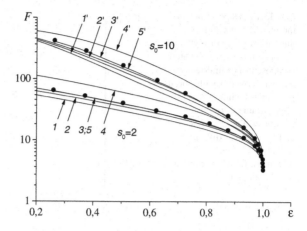

Fig. 13.9 Dependences of the resistance force of cylinder with porous layer on parameter ε at $m = 1, \delta = 0.5$, $s_0 = 2$ (*1–5*) and 10 (*1'–5'*) for different models: (*1–1'*) Happel, (*2–2'*) Kvashnin, (*3–3'*) Kuwabara, (*4–4'*) Cunningham, (*5–5'*) square lattice, and (●) the row of squares [39]

fibers form a square lattice and a row of squares. This figure also makes it possible to clarify the dependence of dimensionless force F, Eq. (13.35), on porosity ε. It is seen that the values of viscous drag found for the Kuwabara and the square lattice models differ insignificantly. At small values of porosity, the effects of the form and boundary conditions of the cell are more pronounced. Based on what was said above, it can be concluded that the proposed analytical models adequately describe the process of liquid flow in the concentrated systems under consideration. In this case, the results of analytical calculations are in good agreement with the data obtained by numerical methods.

Figure 13.10 shows the dependences of the natural logarithm of dimensionless hydrodynamic permeability on the thickness δ of cylindrical porous layer for different models at fixed values $\gamma = 0.5$, $s_0 = 5$, and $m = 2$. All of these dependences are monotonically increasing functions. As in Fig. 13.8, the permeability increases consecutively for the Cunningham, Kuwabara, Kvashnin, and Happel models. Parameter s_0 characterizes the depth of flow penetration into the porous layer; the larger the s_0 value, the greater the depth. At large s_0 values, the flow proceeds not throughout the porous shell but only in the layer with thickness approximately equal to the Brinkman radius. Thus, when the thickness of the porous shell becomes greater than the Brinkman radius, permeability ceases to depend on the radius of rigid core, which is expressed in the presence of smooth, almost horizontal parts of the curves shown in Fig. 13.10.

Figure 13.11 demonstrates the dependences of permeability $L_{11}^{\perp}$ on the ratio of viscosities, m, at $\gamma = 0.8$, $s_0 = 2$, and $\delta = 0.13$. It can be seen that the rise in the viscosity inside the porous layer leads to a decrease in the permeability of membrane. In this case, the permeability first drastically drops, then achieves its limiting value, which corresponds to the higher viscosity inside the porous layer when the flow inside the layer is nearly absent.

Figure 13.12 presents the dependences of the natural logarithm of dimensionless hydrodynamic permeability $L_{11}^{\perp}$ on parameter s_0 for different models at $\gamma = 0.8$, $m = 1$, and $\delta = 0.13$. As was mentioned above, the dimensionless

Fig. 13.10 Dependences of
the natural logarithm of
dimensionless hydrodynamic
permeability $L_{11}^{\perp}$ of a system
composed of cylindrical
particles covered with porous
layer and having
impermeable core on
parameter δ at $m = 2$, $s_0 = 5$,
and $\gamma = 0.5$ for different
models: (1) Happel, (2)
Kvashnin, (3) Kuwabara, and
(4) Cunningham [39]

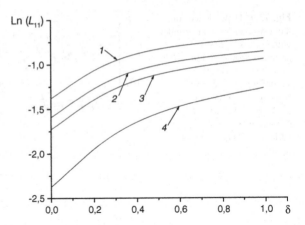

parameter s_0 characterizes the degree of porosity of the Brinkman medium; the
lower the porosity, the larger the s_0 values. At $s_0 \rightarrow \infty$, the particle becomes
absolutely impermeable. It can be seen from the figure that, as the value of s_0
increases, permeability monotonically decreases to its limiting value, as in
Fig. 13.11. As before, the Cunningham model yields lower values of permeability
compared to other models. It should be again emphasized that this structure of a
complex porous medium (membrane) has two additional degrees of freedom as
compared to the set of completely porous particles considered in [5], i.e., the
internal rigidity R, which is the ratio of the radius of the rigid core to that of an
entire particle, and the viscosity m inside the porous layer.

Compared to the set of absolutely rigid particles studied in [1, 34–37], the
model under consideration is characterized by three additional degrees of freedom,
i.e., internal rigidity R, the resistance of the porous layer, s_0, and the viscosity
m inside the porous layer. The introduction of additional degrees of freedom
allows us to describe the effect of the processes of internal contamination and (or)
dissolution of the porous medium (membrane) upon changes in its permeability in
the course of filtration.

Let us now consider significant limiting cases of the solution obtained for the
cylindrical case.

13.3.2.1 Completely Porous Cylindrical Particles

At $\delta = 1$ $(R = 0)$, we have a set of completely porous particles. In this case,
instead of boundary conditions of nonsleeping, the finite velocity condition in the
center of particle is used, according to which $c_3' = c_4' = 0$. Expressions for permeability $L_{11}^{\perp}(\gamma, m, s)$ obtained for different models have the following forms,
respectively,

Fig. 13.11 Dependences of the natural logarithm of dimensionless hydrodynamic permeability $L_{11}^{\perp}$ of a membrane composed of cylindrical particles covered with porous layer on parameter m at $\gamma = 0.8$,

$s_0 = 2$, and $\delta = 0.5$ for

different models: (*1*) Happel, (*2*) Kvashnin, (*3*) Kuwabara, and (*4*) Cunningham [39]

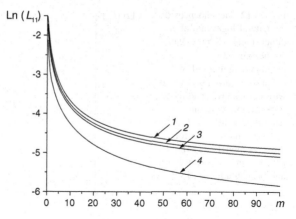

Happel's model:

$$L_{11}^{\perp} = \big[ms\{8(1+\gamma^4) + ms^2(\gamma^4 - 1) + 2ms^2(1+\gamma^4)\ln(1/\gamma)\}I_1(s)$$
$$+ 2\{8 + m(ms^2 - 8) - \gamma^4(8 + m(8 + ms^2)) - 2ms^2(-1 + m + \gamma^4(1+m))\ln(1/\gamma)\}I_2(s)\big]$$
$$/\big[8m^2\gamma^2 s^3(1+\gamma^4)I_1(s) - 16m\gamma^2 s^2(-1 + m + \gamma^4(1+m))I_2(s)\big].$$

(13.54)

Kuwabara's model:

$$L_{11}^{\perp} = \big[2s\{16 + m(-16 + s^2(\gamma^2 - 1)(1 + \gamma^2 + m(\gamma^2 - 3))) - 4ms^2(m-1)\ln(1/\gamma)\}I_0(s)$$
$$+ \{-64 + 64m - 4ms^2(\gamma^4 - 5) - m^2 s^2(3 - 4\gamma^2 + \gamma^4)(4 + s^2)$$
$$+ 4ms^2(-4 + m(4 + s^2))\ln(1/\gamma)\}I_1(s)\big] / \big[16\gamma^2 m^2 s^4 I_1(s) - 32m\gamma^2 s^3(m-1)I_2(s)\big].$$

(13.55)

Kvashnin's model:

$$L_{11}^{\perp} = \big[\{24 - 8\gamma^4 - 8m(3 + \gamma^4) + ms^2(\gamma^2 - 1)(1 + \gamma^2 - 4m) - 2ms^2(-3 + \gamma^4$$
$$+ m(3 + \gamma^4))\ln(1/\gamma)\}sI_0(s) + \{-48 + 16\gamma^4 + 16m(3 + \gamma^4) + 2ms^2(7 + \gamma^4 + 4m(\gamma^2 - 1))$$
$$+ 2m^2 s^4(\gamma^2 - 1) + ms^2(-12 + 4\gamma^4 + 4m(3 + \gamma^4) + ms^2(3 + \gamma^4))\ln(1/\gamma)\}I_1(s)\big]$$
$$/\big[4\gamma^2 m^2 s^4(3 + \gamma^4)I_1(s) - 8\gamma^2 ms^3\{-3 + \gamma^4 + m(3 + \gamma^4)\}I_2(s)\big].$$

(13.56)

Cunningham's model:

$$L_{11}^{\perp} = \big[\{8(m-1) - 8\gamma^4(1+m) - ms^2(\gamma^2 - 1)(1 + \gamma^2 + 2m(\gamma^2 - 1)) - 2ms^2(1 + \gamma^4$$
$$+ m(\gamma^4 - 1))\ln(1/\gamma)\}sI_0(s) + \{16(1 - m + \gamma^4(1+m)) + 2ms^2(\gamma^2 - 1)(3 - 2m$$
$$+ \gamma^2(3 + 2m)) + m^2 s^4(\gamma^2 - 1)^2 + ms^2(4 - m(4 + s^2) + 4\gamma^4 + m\gamma^4(4 + s^2))\ln(1/\gamma)\}I_1(s)\big]$$
$$/\big[4\gamma^2 m^2 s^4(\gamma^4 - 1)I_1(s) - 8\gamma^2 ms^3(1 - m + \gamma^4(1+m))I_2(s)\big].$$

(13.57)

Fig. 13.12 Dependences of the natural logarithm of dimensionless hydrodynamic permeability $L_{11}^{\perp}$ of membrane composed of cylindrical particles covered with porous layer and having impermeable core on parameter s_0 at $m = 1$, $\gamma = 0.8$, and $\delta = 0.5$ for different models: (*1*) Happel, (*2*) Kvashnin, (*3*) Kuwabara, and (*4*) Cunningham [39]

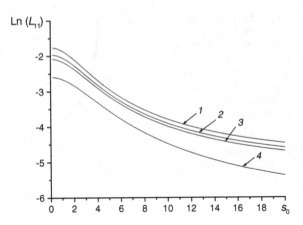

Fig. 13.13 Dependences of the natural logarithm of dimensionless hydrodynamic permeability $L_{11}^{\perp}$ of a membrane composed of completely porous cylindrical particles on parameter γ at $m = 3$ and $s_0 = 4$ for different models: (*1*) Happel, (*2*) Kvashnin, (*3*) Kuwabara, and (*4*) Cunningham [39]

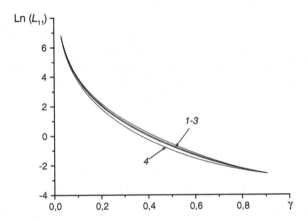

Figure 13.13 presents the dependences of the natural logarithm of dimensionless hydrodynamic permeability $L_{11}^{\perp}$ of a membrane composed of completely porous cylindrical particles on parameter γ for different models at $m = 1$ and $s_0 = 5$. Note that the discrepancy between models in this particular case is less pronounced than in the general case. The figure demonstrates the drop in permeability with an increase in concentration of the solid phase for all models. It is natural that, at $\gamma \to 1$, hydrodynamic permeability $L_{11}^{\perp}$ for all models tends to the same limiting value, i.e., to the hydrodynamic permeability of the continuous Brinkman medium, as follows:

$$L_{11}^{\perp} = 1/s_0^2 \qquad (13.58)$$

13.3.2.2 Solid Nonporous Cylindrical Particles

At $\delta = 0$ or $s \to \infty$ and $m \to \infty$, we have a system of solid impermeable cylindrical particles. Expressions for $L_{11}^{\perp}(\gamma)$ for different models have the following forms,

Happel's model:

$$L_{11}^{\perp} = \frac{-1 + \gamma^4 + 2(1 + \gamma^4)\ln(1/\gamma)}{8\gamma^2(1 + \gamma^4)} \tag{13.59}$$

Kuwabara's model:

$$L_{11}^{\perp} = \frac{-3 + 4\gamma^2 - \gamma^4 + 4\ln(1/\gamma)}{16\gamma^2} \tag{13.60}$$

Kvashnin's model:

$$L_{11}^{\perp} = \frac{2(\gamma^2 - 1) + (3 + \gamma^4)\ln(1/\gamma)}{4\gamma^2(3 + \gamma^4)} \tag{13.61}$$

and *Cunningham's model*:

$$L_{11}^{\perp} = \frac{-1 + \gamma^2 + (1 + \gamma^2)\ln(1/\gamma)}{4\gamma^2(1 + \gamma^2)}, \tag{13.62}$$

which was derived in [1]. Note, that formulas (13.59)–(13.61) coincide with the results reported in [1, 34, 35], respectively.

Figure 13.14 shows the dependences of the behavior of permeability of a membrane composed of solid impermeable cylindrical particles on parameter γ corresponding to different models. It can be seen that curves calculated by the cell models of Happel, Kuwabara, and Kvashnin almost coincide. As regards the Cunningham model, it yields small deviations toward a decrease in permeability, as before for impermeable spherical particles (Fig. 13.7).

13.3.3 Hydrodynamic Permeability of a Porous Medium Modeled as a Swarm of Partially Porous Cylindrical Particles with Impermeable Core (the Flow Is Parallel to the Axes of Cylinders)

In this section, we consider filtration through a membrane composed of impermeable cylinders covered with a porous layer when the flow is directed along the axes of cylinders (Fig. 13.15).

Fig. 13.14 Dependences of the natural logarithm of dimensionless hydrodynamic permeability $L_{11}^{\perp}$ of a membrane composed of solid impermeable cylindrical particles on parameter γ for different models: (*1*) Happel, (*2*) Kvashnin, (*3*) Kuwabara, and (*4*) Cunningham [39]

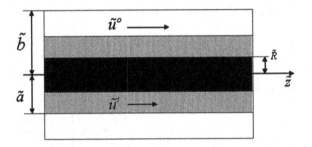

Fig. 13.15 Schematic representation of a cylindrical cell with solid impermeable particle covered with a porous layer upon the longitudinal flow of liquid

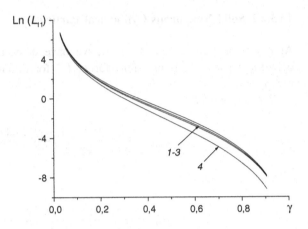

As before, we will consider the flow in a separate cell. In a cylindrical system of coordinates in which the axis is directed along the flow, equations that describe the flow outside and inside the porous layer have, respectively, the following forms:

$$\tilde{\mu}^{o}\frac{1}{\tilde{r}}\frac{\mathrm{d}}{\mathrm{d}\tilde{r}}\left(\tilde{r}\frac{\mathrm{d}\tilde{u}^{o}}{\mathrm{d}\tilde{r}}\right)=\frac{\mathrm{d}\tilde{p}}{\mathrm{d}\tilde{z}}, \quad \tilde{a}<\tilde{r}<\tilde{b}; \tag{13.63}$$

$$\tilde{\mu}^{i}\frac{1}{\tilde{r}}\frac{\mathrm{d}}{\mathrm{d}\tilde{r}}\left(\tilde{r}\frac{\mathrm{d}\tilde{u}^{i}}{\mathrm{d}\tilde{r}}\right)-\frac{\tilde{\mu}^{o}}{\tilde{k}}\tilde{u}^{i}=\frac{\mathrm{d}\tilde{p}}{\mathrm{d}\tilde{z}}, \quad \tilde{R}<\tilde{r}<\tilde{a}, \tag{13.64}$$

where $\tilde{u}^{o}$, $\tilde{u}^{i}$ are flow velocities along the cylinder and, $\mathrm{d}\tilde{p}/\mathrm{d}\tilde{z}$ is the pressure drop, which is considered to be constant and preset.

As we did before, for the surface of a rigid cylinder, we set the sticking condition

$$\tilde{u}^{i}=0, \quad \tilde{r}=\tilde{R}. \tag{13.65}$$

At the interface, we use the following continuity conditions for velocity and tangential stress:

$$\tilde{u}^o = \tilde{u}^i \text{ at } \tilde{r} = \tilde{a} \tag{13.66}$$

$$\tilde{\sigma}^o_{rz} = \tilde{\sigma}^i_{rz} \text{ at } \tilde{r} = \tilde{a}. \tag{13.67}$$

In this case, the Happel, Kuwabara, Kvashnin, and Cunningham boundary conditions all yield the following result:

$$\frac{d\tilde{u}^o}{d\tilde{r}} = 0 \text{ at } \tilde{r} = \tilde{b}. \tag{13.68}$$

Equations (13.63) and (13.64), as well as boundary conditions (13.65)–(13.68) in dimensionless variables (13.20), have the following forms:

$$\frac{d}{dr}\left(r\frac{du^o}{dr}\right) = r\omega, \tag{13.69}$$

$$r\frac{d^2u^i}{dr^2} + \frac{du^i}{dr} - rs^2u^i = r\omega/m, \tag{13.70}$$

$$u^i = 0 \text{ at } r = R, \tag{13.71}$$

$$u^o = u^i \text{ at } r = 1, \tag{13.72}$$

$$\frac{du^o}{dr} = m\frac{du^i}{dr} \text{ at } r = 1, \tag{13.73}$$

$$\frac{du^o}{dr} = 0 \text{ at } r = 1/\gamma, \tag{13.74}$$

where the new designation $\omega = \frac{dp}{dz} \equiv \text{const}$ is introduced.

General solutions to Eqs. (13.69) and (13.70) are as follows:

$$u^o = b_1 + b_2 \ln r + \omega\frac{r^2}{4}, \tag{13.75}$$

$$u^i = c_1 I_0(sr) + c_2 K_0(sr) - \frac{\beta}{ms^2}. \tag{13.76}$$

Substituting expressions (13.75) and (13.76) into the boundary conditions (13.71)–(13.74), we find the values of b_1, b_2, c_1, and c_2 constants

$$b_1 = -\omega\{[msy^2I_1(s) - 2y^2I_2(s) + 2I_0(s)]s^2K_0(Rs) + [msy^2K_1(s)$$
$$+ 2y^2K_2(s) - 2K_0(s)]s^2I_0(Rs) - 4y^2\} \tag{13.77}$$
$$/\{4ms^3y^2(I_1(s)K_0(Rs) + I_0(Rs)K_1(s))\},$$

$$b_2 = -\omega/(2\gamma^2), \tag{13.78}$$

$$c_1 = \frac{\omega\{2\gamma^2 K_1(s) + s(\gamma^2 - 1)K_0(Rs)\}}{2ms^2\gamma^2\{I_1(s)K_0(Rs) + I_0(Rs)K_1(s)\}}, \tag{13.79}$$

$$c_2 = \frac{\omega\{2\gamma^2 I_1(s) - s(\gamma^2 - 1)I_0(Rs)\}}{2ms^2\gamma^2\{I_1(s)K_0(Rs) + I_0(Rs)K_1(s)\}}. \tag{13.80}$$

The dimensionless flow rate of liquid flowing through the cylinder is equal to the following definite integral (see Fig. 13.15):

$$Q = 2\pi \left(\int_R^1 u^i r dr + \int_1^{1/\gamma} u^o r dr \right). \tag{13.81}$$

In the case under consideration, the rate of filtration is equal to $v_f = Q/(\pi/\gamma^2)$ [1]. Using the Darcy law for the flow in a porous medium, we obtain

$$v_f = -L_{11}^{\parallel} dp/dz = -L_{11}^{\parallel}\omega. \tag{13.82}$$

From expressions (13.75)–(13.82), we derive the following formula

$$
\begin{aligned}
L_{11}^{\parallel} =&(-\gamma^2/\omega)\{[\omega(R^2 - 1) + 2ms(c_1 I_1(s) - c_1 RI_1(sR) - c_2 K_1(s) \\
&+ c_2 RK_1(Rs))]/(ms^2) + [8b_2\gamma^2 \ln(1/\gamma) \\
&- (\gamma^2 - 1)(\gamma^2(8b_1 - 4b_2 + \beta) + \beta)]/(8\gamma^4)\}.
\end{aligned}
\tag{13.83}
$$

for the coefficient of hydrodynamic permeability $L_{11}^{\parallel}$ of this medium.

Let us consider the most significant limiting cases of general formula (13.83).

13.3.3.1 Completely Porous Cylindrical Particles

At $R = 0$, we have a porous particle with no rigid impermeable core and the expression for hydrodynamic permeability can be written as

$$L_{11}^{\parallel} = \frac{1}{8}\left(4 - \gamma^2 - \frac{8(\gamma^2 - 2)}{ms^2}\right) + \frac{(\gamma^2 - 1)^2 I_0(s)}{2\gamma^2 msI_1(s)} - \frac{3}{8\gamma^2} + \frac{1}{2\gamma^2}\ln\left(\frac{1}{\gamma}\right) \tag{13.84}$$

Fig. 13.16 Dependences of the natural logarithm of dimensionless hydrodynamic permeability $L_{11}^{\parallel}$ of a membrane composed of cylindrical particles upon the longitudinal flow of liquid on parameter γ at $m = 2$, $s_0 = 1$, and $\delta = 0.5$ for different models: (1) composite (partially porous) particles, (2) completely porous particles, and (3) solid impermeable particles [39]

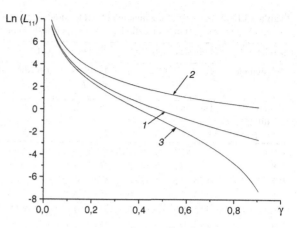

13.3.3.2 Solid Nonporous Cylindrical Particles

At $s_0 \to \infty$ or $m \to \infty$ and $R = 1$, we have a particle without a porous layer that is absolutely impermeable to liquid; then, the expression for hydrodynamic permeability acquires the known form [1]

$$L_{11}^{\parallel} = \frac{1}{8\gamma^2} \left(4 \ln \left(\frac{1}{\gamma} \right) - \gamma^4 + 4\gamma^2 - 3 \right) \tag{13.85}$$

Figure 13.16 demonstrates the dependences of the natural logarithm of dimensionless hydrodynamic permeability $L_{11}^{\parallel}$ of the membrane composed of cylindrical particles on parameter γ at $m = 2$, $s_0 = 1$, and $\delta = 0.5$ for different models. Curves 1–3 are plotted by formulas (13.83)–(13.85). The permeability of the membrane decreases with an increase in γ; i.e., with an increase in the fraction of solid phase. The curve for a particle covered with a porous layer is arranged lower than for completely porous particles and higher than for impermeable solid particles, which is obvious, since the larger the fraction of porous medium, the higher the permeability. As before, at $\gamma \to 0$, permeability of the membrane increases infinitely for all models. At $\gamma \to 1$, porosity for the model with rigid impermeable particles becomes equal to zero and, hence, its permeability also tends to zero (curve 3), which also follows from formula (13.85). At $\gamma \to 1$, the medium formed by completely porous particles (curve 2) is transformed into the Brinkman medium and the permeability tends to its limiting value of $1/(ms^2)$, which follows from formula (13.84). In the case of particles with a porous layer (curve 1), we deal with the flow of liquid in the Brinkman medium with impermeable cylindrical fibers. According to formula (13.83), the hydrodynamic permeability of such medium is equal to

Table 13.1 Values of the coefficients of dimensionless permeability L_{11} for membranes composed of spherical particles or cylinders with impermeable core covered with porous layer[a] [39]

Model	Spherical particles				Cylindrical particles								
Orientation	L_{11}				Perpendicular to the flux $L_{11}^{\perp}$				Parallel $L_{11}^{\parallel}$	Random $L_{11}^{R} = \frac{2}{3}L_{11}^{\perp} + \frac{1}{3}L_{11}^{\parallel}$			
Boundary condition	Ha	Kv	Ku	Cu	Ha	Kv	Ku	Cu	All	Ha	Kv	Ku	Cu
Porosity/No.	1	2	3	4	5	6	7	8	9	10	11	12	13
0.9	1.66	1.55	1.48	1.21	2.54	2.22	2.05	1.57	3.66	2.91	2.70	2.59	2.27
0.8	0.65	0.60	0.57	0.45	0.85	0.73	0.67	0.48	1.14	0.94	0.86	0.82	0.70
0.7	0.36	0.33	0.32	0.25	0.41	0.35	0.32	0.22	0.53	0.45	0.41	0.39	0.33
0.6	0.23	0.21	0.20	0.16	0.23	0.20	0.19	0.13	0.29	0.25	0.23	0.22	0.18
0.5	0.16	0.15	0.14	0.12	0.15	0.13	0.12	0.08	0.18	0.16	0.15	0.14	0.12
0.4	0.12	0.11	0.11	0.09	0.10	0.09	0.08	0.06	0.12	0.11	0.10	0.09	0.08
0.3	0.09	0.09	0.08	0.07	0.07	0.06	0.06	0.04	0.09	0.07	0.07	0.07	0.06
0.2	0.07	0.07	0.07	0.06	0.05	0.05	0.04	0.03	0.07	0.06	0.05	0.05	0.04
0.1	0.06	0.06	0.06	0.05	0.04	0.04	0.03	0.02	0.05	0.04	0.04	0.04	0.03

[a] Abbreviations for models: *Ha* Happel, *Kv* Kvashnin, *Ku* Kuwabara, and *Cu* Cunningham models

$$L_{11}^{\parallel} = \frac{(I_0(Rs) - R^2 I_2(Rs))K_1(Rs) + (K_0(Rs) - R^2 K_2(Rs))}{ms^2(I_1(s)K_0(Rs) + I_0(Rs)K_1(s))} \qquad (13.86)$$

13.4 Comparison of Different Cell Models

In this section, we compare different cell models that describe the flow in a porous media, i.e., we consider the flows through the membranes formed by spherical particles and cylinders with longitudinal, transverse, or chaotic orientations toward the flow.

13.4.1 Composite Particles with a Porous Layer

Table 13.1 lists the values of permeability coefficients L_{11} for membranes composed of rigid spherical particles or cylinders covered with porous layers at different porosities ε and fixed values of $m = 2$, $s_0 = 3$, and $R = 0.4$.

The permeability of a medium formed by spherical particles was calculated by formula (13.38) which was reported in [32]. The permeability of a medium formed by cylinders was calculated using formula (13.38'). For the chaotic orientation of the cylinder, we used an averaging procedure based on the following considerations. When cylinders are oriented perpendicular to the flow, the model yields the same results, which are observed for cylinders with longitudinal or chaotic orientations in the plane normal to the flow. Therefore, when considering the case of chaotic orientation, the averaging was performed so that the weight of the contribution from the cylinders with normal orientation to the flow was twice as large as that of the flow along the cylinders. The thus obtained values of permeability are listed in columns of Table 13.1 referred to as the cases of random packing.

The following conclusions can be drawn from the values of permeability listed in Table 13.1. At a small fraction of solid phase (upper lines in Table 13.1), the permeability of a medium formed by cylinders is higher than for media composed of spherical particles. Moreover, the system of cylinders oriented along the flow is characterized by the largest permeability. The values of permeability for different systems are leveled with an increase in the concentration of solid phase (with a decrease in porosity). This can be explained by the fact that geometric differences between media formed by spherical particles and cylinders become less noticeable at high concentrations of solid phase. At the concentration of solid phase equal to 0.9 ($\varepsilon = 0.1$), the values of dimensionless permeability vary from 0.002 to 0.006. Note also that, unlike weakly concentrated media, at high concentration of solid phase, the values of hydrodynamic permeability appeared to be larger for the set of spherical particles.

13.4.2   Completely Porous Particles

The values of the coefficient of dimensionless permeability L_{11} of membranes formed by porous spherical particles or cylinders at different values of porosity ε and $m = 3$, $s_0 = 2$ are shown in Table 13.2. Table 13.2 is constructed analogously to Table 13.1. The behavior of the permeability coefficients L_{11} for media formed by completely porous and composite particles and cylinders is entirely analogous. Furthermore, the dependence on the type of boundary conditions becomes less marked and almost vanishes at high concentration of solid phase ($\varepsilon < 0.4$). The values of the coefficients of permeability do not tend to zero with a similar tendency of porosity, since, in this case, the liquid can flow through the porous shells.

13.4.3   Rigid Impermeable Particles

The semi-empirical Kozeny–Carman formula [1] results in the following expression for the permeability of isotropic porous media:

Table 13.2 Values of the coefficients of dimensionless permeability L_{11} for membranes composed of fully porous spherical particles or cylinders [39]

Model	Spherical particles				Cylindrical particles					Random			
Orientation	L_{11}				Perpendicular to the flux $L_{11}^{\perp}$				Parallel $L_{11}^{\parallel}$	$L_{11}^{R} = \frac{2}{3}L_{11}^{\perp} + \frac{1}{3}L_{11}^{\parallel}$			
Boundary condition	Ha	Kv	Ku	Cu	Ha	Kv	Ku	Cu	All	Ha	Kv	Ku	Cu
Porosity/No.	1	2	3	4	5	6	7	8	9	10	11	12	13
0.9	3.39	3.30	3.24	3.00	4.43	4.13	3.98	3.53	5.31	4.73	4.53	4.43	4.12
0.8	1.52	1.49	1.47	1.37	1.80	1.70	1.65	1.49	1.99	1.87	1.80	1.76	1.65
0.7	0.95	0.93	0.92	0.88	1.06	1.01	0.99	0.91	1.11	1.08	1.05	1.03	0.98
0.6	0.68	0.67	0.67	0.64	0.73	0.70	0.69	0.66	0.75	0.73	0.72	0.71	0.69
0.5	0.53	0.52	0.52	0.51	0.55	0.54	0.53	0.51	0.55	0.55	0.54	0.54	0.53
0.4	0.43	0.43	0.43	0.42	0.44	0.43	0.43	0.42	0.44	0.44	0.44	0.43	0.43
0.3	0.36	0.36	0.36	0.36	0.37	0.36	0.36	0.36	0.37	0.37	0.36	0.36	0.36
0.2	0.31	0.31	0.31	0.31	0.32	0.31	0.31	0.31	0.32	0.32	0.31	0.31	0.31
0.1	0.28	0.28	0.28	0.28	0.28	0.28	0.28	0.28	0.28	0.28	0.28	0.28	0.28

$$L_{11} = \frac{\varepsilon \tilde{\rho}_H^2}{k_c \tilde{a}^2} \tag{13.87}$$

where k_c is the dimensionless Kozeny constant; $\tilde{\rho}_H$ is the hydraulic radius, which is equal to the ratio of the pore volume to the wetting area; and ε is the porosity.

The calculation of the theoretical values of the Kozeny constant using formula (13.87) yields the following relation:

$$k_c = \frac{\varepsilon \tilde{\rho}_H^2}{\tilde{a}^2 L_{11}} \tag{13.88}$$

For media composed of cylindrical particles, we have

$$\tilde{\rho}_H = \frac{\pi(\tilde{b}^2 - \tilde{a}^2)}{2\pi\tilde{a}} = \frac{\tilde{a}}{2}\left(\frac{1 - \gamma^2}{\gamma^2}\right) = \frac{\varepsilon\tilde{a}}{2(1 - \varepsilon)} \tag{13.89}$$

Substituting expression (13.89) into formula (13.88), we obtain

$$k_c = \frac{\varepsilon^3}{4(1 - \varepsilon)^2 L_{11}'} \tag{13.90}$$

where L_{11}' is found from one of formulas (13.59)–(13.62) or (13.85). For media composed of spherical particles [32], we have

$$\tilde{\rho}_H = \frac{4\pi(\tilde{b}^3 - \tilde{a}^3)/3}{4\pi\tilde{a}^2} = \frac{\tilde{a}}{3}\left(\frac{1 - \gamma^3}{\gamma^3}\right) = \frac{\varepsilon\tilde{a}}{3(1 - \varepsilon)} \qquad (13.91)$$

and the relation for the Kozeny constant instead of (13.80) can be expressed as

$$k_c = \frac{\varepsilon^3}{9(1 - \varepsilon)^2 L_{11}} \qquad (13.92)$$

where L_{11} is calculated using one of formulas (13.49)–(13.52). Table 13.3 presents the theoretical values of the Kozeny constant k_c for membranes formed by spherical particles or cylinders at different values of porosity ε.

According to the current hypothesis, the Kozeny constant is an invariable magnitude, which is independent of both the structure of a membrane and its porosity. Experimental estimations result in values of the Kozeny constant approximately equal to 5. It follows from Table 13.3 that the Happel, Kvashnin, and Kuwabara models yield k_c values that are approximately equal to 5 at porosities of 0.1–0.6 for isotropic media. In this case, the k_c values are nearly identical for media composed of spherical particles or those composed of randomly oriented cylinders. The Cunningham model yields k_c values that differ substantially from those calculated from experimental data.

13.5 Comparison of Theoretical and Experimental Data

The sedimentation of spherical particles made of nylon with glued flexible polyester fibers modeling different porous structures was experimentally studied in [22]. Nylon particles with radius $\tilde{R} = 0.318$ cm to which polyester fibers with radius $\tilde{R}_{\mathrm{cyl}} = 0.065 \cdot \tilde{R}$ and length $\tilde{\delta} = 1.6 \cdot \tilde{R}$ were glued are considered. The number of fibers varied from 1 to 20, which corresponded to changes in the porosity of the medium covering the particle. In the course of the experiment, the sedimentation velocity of particles with glued fibers was measured and the resistance force acting on the composite particle was calculated.

Figure 13.17 demonstrates experimental values of dimensionless force θ acting on composite particle depending on the fraction c of solid phase in the porous layer. Dimensionless force $\theta = \tilde{F}/\tilde{F}_{\mathrm{solid}}$ is equal to the ratio of force acting on the composite particle with general radius $\tilde{a} = \tilde{R} + \tilde{\delta}$ to the Stokes force acting on the rigid particle with radius $\tilde{R}$. For theoretical calculations of dimensionless force θ, we employed formula $\theta = \Omega/R$, where Ω was calculated by formulas (13.76)–(13.78) from [38]. Force $\Omega(R, \delta, s, m)$ depends on four dimensionless parameters introduced previously, e.g.,

Table 13.3 Theoretical values of the Kozeny constant k_c for membranes composed of impermeable spherical particles or cylinders

Model	Spherical particles				Cylindrical particles					Random			
Orientation					Perpendicular to the flux				Parallel				
Boundary condition/No.	Ha 1	Kv 2	Ku 3	Cu 4	Ha 5	Kv 6	Ku 7	Cu 8	All 9	Ha 10	Kv 11	Ku 12	Cu 13
Porosity													
0.1	4.44	4.55	4.66	17.42	5.73	5.95	6.16	22.83	3.08	4.84	4.99	5.14	16.24
0.2	4.42	4.64	4.86	16.87	5.51	5.95	6.36	21.71	3.18	4.73	5.02	5.30	15.53
0.3	4.44	4.79	5.11	16.38	5.36	6.01	6.61	20.66	3.30	4.67	5.11	5.51	14.88
0.4	4.54	5.00	5.43	15.96	5.30	6.16	6.92	19.71	3.46	4.69	5.26	5.77	14.29
0.5	4.74	5.33	5.86	15.66	5.37	6.43	7.34	18.88	3.67	4.80	5.51	6.11	13.81
0.6	5.11	5.83	6.47	15.55	5.62	6.89	7.92	18.26	3.96	5.07	5.91	6.60	13.49
0.7	5.79	6.68	7.42	15.82	6.20	7.68	8.83	18.00	4.41	5.60	6.59	7.36	13.47
0.8	7.22	8.31	9.19	16.96	7.46	9.20	10.5	18.54	5.23	6.72	7.87	8.72	14.11
0.9	11.34	12.8	13.88	21.33	11.03	13.18	14.62	21.88	7.31	9.79	11.22	12.2	17.03

$$\delta = \frac{\tilde{\delta}}{\tilde{a}}, \ R = \frac{\tilde{R}}{\tilde{a}} = 1 - \delta, \ m = \frac{\tilde{\mu}^i}{\tilde{\mu}^o}, \ s = \frac{\tilde{a}}{\sqrt{mk}} \equiv \frac{s_0}{\sqrt{m}}.$$

According to experimental data, $\tilde{a} = 2.6 \cdot \tilde{R}$, and $\tilde{\delta} = 1.6 \cdot \tilde{R}$, and, hence, $R = 5/13$, $\delta = 8/13$. To calculate specific permeability $\tilde{k}$ which is proportional to hydrodynamic permeability $\tilde{L}$ $(\tilde{k} = \tilde{\mu}^o \tilde{L})$ of a porous medium composed of cylindrical fibers arranged perpendicular to the flow, we used the following values calculated by the cell method in [1, 38]:

$$\tilde{L}_{HA} = \frac{-1 + c^2 - (1 + c^2)\ln(c)}{8c(1 + c^2)} \frac{\tilde{R}^2_{cyl}}{\tilde{\mu}^o}, \tag{13.93}$$

$$\tilde{L}_{KU} = \frac{-3 + 4c - c^2 - 2\ln(c)}{16c} \frac{\tilde{R}^2_{cyl}}{\tilde{\mu}^o}, \tag{13.94}$$

$$\tilde{L}_{KV} = \frac{2(c - 1) - 0.5(3 + c^2)\ln(c)}{4c(1 + c^3)} \frac{\tilde{R}^2_{cyl}}{\tilde{\mu}^o}, \tag{13.95}$$

$$\tilde{L}_{CU} = \frac{-1 + c - 0.5(1 + c)\ln(c)}{4c(1 + c)} \frac{\tilde{R}^2_{cyl}}{\tilde{\mu}^o}. \tag{13.96}$$

Coefficients of permeability were calculated by the cell method using boundary conditions: $\tilde{L}_{HA}$, for Happel; $\tilde{L}_{KU}$, for Kuwabara; $\tilde{L}_{KV}$, for Kvashnin; and $\tilde{L}_{CU}$ for Cunningham models.

For the coefficient of permeability of a porous medium composed of cylindrical fibers arranged along the flow, we used expression [1, 38]:

$$\tilde{L}_{along} = -\frac{1}{8c}(c^2 - 4c + 2\ln(c) + 3)\frac{\tilde{R}^2_{cyl}}{\tilde{\mu}^o} \tag{13.97}$$

Coefficient m equal to the ratio of viscosity in the porous layer to that of the bulk liquid was calculated by one of the formulas (13.14)–(13.15).

Figure 13.17 shows dependences $\theta(R, \delta, s, m)$ on the fraction of solid phase at $R = 5/13$, $\delta = 8/13$, $m = 1$, and $s = \tilde{a}/\sqrt{m \tilde{\mu}^o \tilde{L}}$ where permeability $\tilde{L}$ was calculated by one of the formulas (13.93)–(13.97) depending on the chosen model of the porous medium. At zero concentration of solid phase in the porous layer, $\theta = 1$ i.e., the drag force acting on the particle is equal to the Stokes force (see Fig. 13.17). When the fraction of solid phase tends to unity, the θ tends to its limiting value $\tilde{a}/\tilde{R}$ which, in the case under consideration, is equal to 2.6 (see Fig. 13.17). This means that the force acting on the composite particle is equal to the Stokes force acting on the rigid particle with radius $\tilde{a} = 2.6 \cdot \tilde{R}$. An analogous

Fig. 13.17 Dependences of dimensionless force θ on the fraction c of solid phase in porous layer for different models: (•) experimental data from [22], (*1*) Cunningham (Mehta–Morse), (*2*) Kuwabara, (*3*) Kvashnin, and (*4*) Happel; (*5*) model with longitudinal arrangement of fibers [38]

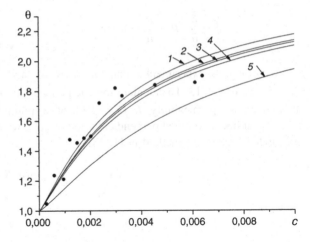

dependence for the Happel model was determined previously in [22]. For each model, root-mean-square deviation σ was calculated as equal to

$$\sigma = \sqrt{\frac{\sum_{i=1}^{n} (\theta_{t_i} - \theta_{e_i})^2}{n-1}}$$

where θ_{ti} and θ_{ei} are theoretical and empirical dependences of $\theta(c)$ function and n is the number of experimental points. The root-mean-square deviation is equal to 0.1168 for Happel model, 0.1015 for Kuwabara, 0.1060 for Kvashnin, 0.0941 for the Cunningham model, and 0.2487 for the model with a longitudinal arrangement of fibers.

It follows from the cited rms deviations and dependences shown in Fig. 13.17 that the Cunningham model better corresponds to experimental data than the Happel, Kuwabara, and Kvashnin models. The model with longitudinal arrangements of fibers with respect to the direction of flow much worse describes experimental data.

When considering other models of a porous medium, the character of dependences of permeability on the fraction of solid phase remains the same as in Fig. 13.17.

The values of rms deviations for different models are listed in the Table 13.4. Columns with numbers 1–4 refer to models with transversal arrangement of fibers with respect to the flow: Ha corresponds to the Happel model, Kv corresponds to Kvashnin, Ku corresponds to Kuwabara; and Cu corresponds to the Cunningham model. The fifth column contains the values of errors for the model with longitudinal arrangements of fibers. Columns 6–13 correspond to the models with random arrangements of fibers where columns six through nine are models containing transversal and longitudinal fibers in 1:1 ratio, while 10–13 columns refer

Table 13.4 Values of root-mean-square deviation for different models

Arrangement	Perpendicular				Along	Random (1/2 and 1/2)				Random (2/3 and 1/3)			
No. Model	1 Ha	2 Kv	3 Ku	4 Cu	5 Along	6 Ha	7 Kv	8 Ku	9 Cu	10 Ha	11 Kv	12 Ku	13 Cu
1	0.1168	0.1060	0.1015	0.0941	0.2487	0.1905	0.1849	0.1821	0.1735	0.1673	0.1594	0.1554	0.1434
$1 + 2.5c$	0.1161	0.1055	0.1011	0.0942	0.2505	0.1921	0.1865	0.1836	0.1750	0.1688	0.1608	0.1567	0.1446
$1/(1 - c)^{2.5}$	0.1161	0.1055	0.1011	0.0942	0.2505	0.1921	0.1865	0.1837	0.1750	0.1688	0.1608	0.1568	0.1446

to the same mixed arrangements of fibers in 2:1 ratio. Flows with constant (fourth line) and variable (5th–6th lines) viscosities are considered. The Cunningham model ($\sigma = 0.0941$) describing the flow in a porous medium composed of cylindrical fibers arranged perpendicular to the flow is characterized by the minimal error among all of the models considered. Maximal error ($\sigma = 0.2487$) is registered in the model with longitudinal arrangements of fibers. Note that the porous layer with a very poor content of solid phase was considered and, therefore, the viscosity ratios m calculated by formulas (13.4)–(13.5) slightly differ from unity and, correspondingly, errors for models with different viscosities nearly coincide with one another.

As was mentioned above, it should be assumed that, for highly porous media, viscosities of liquid in the bulk and the layer are equal.

13.6 Observations on the Cell Models

The studies performed make it possible to draw the following conclusions.

(1) Cell models of the flow in porous media formed by partially porous spherical particles or cylinders adequately describe the filtration of highly concentrated media (porosity is equal to 0.6 or less), when using the Happel, Kvashnin, and Kuwabara boundary conditions. Note that, as was theoretically shown in a recently published work [10], the use of the Kuwabara boundary condition, in contrast to other types of boundary conditions, does not lead to contradictions when studying electrokinetic phenomena.

(2) The Cunningham boundary condition poorly describes the influence of neighboring particles on the flow in a chosen cell and, hence, cannot be recommended for application.

(3) It is clear that the dynamics of flows in channels bounded by solid and/or porous walls may be adequately described, provided that the physics of nonequilibrium processes that take place at the liquid–solid interface is clear. It is often the case that the classical boundary condition of sticking (non-slipping), which has been used for more than a century, cannot adequately describe experimental data on liquid flows in micro and nanosized channels. Therefore, the study of the character of liquid flows near solid surfaces has now become not only of theoretical, but also of great practical significance. In particular, when extracting oil reserves that are residual or difficult to extract from porous rocks, various polymers (polyelectrolytes), ionic surfactants, or gases are added to displacing fluids. This causes a noticeable change in the character of the oil flow in micropores due to a reduction in the viscous friction in boundary layers. As a result, the coefficient of oil displacement from a porous medium may be significantly increased. The data on the hydrodynamic forces that act on molecularly smooth hydrophilic surfaces are adequately described by the Reynolds lubrication theory with the use of the

sticking conditions. However, molecularly smooth surfaces are extremely rare. Commonly, surfaces are rough, porous, or rough and porous simultaneously. In practice, we encounter liquid flows in channels or pores with such surfaces. They can be exemplified by flows in capillaries, chromatographic columns, processes of wetting and spreading of liquids on porous surfaces, and tangential flows of solutions between planar walls of apparatuses covered with membranes, solid particles, proteins, and monomolecular or bimolecular layers of polyelectrolytes. In all these cases, a liquid partly flows either inside a porous layer or near a rough surface. The effective velocity of a liquid flow may be reduced in these situations. At the same time, the covering of surfaces with hydrophobizing layers causes an increase in the liquid flow velocity, which is related to a slip on an interfacial surface [42, 43]. The degree of deviation from complete sticking is commonly characterized by a parameter that is referred to as the slip length, which represents a distance such that the extrapolation of the weighted velocity profile to which yields the zero flow velocity. In Ref. [44] three different mathematical models have been proposed for describing liquid flow in a long cylindrical capillary, the internal surface of which is covered with an adsorbed porous layer and experimental data on an increase in the liquid flow rate in such a capillary [45] have been explained in terms of these models. In the first model, the liquid flow in the porous layer is described by the Brinkman equation; according to the second one, the presence of the porous layer is taken into account using the Navier slip boundary conditions; and in the third model the Navier condition is imposed on the porous layer—liquid interface, with the flow inside the porous layer being excluded. The theoretical predictions are compared with the experimental data that was obtained for liquid flow rates in porous capillaries. The validity and appropriateness of the application of the proposed models are discussed. The selection of a model is governed by experimental conditions and the features of a system under examination. The results of [44] can be used for further expansion of the cell method.

(4) Paper [46] concerns the flow of an incompressible, viscous fluid past a porous spherical particle enclosing a solid core, using also the particle-in-cell method. The Brinkman equation in the porous region and the Stokes equation for clear fluid are used. At the fluid–porous interface, the stress jump boundary condition for the tangential stresses along with continuity of normal stress and velocity components are employed. No-slip and impenetrability boundary conditions on the solid spherical core have been used. The hydrodynamic drag force experienced by a porous spherical particle enclosing a solid core and the permeability of a membrane built up by solid particles with a porous shell are evaluated. It is found that the hydrodynamic drag force and dimensionless hydrodynamic permeability depends not only on the porous shell thickness, particle volume fraction γ and viscosities of porous and fluid media, but also on the stress jump coefficient. Four known boundary conditions on the hypothetical surface are also considered and compared: Happel's, Kuwabara's, Kvashnin's, and Cunningham's (Mehta–

Morse's) conditions. Some previous results for the hydrodynamic drag force and dimensionless hydrodynamic permeability reported in the present chapter have been verified.

(5) A hydrodynamic permeability of membranes built up by porous cylindrical particles with impermeable core is investigated in [47]. Different versions of the cell method are used to calculate the hydrodynamic permeability of membranes. A possible jump of a shear stress at the fluid-membrane interface, and its impact on the hydrodynamic permeability is also investigated.

(6) Tensor of the permeability coefficients can be constructed in the principal axes in the following forms (see Tables 13.1 and 13.2):

for isotropic porous medium consisting of spherical particles randomly distributed in the space,

$$\tilde{k}_{ij} = \begin{pmatrix} \tilde{\mu}^o \tilde{L}_{11} & 0 & 0 \\ 0 & \tilde{\mu}^o \tilde{L}_{11} & 0 \\ 0 & 0 & \tilde{\mu}^o \tilde{L}_{11} \end{pmatrix} \tag{13.98}$$

for isotropic porous medium consisting of cylindrical particles randomly distributed in the space,

$$\tilde{k}_{ij} = \begin{pmatrix} \tilde{\mu}^o \tilde{L}_{11}^R & 0 & 0 \\ 0 & \tilde{\mu}^o \tilde{L}_{11}^R & 0 \\ 0 & 0 & \tilde{\mu}^o \tilde{L}_{11}^R \end{pmatrix} \tag{13.99}$$

for transversal isotropic porous medium consisting of cylindrical particles with parallel axes in the space,

$$\tilde{k}_{ij} = \begin{pmatrix} \tilde{\mu}^o \tilde{L}_{11}^\perp & 0 & 0 \\ 0 & \tilde{\mu}^o \tilde{L}_{11}^\perp & 0 \\ 0 & 0 & \tilde{\mu}^o \tilde{L}_{11}^\| \end{pmatrix} \tag{13.100}$$

Note that an alternative to the cell method in the case of a porous transversely isotropic medium is a method of representing a porous medium as a set of straight and parallel capillaries-pores having a given distribution $f(r)$ for the dimensionless radii. At the same for getting dimensionless quantities, all radii are divided by some characteristic value, for example, the average pore radius a_0, so dimensionless minimum and maximum pore radii are equal to $r_{min} = a_{min}/a_0$ and $r_{max} = a_{max}/a_0$, correspondingly. It is usually assumed that liquid flow in each capillary opened for filtration is governed by the Hagen-Poiseuille law. Then the expression for the specific permeability in the direction of the capillary axes is obtained by integrating over all elementary fluxes through all opened pores [48]:

$$k_D = \frac{a_0^2 \varepsilon}{8} \cdot \int\limits_{r_{\min}}^{r_{\max}} r^4 f(r) \mathrm{d}r \bigg/ \int\limits_{r_{\min}}^{r_{\max}} r^2 f(r) \mathrm{d}r, \qquad (13.101)$$

Differential distribution function $f(r)$ can be experimentally determined by the MSCP. If the membrane is hydrophobic with respect to the liquid to be filtered, then at a given pressure drop ΔP only part of its pores with radii $a > a_{cr}$ are open for filtration. The critical pore radius a_{cr} is determined by the Laplace equation:

$$\Delta P = \frac{2\sigma}{a_{cr}} \cos\theta \qquad (13.102)$$

where σ—the surface tension at the interface "displacing liquid—displaced liquid (gas)," θ—the contact angle of wetting the capillary walls by displacing liquid. In the case under consideration, the lower limit of the first integral in (13.101) must be replaced by $r_{cr} = a_{cr}/a_0$.

The situation opposite to the above-described arises when initially all the pores of the membrane are opened for filtering suspensions and a part of them is clogged in time by suspended particles. This case has the important practical application in the membrane microfiltration, discussed in detail below, and also during penetration of proppant particles inside hydraulic fractures in the process of hydrocarbons extraction, that is simulated in Chap. 8.

13.7 Mathematical Modeling of Microfiltration of Polydisperse Suspension Through Heterogeneous Membranes

The process of microfiltration in the dead-end regime, when the flow of a polydisperse suspension, which has the log-normal density distribution of particle diameters, is perpendicular to the membrane surface, has been considered. The probabilistic sieving mechanism is used to describe the process of filtration through a heterogeneous membrane that has a bi-lognormal density of distribution of pores diameters. The integral equation for determining the specific productivity of the membrane in time is solved numerically for five different positions of the maximum of the distribution density of particles diameters relative to the two maximums of the distribution density of membrane pores size. The results are illustrated by plots to reflect time changes in specific productivity, rejection coefficient of the membrane and in pore size distribution during pore blocking.

Fig. 13.18 Schematic representation of the microfiltration process: (*1*) a particle has adhered to the membrane surface, (*2*) a particle has clogged the membrane pore, (*3*) a small particle passes through a large pore, and (*4*) a particle will plug the membrane pore

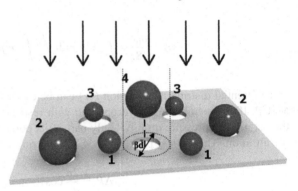

13.7.1 Preliminary Remarks

The membrane microfiltration technique is widely used to purify liquids by removal of suspended particles of 0.1–20 μm size [49]. For example, it is known that ordinary drinking water contains up to 100,000 of these microparticles in 1 cm^3, while the limit value is 500 for the needs of microelectronics and even two particles in 1 cm^3 in some other cases [50]. The process of microfiltration separation is characterized by a rise in hydraulic resistance of the membrane (drop in the liquid flux) with time due to clogging membrane pores by retained particles and the formation of a layer from deposited colloidal particles [49]. In the case of running the process at a constant pressure, the membrane productivity decreases; in order to maintain a constant productivity, it is necessary to increase the transmembrane pressure. The experimental results and analytical estimates reported to date show that the effect of osmotic pressure during microfiltration is negligible compared with the flux reduction due to the resistance of the additional surface layers of dispersed material with a thickness of about 1 μm, deposited at relatively high concentrations of dispersed particles in the feed solution, and membrane pore blockage, which plays a significant role in the case of dilute dispersions [51, 52]. A microphotography study has shown that individual particles do not enter the pore structure [52]. Even in the case of ultrafiltration of proteins, the number of their molecules penetrating into the membrane pores is negligible [53]. Such behavior of particles suggests that the membrane operates mainly by the sieving mechanism at a low concentration of the dispersed phase.

The sieving mechanism is characterized by the following conditions: if the diameter of a particle that approaches the pore mouth is smaller than the pore diameter, this particle passes into the permeate, but if the particle diameter exceeds the diameter of the pore, it is retained and the pore is blocked and eliminated from further separation process. In this case, the hydrodynamic resistance of the membrane increases. The "approach to the pore" concept is defined in terms of the "pore influence area" (Fig. 13.18). In this respect, for a single pore with a diameter $\tilde{d}$ and a cross-sectional area $\pi\tilde{d}^2/4$, all particles over the membrane in a cylinder with a base area $\beta\pi\tilde{d}^2/4$ either pass through this pore or plug it depending on the pore diameter $\tilde{d}$. If a single particle is outside the cylinder, it is assumed that the pore in question does not affect in any way the behavior of this particle. The parameter $\beta > 1$ takes into account the hydrodynamics of the process and the specific surface interaction forces between the particle and the membrane pore.

The sieving mechanism of pore blocking was observed as early as more than 30 years ago [54]. However, for a long time it was just the attempts to approximate the experimental dependence of the productivity of microfilters upon time by specially selected functions that were being made [55]. In this case, the physical meaning of fitting constants stayed in the background, not enabling the empirical relationships to be extended to calculation of processes with new membranes or under other conditions (e.g., pressure). Comparative analysis of the effect of filtration mechanisms, such as sieving and silting the working surface, on membrane fouling was given in [56].

Hubble [57] who studied the separation of proteins on microfiltration membranes considered filtration in terms of the bi-parametrical model of unblocked and partially blocked membrane pores. It was assumed that the ratio of the number of unblocked and partially blocked pores depends linearly on the ratio of the observed flux to a certain pressure invariant flux. Note that the time dependence for the rate of the dispersion flux drop across the membrane also was not considered in [57].

A log-normal pore size distribution was observed in a number of studies, as well as the effect of membrane contamination during the filtration process on change in this distribution [58, 59]. However, neither the continuous fall in productivity nor the change in pore size distribution with time has been analyzed in these studies. A stochastic model of deep bed filtration was proposed and developed in [60, 61]. Later, this model was used in the form complemented with particle diameter and pore size distribution functions. In Refs. [62–64], it was noted that the membrane rejection factor has a log-normal distribution, depending on the molecular weight of the filtered material, a fact that can be interpreted as a consequence of the log-normal pore size distribution of the membrane.

We consider here the probabilistic sieving model of dead-end microfiltration, using the common approach proposed by Kutepov and Sokolov [65] and further developed in [66–70]. In those studies, particular cases of membranes with one or two types of pores that allow for the exact solution of the problem were considered, experimental data on microfiltration of tap water through Russian MMF (fluoroplastic) membranes were theoretically described, and the similarity property of microfiltration processes was revealed. In addition, approximate formulas for calculating the productivity and rejection of the membranes were derived with some weak restrictions on the pore diameter distribution law. The approximate formulas were confirmed by both numerical calculations and experimental results with the use of track-etched membranes.

13.7.2 Theoretical Model

Let us consider the dead-end microfiltration process when the flow of a polydisperse colloid solution is perpendicular to the membrane surface (Fig. 13.18). It is convenient to make all the diameters of the dispersion particles and membrane pores dimensionless using a characteristic length $(\tilde{l}*)$, which can be the smallest particle diameter or membrane pore diameter. Let the colloid solution have a certain preset differential distribution function $f_p(\tilde{D})$ of particle diameters. Let the probability density of the diameter distribution of the membrane pores $f_m(\tilde{d})$ be also known. Note that either log-normal or bi-lognormal distribution function is commonly used for this purpose [58, 59]. We believe that the suspension has a constant dispersion composition, i.e., the function $f_p(\tilde{D})$ is not dependent on time. Let $\tilde{D}_{min}$ and $\tilde{D}_{max}$ denote respectively the minimum and the maximum diameters of particles in the solution; $c_0 = $ const be the number density of dispersed particles; $\tilde{d}_{min}$ and $\tilde{d}_{max}$ be the smallest and the largest membrane pore diameters, correspondingly. Below, we will denote all the dimensionless lengths and time without the upper tilde.

The diffusion coefficient of a particle can be defined as $\sim k_B T/(3\pi\mu\tilde{D})$ where k_B is the Boltzmann constant, T is the absolute temperature in kelvins, and μ is the viscosity of the suspension. Then, the diffusion coefficient for particles of ~ 1 μm diameter will be $\sim 10^{-13}$ m^2/s. In fact, the diffusion coefficient can be even smaller because of a considerable reduction in the hydrodynamic mobility near the wall [71–73]. We estimate the characteristic distance between the pores at $(S/N_m)^{1/2} \sim (\pi\tilde{d}^2/4\varepsilon)^{1/2}$, where S is the surface area of the membrane and N_m is the total number of pores on the membrane surface. For track-etched membranes as an example, the porosity ε is ~ 3 %; $\tilde{d} \sim 1$ μm; and, consequently, the characteristic distance is $\sim 10^{-5}$ m. Thus, for a typical filtration rate of $\sim 10^{-5}$ m/s and a Peclet number of $\sim 10^3 \gg 1$, the diffusion of particles and the phenomenon of concentration polarization can be ignored. In this sense, microfiltration differs from ultrafiltration in which the concentration polarization is substantial [74, 75] and the true rejection coefficient differs from its apparent value [76, 77].

It is natural to assume that the fluid motion in every pore of the membrane obeys the Hagen–Poiseuille law; i.e., $q = \frac{\pi\tilde{l}_*^4\tilde{d}^4}{128\mu\tilde{L}}\Delta p$, where q is the flux of the fluid through a pore with a diameter $\tilde{d}$, $\tilde{L}$ is the membrane pore length, and Δp is the pressure difference across the membrane. Summing up the fluxes of fluid in all the pores of the fresh membrane, we obtain the following equation for the initial productivity:

$$J_0 = \pi N_m \tilde{l}_*^4 \Delta p / 128\mu\tilde{L}SA \qquad (13.103)$$

where the parameter A is related to the size distribution function of membrane pores as follows [67]:

$$A = \left[\int\limits_{d_{min}}^{d_{max}} x^4 f_m(x) \mathrm{d}x \right]^{-1} \tag{13.104}$$

At time $\tilde{t}$ with allowance for pore blocking, the membrane productivity is given by

$$J(\tilde{t}) = J_0 A \int\limits_{d_{min}}^{d_{max}} x^4 f_m(x) P(x, \tilde{t}) \mathrm{d}x \tag{13.105}$$

Equation (13.105) includes the probability $P(d, \tilde{t})$ that the single pore with a diameter d will not be blocked at time $\tilde{t}$ when an ensemble of particles with diameters $D > d$ approaches it. Since the probability densities $f_p(D)$ and $f_m(d)$ are integrally normalized to unity, the number density of particles with a diameter $D > d$ in the suspension is $c_d = c_0 \int_d^{D_{max}} f_p(D) \mathrm{d}D$. Assuming that the diffusion flux is negligible, we define the amount of these particles to approach the working surface of the membrane at a time $\tilde{t}$ as follows:

$$n(d, \tilde{t}) = c_0 S \int\limits_0^{\tilde{t}} J(u) \mathrm{d}u \int\limits_d^{D_{max}} f_p(D) \mathrm{d}D \tag{13.106}$$

Then, as shown in [67–70], the above probability is given by the expression

$$P(d, \tilde{t}) = \left(1 - \frac{\pi \beta \tilde{l}_*^2 d^2}{4S} \right)^{n(d, \tilde{t})} \approx \exp\left[-(\pi \beta \tilde{l}_*^2 d^2 c_0 / 4) \times \int\limits_0^{\tilde{t}} J(u) \mathrm{d}u \int\limits_d^{D_{max}} f_p(D) \mathrm{d}D \right] \tag{13.107}$$

provided that the obvious conditions $\pi \tilde{l}_*^2 d^2 / 4 \ll S$ and $n(d, \tilde{t}) \gg 1$ are fulfilled. Here, the term $1 - \frac{\pi \beta \tilde{l}_*^2 d^2}{4S}$ stands for the probability that the membrane pore is not blocked when a single particle with a diameter $D > d$ approaches it. Introducing the dimensionless flux $j = J/J_0$ and the dimensionless time $t = \tilde{t}/\tilde{t}_*$, where $\tilde{t}_* = 4/\pi \beta \tilde{l}_*^2 c_0 J_0$ is the characteristic time of the filtration process [68], we get the integral equation for the dimensionless productivity of the membrane in the following form:

$$j(t) = A \int_{d_{\min}}^{d_{\max}} x^4 f_m(x) \exp\left[-\int_0^t j(\zeta)d\zeta \times x^2 \times H(D_{\max} - x) \int_x^{D_{\max}} f_p(y)dy \right] dx$$

$$(13.108)$$

where $H(x)$ is the Heaviside function.

The expression for the membrane rejection coefficient as a function of time is then given by [66–70]:

$$\phi(t) = 1 - \left(N_m \beta \pi \tilde{l}_*^2 / 4S \right) \times \int_{D_{\min}}^{\min(d_{\max}, D_{\max})} \left(\int_y^{d_{\max}} x^2 f_m(x) P(x, t) dx \right) f_p(y) dy$$

$$(13.109)$$

It is noteworthy to mention once again the features of separation mechanism that distinguish microfiltration from reverse osmosis [78] and ultrafiltration. Figure 13.18 illustrates this mechanism in the case of microfiltration. Particles adhering to the membrane surface and particles blocking the membrane pores are rejected by the membrane. Only particles from the pore influence area, which have a diameter smaller than the pore diameter pass through the pore. All of these elementary events are taken into account in Eq. (13.107). Note that the given model does not suggest pore blocking by large particles unless their center falls on the area occupied by the pore influence region, although a geometrically large particle may in principle close a small pore by its "shadow." Thus, finding the productivity $J(t)$ and the retention $\phi(t)$ of the membrane at time t is reduced to solving the integral Eq. (13.108).

13.7.3   Numerical Calculations

To draw the time-dependent curves for the dimensionless productivity j and retention ϕ and to determine the evolution of the distribution of unblocked pores, numerical calculations were performed (Figs. 13.19–13.22). In the calculations, all the lengths were made dimensionless by $l_* = 10^{-8}$ m. Let the suspension have a

log-normal diameter distribution function of suspended particles $f_p(D) = \dfrac{\exp\left(-\ln(D - e_p)^2 / 2\sigma_p^2 \right)}{\sqrt{2\pi}\sigma_p D}$ with the average particle diameter $E_p = \exp\left(\sigma_p^2/2 + e_p \right)$ and a variance $\Sigma_p = \exp\left(\sigma_p^2 + 2e_p \right) \left[\exp\left(\sigma_p^2 - 1 \right) \right]$. Let us also assume that the

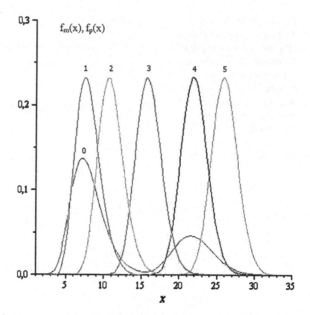

Fig. 13.19 Plots of (*0*) the bi-lognormal distribution density of membrane pore diameters and (*1–5*) log-normal distribution densities of particle diameters of a suspension [70]

membrane pore diameters are distributed according to the bi-lognormal law $f_m(d) = \alpha f_m^1(d) + (1 - \alpha)f_m^2(d)$ where f_m^1, f_m^2 are the log-normal distribution functions with average pore diameters $E_{m1} = \exp(\sigma_{m1}^2/2 + e_{m1})$ and $E_{m2} = \exp(\sigma_{m2}^2/2 + e_{m2})$ and variances $\Sigma_{m1} = \exp(\sigma_{m1}^2 + 2e_{m1})[\exp(\sigma_{m1}^2 - 1)]$, $\Sigma_{m2} = \exp(\sigma_{m2}^2 + 2e_{m2})[\exp(\sigma_{m2}^2 - 1)]$ respectively, and α is the fraction of small pores.

We used the following parameters of bi-lognormal distribution of membrane pores: $\alpha = 0.7$, $E_{m1} = 8$, $D_{m1} = 5$, $E_{m2} = 22$, $D_{m2} = 7$, $d_{\min} = 1$, $d_{\max} = 35$ $(A = 1.555 \cdot 10^{-4})$ and log-normal size distributions of suspended particles (Fig. 13.19):

1) $E_p = 8$, $D_p = 3$, $D_{\min} = 1$, $D_{\max} = 11$;   2) $E_p = 11$, $D_p = 3$, $D_{\min} = 6$, $D_{\max} = 16$;
3) $E_p = 16$, $D_p = 3$, $D_{\min} = 11$, $D_{\max} = 21$; 4) $E_p = 22$, $D_p = 3$, $D_{\min} = 17$, $D_{\max} = 27$;
5) $E_p = 26$, $D_p = 3$, $D_{\min} = 21$, $D_{\max} = 31$.

Note that in the numerical solution of Eq. (13.108), explicit finite difference schemes (Runge–Kutta and Euler methods), an implicit scheme with simple iteration searching, and an implicit scheme with searching by dichotomy were used. At the maximum position of the distribution function for particles near large pores of the membrane, numerical instability was observed when conditionally stable explicit finite difference schemes were used. To implement the computation process in these cases, unconditionally stable implicit finite difference schemes were used.

Fig. 13.20 Plots of the dimensionless productivity j for the bi-lognormal size distribution of membrane pores and the log-normal size distribution of particles in the suspension, shown in Fig. 13.19, versus the dimensionless time t [70]

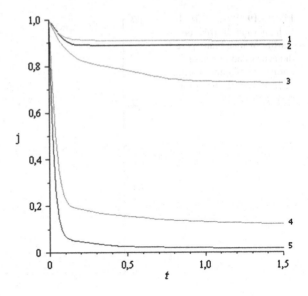

In order to solve Eq. (13.108) for distributions of particles 3 and 4 (Fig. 13.19), it is necessary to employ the implicit scheme involving simple iterative search. For case 5 of the particles distribution, the solution was obtained with the use of an implicit scheme with searching by dichotomy, since the search by simple iteration is only conditionally stable. Although explicit difference schemes have significant shortcomings due to conditional stability, their use is justified because it enables a solution to be found much faster than in the case of implicit schemes.

Figure 13.20 shows that the dimensionless flux through the membrane rapidly decreases at the beginning of the process and then changes little with time. Figure 13.21 shows monotonically increasing membrane retention ϕ plotted against the dimensionless time t for the density distributions presented in the same Fig. 13.19.

It is seen that unlike productivity, this quantity varies unpredictably from one distribution of particles to another, thereby reflecting the complex time behavior of the clogging process associated with an overlap of different distribution densities of particle sizes and membrane pore diameters (Fig. 13.19). The big difference between curves *3* and *4* in the plot of the membrane productivity j is explained in terms of the fact that small pores are basically blocked in cases 1 and 2 (Fig. 13.22a, b), since primarily small particles contained in the suspension are not able to block large pores. However, the opposite picture is observed in the case of curves *4* and *5* in Fig. 13.20 (see Fig. 13.22d, e), primarily large particles in the suspension first block the large pores, which make a significant contribution to the total productivity. Case 3 (Fig. 13.22c) refers to the intermediate situation when the suspension contains particles with a middle diameter falling in the interval between the maximums of the size distribution function.

Fig. 13.21 Plots of the retention ϕ for the bi-lognormal size distribution of membrane pores and the log-normal size distribution of particles in the suspension, shown in Fig. 13.19, versus the dimensionless time t [70]

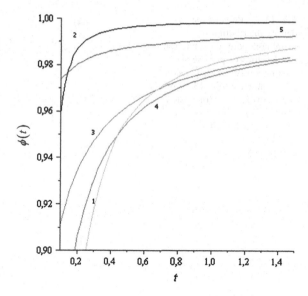

Figure 13.22a–e characterize the time change in the diameter distribution of unblocked membrane pores for the types of the particle diameter distribution function shown in Fig. 13.19. It can be seen that the fraction of pores with smaller diameters drops with time for cases Fig. 13.22a–c and increases with a simultaneous increase in the distance between the maximums for cases Fig. 13.22d, e. Moreover, the fraction of large pores that remain unblocked is greater than that of the small ones in Fig. 13.22d, while Fig. 13.22e shows the opposite behavior. Thus, the time behavior of the density distribution of unblocked pores, as shown in Fig. 13.22d, e, can be interpreted as the transformation of the bimodal membrane into a unimodal microfiltration and a unimodal ultrafiltration membrane, respectively. Note that the use of the log-normal law for describing the size distribution of dispersed particles fully complies with the physical thoughts on the random nature of such a distribution [79].

13.8   Comparison Between the Cell and Probabilistic Models

Despite the difference between the models which are considered here, there are common points. The main of them is that both kinds of models are based on using the Navier-Stokes equations. Second one is that all models apply an averaging procedure over particles or pores assemble. And third—all models are in a good agreement with available experimental data [22, 38, 66, 68, 69].

Fig. 13.22 Plots of the
density distribution of
membrane pore diameters for
the densities of particle size
distribution of the suspension
$f_p(x)$, shown in Fig. 13.19 by
curves (**a**) *1*, (**b**) 2, (**c**) 3, (**d**) *4*
and (**e**) *5*, at different points
of dimensionless time $t = (1)$
0, (*2*) 0.07, (*3*) 0.14, and
(*4*) 0.7 [70]

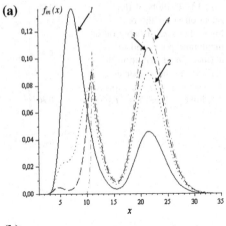

(**a**)

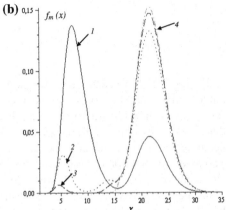

(**b**)

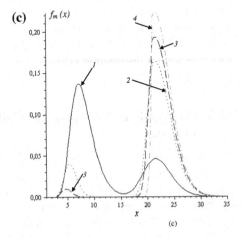

(**c**)

Fig. 13.22   continued

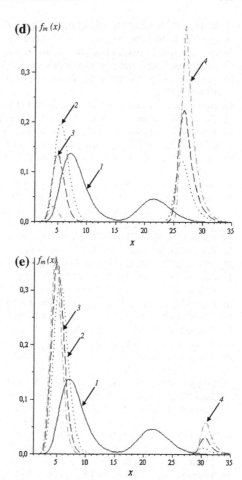

Acknowledgment   The support of Russian Science Foundation (RSF) is acknowledged (grant No 14-19-01045).

References

1. Happel J, Brenner H (1965) Low Reynolds number hydrodynamics with special applications to particulate media. Prentice-Hall, Englewood Cliffs (reprinted by Wolters-Nordhoff, 1973); paperback edition, (Martinus Nijhoff; Kluwer Academic Publishers, 1983)
2. Barenblatt GI, Entov VM, Ryzhik VM (1990) Theory of fluid flows through natural rocks. Kluwer Academic Publishers, Dordrecht
3. Whitaker S (1986) Transp Porous Med 1:3
4. Auriault J-L (2009) Transp Porous Med 79:215
5. Brinkman HC (1947) Appl Sci Res A1:27
6. Starov VM, Zhdanov VG (2008) Adv Colloid Interface Sci 137:2

7. McHale G, Newton MI, Shirtcliffe NJ (2010) Soft Matter 6:714
8. Qin Yu, Kaloni PN, Angew Z (1993) Math Mech 73:77
9. Greenkorn RA (1983) Flow phenomena in porous media: fundamentals and applications in petroleum, water, and food production. Marcel Dekker, New York
10. Zholkovskij EK, Masliyah JH, Shilov VN, Bhattacharjee S (2007) Adv Colloid Interface Sci 134–135:279
11. Kotov AA, Solomentsev YuE, Starov VM (1991) Prog Colloid Polym Sci 84:293
12. Kotov AA, Solomentsev YuE, Starov VM (1991) Colloid J 53(6):867
13. Kotov AA, Solomentsev YuE, Starov VM (1992) Int J Multiph Flow 18(5):739
14. Churaev N, Sergeeva I, Derjaguin B (1981) J Colloid Interface Sci 84:451
15. Churaev NV, Kotov AA, Solomentsev YuE, Starov VM (1991) Prog Colloid Polym Sci 84:290
16. Solomentsev YuE, Starov VM (1993) Colloid J 54(4):591
17. Starov VM, Solomentsev YuE (1993) J Colloid Interface Sci 158:159
18. Starov VM, Solomentsev YuE (1993) J Colloid Interface Sci 158:166
19. Varoqui R, Dejardin P (1977) J Chem Phys 66:4395
20. Hiller J, Hoffman H (1953) J Comp Physiol 42:203
21. Parsons D, Subjeck J (1972) Biochem Biophys Acta 55:440
22. Masliyah J, Neale G, Malysa K, van de Ven T (1987) Chem Eng Sci 42:245
23. Garvey M, Tadros Th, Vincent B (1975) J Colloid Interface Sci 55:440
24. Pefferkorn E, Dejardin P, Varoqui R (1975) J Colloid Interface Sci 63:353
25. Idol WK, Anderson JL (1986) J Membr Sci 28(3):269
26. Anderson JL, Kim J (1987) J Chem Phys 86:5163
27. Perepelkin PV, Starov VM, Filippov AN (1992) Colloid J 54(2):139
28. Vasin SI, Starov VM, Filippov AN (1996) Colloid J 58(3):291
29. Vasin SI, Filippov AN (2004) Colloid J 66(3):261
30. Vasin SI, Starov VM, Filippov AN (1996) Colloid J 58(3):282
31. Filippov AN, Vasin SI, Starov VM (2006) Colloids Surf A: Physicochem Eng Aspects 282–283:272
32. Vasin SI, Filippov AN, Starov VM (2008) Adv Colloid Interface Sci 139:83
33. Koplic J, Levine H, Zee A (1983) Phys Fluids 26(10):2864
34. Kuwabara S, Rhys J (1959) Soc Jpn 14:527
35. Kvashnin AG (1979) Izv Akad Nauk SSSR (Proceedings of the Academy of Sciences of the USSR), Mekh Zhidk Gaza (Mechanics of Liquid and Gas) 4, 154 (in Russian)
36. Mehta G, Morse T (1975) J Chem Phys 63(5):1877
37. Cunningham E (1910) Proc R Soc (London) A83:357
38. Vasin SI, Filippov AN (2009) Colloid J 71(1):31
39. Vasin SI, Filippov AN (2009) Colloid J 71(2):141
40. de Groot S, Mazur P (1962) Non-equilibrium thermodynamics. North-Holland, Amsterdam
41. Kirsh VA (2006) Colloid J 68(2):173
42. Sergeeva IP, Semenov DA, Sobolev VD, Churaev NV (2008) Colloid J 70(5):616
43. Churaev NV, Sobolev VD, Somov AN (1984) J Colloid Interface Sci 97:574
44. Filippov AN, Khanukaeva DYu, Vasin SI, Sobolev VD, Starov VM (2013) Colloid J 75(2):214
45. Kiseleva OA, Sobolev VD, Semenov DA, Ershov AP, Sergeeva IP, Churaev NV (2009) Colloid J 71(1):76
46. Yadav PK, Tiwari A, Deo S, Filippov A, Vasin S (2010) Acta Mech 215:193
47. Deo S, Filippov A, Tiwari A, Vasin S, Starov V (2011) Adv Colloid Interface Sci 164:21
48. Ivanov VI (2011) Vestnik Nizhegorodskogo universiteta imeni N.I.Lobachevskogo. Mehanika zhidkosti i gaza (Vestnik of Lobachevsky State University of Nizhny Novgorod, Mechanics of Liquid and Gas) 4(2), 438 (in Russian)
49. Brock TD (1983) Membrane filtration: a user's guide and reference manual. Science Tech, Madison
50. Kemmer FN (ed) (1987) The NALCO water handbook, 2nd edn. McGraw-Hill, New York

51. Cohen RD, Probstein RF (1986) J Colloid Interface Sci 114:194
52. Schulz G, Ripperger S (1989) J Membrane Sci 40:173
53. Suki A, Fane AG, Fell CJD (1986) J Membrane Sci 27:181
54. Howell JA, Velicangil O, Lee MS, Herrera-Zeppelin AL (1981) Ann NY Acad Sci 369:355
55. Yasminov AA, Grekov AV, Gaidukova IP et al (1988) Vysokochist. Veshchestva (High-Purity Substances) 1, 110 (in Russian)
56. Persson KM, Nilsson JL (1991) Desalination 80(2–3):123
57. Hubble J (1989) Fouling and cleaning in food processing.In: Kessler HG, Lund DB (eds) Third international conference on fouling and cleaning in food processing, Prien, Bavaria, June 1989. p 239
58. Aimar P, Meireles M, Sanchez V (1990) J Membrane Sci 54(3):321
59. Meireles M, Aimar P, Sanchez V (1991) J Membrane Sci 56(1):13
60. Fan LT, Nassan R, Hwang SH, Chou ST (1985) AIChE J 1781
61. Fan LT, Hwang SH, Chou ST, Nassan R (1985) Chem Eng Commun 35:101
62. Polotskii AE, Cherkasov AN (1983) Colloid J 43:467
63. Cherkasov AN (1985) Colloid J 47:363
64. Cherkasov AN, Vlasova OL, Tsareva SV et al (1990) Colloid J 52:323
65. Kutepov AM, Sokolov MV (1986) Teor Osn Khim Tekhnol (Theoretical Foundations of Chemical Engineering) 19:123
66. Filippov AN, Starov VM, Gleizer SV, Yasminov AA (1990) Khim Tekhnol Vody (Water Chemistry and Technology) 12:483 (in Russian)
67. Torkunov AM, Filippov AN, Starov VM (1992) Colloid J 54:126
68. Filippov AN, Starov VM, Lloyd DR et al (1994) J Membr Sci 89:199
69. Starov V, Lloyd D, Filippov A, Glaser S (2001) Sep Purif Technol 26:51
70. Filippov AN, Iksanov RKh (2012) Petrol Chem 52(7):520
71. Brenner H (1961) Chem Eng Sci 16:242
72. Goldmann AJ, Cox RG, Brenner H (1967) Chem Eng Sci 22:637
73. Goldmann AJ, Cox RG, Brenner H (1967) Chem Eng Sci 22:653
74. Liu MK, Williams FA (1970) Int J Heat Mass Transfer 13:1441
75. Trettin DR, Doshi MK (1980) Ind Eng Chem Fundam 19:189
76. Blatt WF, Dravid A, Michaels AS, Nelsen L(1970) Membrane science and technology. In: Flinn JE (ed). Plenum, New York, p 47
77. Opong WS, Zydney AI (1991) AIChE J 37:1497
78. Starov VM, Churaev NV (1993) Adv Colloid Interface Sci 43:145
79. Kolmogorov AN (1941) Dokl Akad Nauk SSSR 31:99

Conclusion

Figure 1 illustrates a cross section of the stem of horsetail. One can see the perfection in the porous structure of plants, created by nature. At the cut it is seen that, first, it has *radial* symmetry, and, second, the pores of approximately the same dimensions (D) are arranged regularly in a circle in the form of clusters of the same radius R. Where the function D of R is complex but perhaps not random and rather *reasonable*, it is probably due to the optimization of moisture transport in the stem. This figure illustrates that NATURE optimized pore structure of plants in the process of evolution. Not only plants, but also the skin (see Chap. 12) and bones of animals, etc. In this regard, it is clear why bionics—applied science which uses principles of organization, properties, functions, and structures of nature in technical devices and systems, that is, the study of living forms in nature and their industrial counterparts, has appeared. For example, when creating artificial skin and bones, engineers seek to emulate relevant natural analogs.

Fig. 1 Cross section of the stem of horsetail

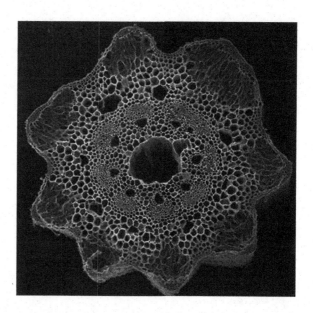

Y. M. Volfkovich et al., *Structural Properties of Porous Materials and Powders Used in Different Fields of Science and Technology*, Engineering Materials and Processes, DOI: 10.1007/978-1-4471-6377-0, © Springer-Verlag London 2014

In this book, the authors examined the porous structure of various synthetic and natural materials and its impact on the functional properties of the latter, as well as simulation of viscous flow in porous objects. As you can see, the optimal organization of the porous structure of these materials can be described by the universal laws of mass and heat transfer and capillary equilibrium in porous media.

Summary

Porous materials and powders of different pore sizes are used in many areas of industry, geology, agriculture (crops and livestock), pharmacy and science. These areas include (i) a variety of devices and supplies—filters and membranes, chemical power sources (batteries and supercapacitors), sintering, paper, pharmaceutical powders; (ii) thermal insulation and building materials; (iii) oil-bearing geological, gas-bearing and water-bearing rocks; (iv) biological objects—food materials, soils and roots of plants, mammalian skin, and others.

The book describes modern methods for measuring the characteristics of the porous structure, such as pore size distribution, porosity, specific surface area, wettability by liquids, and others. Characteristics of the porous structure of different materials, products, powders, and natural objects are given as well. Furthermore, it is shown how these characteristics affect the functional properties of porous objects and materials: electrical conductivity, gas and moisture permeability, thermal conductivity, strength, and selectivity of filters and membranes towards different chemicals, electric capacity of batteries and supercapacitors, oil saturation of geological rocks, productivity of soil, absorption of pharmaceuticals by humans and animals, etc.

In this book, on the basis of modern physics, chemistry, and mathematics important examples of mathematical modeling of mass transfer occurring in different porous materials and products, as well as modern models of the hydrodynamic permeability of structured complex-porous media, are presented. Mathematical models validated experimentally allow optimizing these processes, as well as the porous structure of different products. It is shown that the principles and approaches of mathematical modeling are united and thus are applicable to all porous objects such as artificial and created by nature during the process of evolution, i.e., inorganic and biological.

Y. M. Volfkovich et al., *Structural Properties of Porous Materials and Powders Used in Different Fields of Science and Technology*, Engineering Materials and Processes, DOI: 10.1007/978-1-4471-6377-0, © Springer-Verlag London 2014

This book allows the reader (a) to understand the basic regularities of heat and mass transfer and adsorption occurring in qualitatively different porous materials and products and (b) to optimize the functional properties of porous and powdered products and materials.

The book is written in popular language and is intended for a wide audience, specializing in different fields of science and engineering: engineers, geologists, geophysicists, oil and gas producers, agronomists and other agricultural workers, physiologists, pharmacists, researchers, teachers, and students.

About the Authors

Dr.Sci. Yury M. Volfkovich at present is Principal Scientist at A.N. Frumkin Institute of Physical Chemistry and Electrochemistry, Russian Academy of Sciences, Moscow. His main areas of work during 50 years have been porosimetric methods, fuel cells, batteries, supercapacitors, and electrochemical water purification. During 1970–1995 he participated in development of fuel cells for Soviet and Russian space programs. During 2003–2004, he worked as guest scientist at Axion Power Int. Co. and Porotech Ltd. (Toronto, Canada) as head of laboratories of electrochemistry and porosimetry. He developed a new porosimetric method—Method of Standard Contact Porosimetry and corresponding Porosimeter. At present this porosimeter is producing by Porotech Ltd. He is author of the two monographs: I.G. Gurevich, Yu.M. Volfkovich, V.S. Bagotsky, "Macrokinetics of Porous Electrodes", Minsk, 1974 (in Russian) and Yu. M. Volfkovich, T.M. Sedyuk, Electrochemical capacitors. In book: Reference book on Power Sources, Moscow, 2003 (in Russian). He is the author of more than 400 papers in scientific journals (Colloid and Surfaces, Advanced Materials, Electrochimica Acta, J. of Electrochemical Society, J. of Power Sources, J. of Applied Electrochemistry, Polymer Communications, Membrane Science, Russian Journal of Electrochemistry, etc.), and 30 patents.

Dr.Sci. Anatoly N. Filippov is Professor (Full) in the Higher Mathematics Department at Gubkin Russian State University of Oil and Gas from February 2010. From 1987 to January 2010 he worked at Moscow State University of Food Production successively as Assistant, Associate, and Full Professor, Head of Pure and Applied Mathematics Department. Areas of his scientific interests include physicochemical mechanics, modeling of membrane processes, viscous flow in porous medium, colloidal interactions, and nanotechnology. Professor Filippov published more than 250 papers in such journals as Colloid J., Advances in Colloid and Interface Science, Colloids and Surfaces, J. Membrane Sci., Separation and Purification, Acta Mechanica, Russian Journal of Electrochemistry, Petroleum Chemistry, etc. He is co-author of two monographs: S.N. Goryachev, A.N. Filippov "Theoretical aspects of electromigration of low-molecular components during fur manufacturing", Furs of the World Publishing, Moscow, 1999 (in Russian); K.I. Popov, I.V. Gmoshinsky, A.N. Filippov, A.V. Zherdev, S.A. Khotimchenko, V.A. Tutelian "Food nanotechnologies: prospective and

Y. M. Volfkovich et al., *Structural Properties of Porous Materials and Powders Used in Different Fields of Science and Technology*, Engineering Materials and Processes, DOI: 10.1007/978-1-4471-6377-0, © Springer-Verlag London 2014

problems", MSUFP Publishing Complex, Moscow, 2010 (in Russian) and the Chapter "Process of moisture sorption by hydrophilic biopolymers and biopolymer mixtures" (V.V. Ugrozov, A.N. Filippov, Yu.I. Sidorenko, N.N. Shebershneva) in the book "Theoretical Backgrounds of Food Technologies", Moscow: KolosS Publishing, Book 2, 2009 (in Russian), and also of three patents.

Dr.Sci. Vladimir S. Bagotsky passed away on November 12, 2012 at the age of 92. Between 1949 and 1965 Vladimir Bagotsky worked at the All-Union Research Institute of Power Sources in Moscow. He was head of the laboratory which developed batteries for spacecrafts (mostly high capacity silver-zinc and mercury-zinc). From 1960, V. Bagotsky became a leader of fuel cell development in the Soviet Union, and from 1980 he supervised the Russian R&D related to lithium batteries. From 1965 to 1985 before moving to the USA, Bagotsky worked at the Institute of Electrochemistry of the Russian Academy of Sciences as head of the power sources division. Basic aspects of electrochemical energy conversion addressed by Bagotsky and his co-workers in this Institute are related to electrocatalysis, electrode kinetics on porous electrodes, and electrochemical intercalation. During his scientific career V.S.Bagotsky authored more than 400 papers in scientific journals (Electrochimica Acta, J. of Electroanalyt.Chemistry, J. of Power Sources, Russian Journal of Electrochemistry, etc.) and six monographs. The first was "Kinetics of Electrode Processes" (1952) published together with A.N. Frumkin, Z.A. Iofa and B.N. Kabanov in Russian and available in English as NBS translation. Already in the twenty-first century, during his USA period, he published "Fundamentals of Electrochemistry" (Wiley, 2006) and the treatise "Fuel Cells: Problems and Solutions" (Wiley, 2009)." Till his last days Professor Bagotsky has been working on the present book.

Printed in the United States
By Bookmasters